# AIRCREW TRAINING AND ASSESSMENT

**HUMAN FACTORS IN TRANSPORTATION**
**A Series of Volumes Edited by**
**Barry H. Kantowitz**

---

Barfield/Dingus • *Human Factors in Intelligent Transportation Systems*

Billings • *Aviation Automation: The Search for a Human-Centered Approach*

Garland/Wise/Hopkin • *Handbook of Aviation Human Factors*

Hancock/Desmond • *Stress, Workload, and Fatigue*

Noy • *Ergonomics and Safety of Intelligent Driver Interfaces*

O'Neil/Andrews • *Aircrew Training and Assessment*

Parasuraman/Mouloua • *Automation and Human Performance: Theory and Applications*

# AIRCREW TRAINING AND ASSESSMENT

Edited by

Harold F. O'Neil, Jr.
*University of Southern California/CRESST*

Dee H. Andrews
*Warfighter Training Research Division*
*U.S. Air Force Research Laboratory*

LEA
2000 LAWRENCE ERLBAUM ASSOCIATES, PUBLISHERS
Mahwah, New Jersey London

Lawrence Erlbaum Associates, Inc., Publishers
10 Industrial Avenue
Mahwah, New Jersey 07430

Cover design by Kathryn Houghtaling Lacey

**Library of Congress Cataloging-in-Publication Data**

Aircrew training and assessment / edited by Harold F. O'Neil, Jr., Dee H. Andrews.
p. cm.
Includes bibliographical references and index.
ISBN 0-8058-2977-6 (cloth : alk. paper)
1. United States. Air Force—Airmen—Training of. 2. Flight crews—United States. I. O'Neil, Harold F., 1943- II. Andrews, Dee H.

UG638 .A63 2001
358.4'15—dc21 00-037150

Books published by Lawrence Erlbaum Associates are printed on acid-free paper, and their bindings are chosen for strength and durability.

Printed in the United States of America
10 9 8 7 6 5 4 3 2 1

# Contents

# Preface

*Aircrew Training and Assessment* is designed for professionals in the aviation psychology, assessment and evaluation, aircraft simulation engineering, human factors, vocational, technical, educational psychology, and educational technology communities. It explores the state of the art in the training and assessment of aircrews, covering military and civilian environments and including both individual and team training R&D. While there has been considerable research performed over the years examining aircrew procedural training, the examination of military mission (tactical) training has not received nearly as much research attention, mainly because it has been difficult to examine multiship tactical training (due to a lack of networked simulators for research or training) until the late 1970s.

An aircrew consists of one or more persons who are responsible for achieving a mission goal through use of an aircraft. Depending on one's point of view, an aircrew can be as small as one pilot flying a single-seat aircraft, or as large as a full crew operating an airliner, or all of the people on board an aircraft plus others who provide assistance to the on-board personnel during the flight (e.g., air traffic controllers). This book contains chapters that highlight these various definitions of the term *aircrew.* We believe the important point is that it takes more than just the pilot to *safely* complete a mission with an aircraft.

Despite advances in aircrew selection and human factors engineering techniques, the need for better aircrew training is still readily apparent. In military aircraft the missions keep getting more complex. For example, military aircrews must now fly with joint forces (allied aircraft) more than they ever have in the past, new weapons that require new skills are added to the inventory all the time, new aids to flying are added such as night vision goggles, and new tactics must be learned as more capable aircraft are added to the inventory. All of these new operational requirements necessitate more effective and efficient training. Such training needs to be assessed for its utility in solving practical flying problems. Commercial jetliner aircrews also have an increased need for better aircrew training methods. The fact that "human error" still accounts for the majority of aircraft accidents, both civilian and military, shows that better selection and equipment alone will not be enough to improve aircrew performance to levels desired by operators, companies, regulatory agents, and customers.

With the advent of new technologies, especially in the area of microcomputers, the expansion and utilization of simulators for flight training is burgeoning because the cost of training a pilot is high and safety is of the utmost importance in actual aircraft. Flight simulators are today's most modern training simulators. Generally, flight simulators have extensive physical fidelity and can present numerous environmental situations (e.g., emergencies). Simulators are also used extensively for both initial and follow-on training. For example, simulators are the only means used by civilian airlines to train experienced pilots who are upgrading from one airplane to another. Thus, the first time a pilot trained solely on the upgrade simulator flies the new airplane, it has passengers on it. Such training has been certified as safe and effective by the Federal Aviation Administration (FAA).

More complex military training requirements are continually evolving at the same time that training resources are becoming scarce. For example, the Air Force flight training budget has been reduced, whereas the training requirements for pilots have increased. This combination of doing more with less has heightened interest in effectively using innovative training techniques and technologies. For example, the utility of using simulation to increase the scope and realism of training at an affordable cost has been investigated.

In this book, a succinct review and description of training simulators in the area of aircrew training is presented along with an exploration and discussion of related assessment issues and concerns. For example, in aircrew training what are the characteristics and functions of training simulators? What are the advantages and disadvantages of simulators? A description of the diverse applications and uses of training simulators is also presented. And the notion of cost-effectiveness of training simulators is addressed.

Our book explores the foregoing issues in two major sections: Models and Tools for Training of Aircrews and Models and Tools for Assessment of Aircrew Training. The volume is unique in the following montage of factors: (a) a focus on aircrews; (b) use of simulation as the major training and assessment tool; (c) a focus on competencies of teams as well as individuals; and (d) contexts in Air Force schools as well as job sites. In many cases, there is a significant body of research and experience that is discussed. In other cases, the training community is just starting to articulate the key training and assessment issues. The data that are reported were collected in various settings, for instance, military aviation laboratories and industrial settings. The chapter authors have extensive backgrounds in the topics they will address and they highlight topics where considerable training and assessment R&D needs to be performed. These authors are the researchers that, in many cases, have defined the state of the art by their work. We expect that readers of this book should learn much that can be of value as they consider the complex and challenging nature of manned flight.

## ACKNOWLEDGMENTS

Support for this work was provided in part by the Air Force Research Laboratory. However, the views, opinions, and findings contained in this book are the authors' and should not be construed as an official position, policy, or decision unless so designated by other, official documentation.

*Harold F. O'Neil, Jr.*
*Dee H. Andrews*

# Book Introduction and Overview of Military Aircrew Training

Dee H. Andrews
Justine N. Good
*Air Force Research Laboratory, Mesa, AZ*

Safely and successfully piloting an aircraft is one of the most difficult tasks a person can perform. Although pilots talk about having the "right stuff," defined as an almost mystical set of skills and abilities that good pilots naturally possess, experience has shown that almost anyone can learn to fly safely with the proper training. We believe it is true that some pilots do have a natural head start in being proficient at piloting due to some innate set of cognitive and motor skills. However, we also believe that quality instruction can make it possible for the vast majority of the population to fly safely.

This book covers a wide range of topics concerning the training and assessment of aircrews. We prefer the term *aircrews* rather than *pilots* because many piloting tasks require more than one person (a pilot) to be successfully performed. An aircrew can mean one pilot or it can mean an entire set of people needed to fly the aircraft. People other than the pilot can include copilots, navigators, flight engineers, and loadmasters who must make sure an aircraft's cargo is properly loaded and secured. In military aircraft there may be a whole range of aircrew members besides those just listed, such as special sensor operators and weapon system operators (e.g., machine gunners). Even when a single pilot is alone in an aircraft, such as a fighter, it is possible to think of other pilots in other aircraft as members of one aircrew. For example, fighter pilots usually fly in multiple-ship formations (2, 4, or more) for mutual support when they go into battle. Although they are in separate aircraft, they must function as one unit, just as multiple aircrew members must function as one aircrew on a multiplace aircraft. Therefore, their flying and communication skills must be honed to function with high synchronicity.

We purposely sought a wide range of chapters for this book from a variety of experts on aircrew training and assessment. However, we did not attempt to address every aspect of aircrew training; such an endeavor would require many volumes. We solicited chapters that describe models of aircrew learning and instruction, chapters that describe technologies (such as simulation) that are vital to conveying aircrew instruction, and chapters that examine techniques and issues involved in assessing aircrew training. In these chapters, there are many well known references to help readers understand the "nuts and bolts" of aircrew training and assessment, so generally we did not concentrate on those fundamentals. Rather, we attempted to explore some of the more difficult problems in aircrew training that are addressed by newer models, technologies, and assessment techniques. Some of the chapters are aimed at readers who are new to the topic of aircrew training, whereas others are targeted at people who may have been involved in aviation training for years and are interested in concepts and technologies that only recently were developed. Because many of the chapters deal with military aircrews, a brief overview of training military aircrews appears at the end of this section (http://www.aetc.randolph.af.mil/ho). See chapter 1 by Taylor and Emanuel for a similar discussion on the training of civilian aircrews.

This book is divided into two main sections: the training of aircrews and the assessment of aircrew training. In each section, we present various models and tools for either training or assessment. Although we built the book around these two conceptual domains, both topics are so closely related that they appear on a frequent basis in each chapter. Many of the chapters could fit neatly into either section. However, we placed each chapter in the section that we felt represented the chapter's major focus, either training or assessment. We provide a brief overview for each of the chapters, in the order in which they appear.

## Models and Tools for Training Aircrews

This section has an eclectic mix of chapters that examine a variety of aircrew training issues from multiple perspectives. The first chapter, by Taylor and Emanuel, documents how pilots in civil aviation gain their skills. Civil aviation includes all nonmilitary training such as commercial airline training, corporate aviation, and general aviation for private pilots. Taylor and Emanuel focus on the increased demand for civil aviation pilots at the same time that a major traditional source of pilots, military aviation, has cut back on the available supply. They also discuss the regulation of civil aircrew training and the impact of cockpit automation on training.

Advanced simulation for civil aviation is discussed, along with the training of cockpit resource management skills.

Andrews and Thurman (chapter 2) review the application of the *instructional systems development* (ISD) approach to developing and evaluating aircrew training. They describe how instructional systems development was once strictly a behaviorally oriented approach to instructional development that evolved to a more cognitive perspective. This chapter, which is a primer for the use of instructional systems development in aircrew training, discusses how recent advances in cognitive psychology were incorporated in the traditional instructional systems development model. This new approach, called the *Systems Approach to Training* in the Air Force, has proven to be especially useful for developing multifaceted training for the complex aircrew world.

Jensen, Boyle, and Fuller (chapter 3) discuss an important perspective about the need for systematic user involvement in the development of aircrew training systems. Poor results in aircrew training can often be traced to the development of training systems that do not include enough user input. It is tempting for experienced aircrew training developers to make assumptions about user needs and desires without consulting the eventual users of the training systems. This chapter describes some pitfalls and then some techniques to increase user involvement.

In chapter 4, Bowers, Jentsch, and Salas present a comprehensive approach to identifying *Cockpit Resource Management* (CRM) training needs, focusing on aircrew competencies. CRM training attempts to provide aircrews with skills for quality and timely communication and coordination at all times on the flight deck. It has been adopted by civil and military aviation training institutions throughout the world.

Thurman and Russo (chapter 5) discuss the use of commercial, off-the-shelf virtual reality technology for aircrew training and review current virtual reality technologies and their capabilities and limitations. They also explore the use of virtual reality as it relates to the ISD process in aircrew training.

O'Neil, Mayer, Herl, Niemi, Olin, and Thurman continue, in chapter 6, the theme of the use of virtual reality for aviation training by presenting the results of an empirical study that investigated a specific instructional strategy to improve aviation training. Note that the term *aviation training* is used to describe the topic of this chapter instead of *aircrew training*. That is because the type of instructional strategy (visual vs. auditory) examined in the study should have use in multiple aviation domains (e.g., aircrew, maintenance, etc.).

Nullmeyer and Spiker (chapter 7) discuss an important new tool for training military aircrews to successfully perform their missions. Mission rehearsal provides aircrews with realistic simulations that can replicate quite accurately the mission environment crews will face. The authors have considerable experience in working with aircrews who have performed mission rehearsals in simulators, and their insights provide principles for effective mission rehearsals.

A very interesting simulation-based approach to aircrew training called *above real-time training* is presented by Crane and Guckenberger in chapter 8. This approach takes advantage of a computer-based simulator's capability to allow a trainee to fly a simulated aircraft at faster speeds than could be achieved in the actual aircraft. The authors discuss the technical aspects of this training approach and present research findings from studies that examined above real-time training.

The final chapter of the *Models and Tools for Training of Aircrews* section is chapter 9, written by Klein. He conducted extensive research into complex decision-making processes, obviously an important set of skills for any aircrew to possess. Klein presents important concepts and empirical results about how aircrew training can better prepare aircrews to work in complex decision-making environments.

### Models and Tools for Assessment of Aircrew Training

This section begins with a discussion by Alessi (chapter 10) concerning a general model for simulation-based training design. Of all the types of training that use simulation as a medium of instruction, perhaps aircrew training makes the most profound use. Alessi argues the need for a theory of learning and assessment for simulation design. He then provides a conceptual model framework for such a theory. His ideas can readily be applied to any type of simulation-based learning, but as is quickly obvious in this book, aircrew training simulation is potentially the major beneficiary of such a simulation design.

Dwyer and Salas (chapter 11) detail some key principles of performance measurement for assessing aircrew training effectiveness. Frequently when aircrew training results are disappointing, the root cause of the dissatisfaction is poor performance measurement. When the training fails for some reason, trainers are often at a loss to diagnose the difficulty because their performance measurement system is lacking or, in too many cases, is absent altogether. This chapter is based on both empirical research and the authors' rich experience in this important topic.

Chapter 12 examines the use of simulation to assess teamwork skills in a domain other than aircrew training. However, these research results hold promise for the development of a real-time measurement system for teamwork skills. O'Neil, Wang, Chung, and Herl present findings from some empirical studies that examined teamwork processes and teamwork skills using two different computer-based group tasks, and we believe the lessons learned will have positive transfer for the assessment of aircrew training.

Niall and Pierce (chapter 13) discuss the role of visual requirements in simulation-based training for aircrews. Presenting visual stimuli and cues that are realistic has long been a major goal for simulation developers. This challenge is especially difficult in the highly dynamic realm of aircrew simulation. It is vital that simulation developers understand the human visual system and its requirements for visual simulation if high fidelity simulation is to be achieved. This chapter depicts the current understanding of those visual requirements.

Moor, Andrews, Nullmeyer, and Burright (chapter 14) look at an important, but seldom researched, aspect of training assessment—benefit–cost analysis and the cost-effectiveness of simulation for aircrew training systems. Next to actual flying, simulation is usually the most expensive part of aircrew training. Yet, when key decision makers desire information about the benefit–cost and cost-effectiveness of this expensive training component, the training community too often cannot provide a satisfactory answer. This chapter examines fundamental principles of benefit–cost and cost-effectiveness analysis for simulation, and then discusses two examples where these principles are applied to actual simulation approaches to training.

As mentioned earlier, aircrews include more members than just those who fly the airplane. Some military aircraft, such as the Airborne Warning and Control System (AWACS), carry large teams of aircrew members who operate radar command and control aircraft. In the final chapter, Schiflett and Elliot discuss the challenges these types of aircrew members face in their training and some approaches in use to explore and alleviate these training issues, including the use of synthetic team training.

## OVERVIEW OF MILITARY AIRCREW TRAINING

This section describes current approaches in the military, especially the U.S. Air Force, to training fixed-wing aircraft aircrews. For brevity's sake, the focus here is on pilot training, and not the training of other aircrew positions (e.g., navigator, electronic warfare, machine gunners). However, the training of those other military aircrew positions is challenging enough that entire chapters could be devoted to each of them. The training sequence for U.S. Air Force pilots is described, taking a pilot from novice to undergraduate trainee to graduate with wings to operational pilot. Although there are some differences between the ways the Air Force and the Navy train their fixed-wing aircraft pilots, there are many similarities. Therefore, this discussion of Air Force pilot training gives readers a good general idea of how Navy pilots are trained.

**Air Education and Training Command (AETC)**

Headquartered at Randolph Air Force Base, TX, Air Education and Training Command (AETC) is responsible for Air Force flying training. AETC has a large number of supporting organizations that provide assistance to the training units, including organizations that are expert in instructional design, media development, instructor pilot training, and training evaluation. AETC Headquarters sets policies, standards, and directives, and waivers and deviations must be approved through the chain of command. The structure of AETC includes the 19th Air Force. Under the 19th Air Force, 11 flying wings conduct various types of flight training for both undergraduate pilots (those who are in training to receive their wings) and graduate pilots at Formal Training Units (FTUs) in the specific aircraft that they will fly when they are assigned to operational units.

**Initial Flight Screening**

Regardless of the source of undergraduate pilot candidates, they all must pass a series of rigorous aptitude and psychomotor tests as well as show some proficiency in flying introductory propeller-driven aircraft. The typical progression of a pilot-training student is first through a flight screening program (unless the student possesses a private pilot certificate), then through Undergraduate Pilot Training. The flight screening programs were implemented to give flying experience to those individuals who do not possess a private pilot license and to increase the success rate of officers entering pilot training.

**Undergraduate Pilot Training (UPT)**

In 1959, the Cessna T-37 (Fig. I-1), a twin-engine subsonic aircraft in which the instructor and student sit side-by-side, was integrated into regular student training. In 1961, Air Training Command (ATC, later called Air Education and Training Command, or AETC) received its first T-38 Talon (Fig. I-2), a twin-engine supersonic aircraft (the first ever to be used in pilot training) with tandem seating. The T-37/T-38 combination of aircraft for training continues through the turn of the century.

**Specialized Undergraduate Pilot Training (SUPT)**

In January 1992, AETC received its first T-1A Jayhawk trainer aircraft at Reese AFB, TX. The T-1A marked the first time the USAF purchased a commercially available aircraft and modified it to accomplish military flight training. The T-1A is a medium-range, twin-engine, business-type jet trainer version of the Beechcraft 400A. The key modification was the

FIG. I.1. T-37 aircraft. USAF photo by Don Fike.

FIG. I.2. T-38 aircraft. USAF photo by Don Fike.

addition of a jump seat between and slightly behind the pilot's and copilot's seats. The two trainees can sit in any of the three seats depending on the goal of the instructor; the instructor always occupies one of the primary positions at the flight controls with student pilots.

With the T-1A, AETC turned away from the "universally assignable" pilot theory (i.e., any graduate could be assigned to fly any aircraft) and implemented the more efficient Specialized Undergraduate Pilot Training (SUPT) program, a program specifically designed to train students in different tracks earlier in the training process. The rationale for creating a specialized track after primary training was to allow instruction to be specially modified for the general types of missions that trainees would eventually be flying when they moved beyond SUPT. Although trainees were all thrilled to fly the high-performance T-38 (Fig. I-2), the Air Force decided that it was more important for trainees who would eventually fly in multiplace aircraft to start getting that experience early in their flying careers. Additionally, the USAF selected this approach to reduce the number of sorties flown in the more expensive operational aircraft used at FTUs (schools for graduate pilots).

Until recently, after the students completed their initial primary training course in the T-37, they were ranked and allowed to select between the airlift-tanker (AT) track in the T-1 and the fighter-bomber (FB) track in the T-38. Clearly, the decision about which track to place a trainee in is a critical one. Highly rated trainees from the T-37 course were asked to give their preference for which specialized track they would prefer to follow. The availability of specialized tracks (T-38, T1, helicopter) was determined by USAF needs at the time of the student's selection.

To help eliminate "mismatches" in track selection, in 1998 the merit-based assignment selection system was discontinued in favor of the more traditional assignment system. However, the new system combines elements of the old with the new. USAF SUPT students make their desires formally known at the end of T-37 training, but the instructor pilots (IPs) and the flight commander ultimately determine which specialized track and final aircraft assignment the student receives. The determination is made after consideration of various factors including the student's personal desires and daily flight and academic performance, as well as the needs of the Air Force.

### Joint Specialized Undergraduate Pilot Training (JSUPT)

Traditionally, flight training was conducted by each service individually to meet the specific requirements of its mission. With the large reductions in the defense budget following the end of the cold war, the services looked to consolidate training where possible. In the early 1990s, the Department of Defense decided to institute Joint Specialized Undergraduate Pilot Training (JSUPT) for the USAF and US Navy.

### The Curriculum

Currently Air Force SUPT takes 52 weeks to complete. Although it is possible for a trainee to take longer than that, for various reasons such as transferring to joint training bases or academic or medical problems, the large majority of USAF pilot trainees receive their wings (graduate) in 1 year. The Air Force examined the feasibility of self-pacing instruction;

however, such an instructional approach has never been adopted, and group-pacing has been the norm for many years. One of the main reasons for the group-paced approach is to make sure the personnel system can accurately predict when pilots will be available to fill specific training courses in formal training units beyond SUPT and in operational assignments beyond that. The Navy has a modified self-paced approach that allows some flexibility as to when undergraduate pilots receive their wings.

The 52-week SUPT program is divided into three distinct phases. Phase I consists of in-processing, indoctrination, and basic academics (ground school) and lasts 1 month. Phase II is primary and intermediate flight training conducted in the T-37 and lasts approximately 5 months. Phase III is the specialized training track lasting about 6 months. This training includes flying training to teach the principles and techniques used in operating high-speed jet aircraft, ground training to supplement and reinforce flying training, and officer development training to strengthen the graduate's leadership skills, officer qualities, and understanding of the role of the military pilot as an officer and supervisor.

The basic methodology for USAF training is to first instruct the concepts in an academic training environment, next demonstrate the concepts to the student in a training medium either on the ground or in the air, then provide the student an opportunity to practice the concepts. Before they sit in an airplane, the students learn all aspects of their aircraft systems and operation.

Phase I training is conducted solely in an academic training environment using platform instruction supplemented by individual Computer Aided Instruction (CAI), video and slide presentations, and hands-on practice with some concepts. Academic training continues for the students throughout SUPT. Basic aircraft systems and aviation knowledge tailored to specific training aircraft are presented in each new phase of training; however, general ground training instruction such as flying fundamentals, aerodynamics, navigation, aviation meteorology, mission planning, local flying procedures, and aircraft mishap prevention are primarily presented during Phase I. At this time, the students also receive critical life-saving training in aerospace physiology. There they learn about the effects of, and precautions associated with, high altitude flight. They also learn how to get out of an aircraft in case of an emergency—instruction that includes basic parachute training. In total, each student completes more than 200 courses totaling more than 300 hours of academic instruction.

Another significant part of instruction during flight training is the pre- and postmission briefing. Generally, these briefings are conducted on a one-on-one basis between the instructor and the student, although many times other students monitor the brief–debrief to gain additional knowledge. During the brief, the instructor lays out the objectives for the mission and evaluates the student's preparation, including general knowledge for the flight. The mission is then flown, either in the simulator or aircraft. Following the sortie, the instructor debriefs the mission, evaluating and critiquing the student's performance, providing feedback to the student about his or her performance during the mission, and providing guidance for areas of study in preparation for the student's next sortie. If a student fails a mission, he or she is given specific training to prepare for the repeat ride. Many times the student is encouraged to "chair fly" the mission, a term used for mental rehearsal for the next mission. The brief–debrief is a critical area of training that continues throughout a military pilot's career.

In USAF SUPT, the formal military briefing conducted at the beginning of the training day with both students and instructors is also an opportunity for learning and training for the students. Not only does the morning brief force the students to conduct themselves with a military bearing, but it puts them in a high-pressure situation with the Stand-up Emergency Procedure of the day. The emergency procedure, or EP, is a hypothetical emergency situation in the aircraft that is presented to the student on the ground. The student must then verbally walk through the appropriate steps to resolve the problem and "recover the aircraft" to the ground safely. The instructor presents EPs to the student for every mission, but the Stand-up EP is presented differently. The simulated situation is presented to the entire group of students, then one or two students are singled out of the group to literally stand up and talk the instructors and other students through the situation.

Once the students have a sufficient amount of "book knowledge," they proceed to the next phase of training in a specific aircraft. Again, using a building block approach to training, the student accomplishes a significant amount of initial training in a simulator. During all the flying phases of their training, the students receive instruction in simulators that are working duplicates of the aircraft.

Once the prerequisite ground and simulator training has been completed, the student receives flight training in actual aircraft. The categories of instruction for SUPT are generally the same for every aircraft: contact, instruments, navigation, and formation. The specific amount of training received in each area in the aircraft depends on the desired training objectives for the phase of training and student.

In their first category of instruction, contact, students learn how to fly primarily using outside references. In contact, they learn basic flight maneuvers and characteristics, including aerobatics. The next category of instruction teaches the students how to fly without outside references. In the instrument and navigation categories, students learn how to tell where they are in the sky and how to get where they need to go without looking at outside references. They learn how to navigate from one place to another and land at a base other than the one from which they took off. The final category of instruction is also one of the toughest: formation. In this category, students learn how to fly as close as 3 feet from another aircraft or farther apart as the specific mission and/or aircraft training dictates.

A student's performance is evaluated at specific points in his or her training with *checkrides*. These are essentially quality control evaluations conducted by a standardization–evaluation flight of instructors. Checkrides are also a normal way of life for operational aircrew members, so the student is introduced to this high-pressure evaluation early in his or her career. Checkrides in SUPT are essentially a final exam in a category of training in a specific aircraft.

In all phases of flight training, regardless of aircraft, the main training objectives for flight are essentially the same. They include multiple performances, conditions, and standards. The objective topics are:

1. basic control/inflight planning/clearing/cross-check
2. mission planning/ground operations
3. takeoff/transition to instruments
4. local area procedures
5. inflight checks
6. visual flight rules and pattern procedures
7. aerobatic maneuvers
8. stall recognition and recovery/slow flight
9. stall recognition
10. spins/stability demonstration
11. communications
12. instrument maneuvers
13. high/low instrument navigation
14. pilotage/dead reckoning navigation
15. descent and approach
16. formation takeoff
17. close formation
18. offset trail
19. formation approach and landing
20. emergency procedures/general knowledge
21. situational awareness.

An important development on the near horizon will be the introduction to both Air Force and Navy JSUPT training programs of the Joint Primary Aircraft Training System (JPATS). The aircraft is a propeller-driven "Raytheon's Beech/Pilatus PC-9 Mk II based on the Swiss Pilatus PC-9 aircraft, modified to include a strengthened fuselage, upgraded engine, more fuel, pressurized cockpit, larger, bird-resistant canopy, and new digital avionics" (Air Force Almanac Commentary, 1999). The JPATS aircraft will replace the Air Force's T-37 and the Navy's T-34 aircraft, both of which have been used as primary aircraft for many years.

In the past, all parts of the training system in undergraduate and graduate training were developed and run by military personnel. That included courseware development, academic instruction, simulator maintenance, and in-flight instruction. Over the years, the military determined that it was more cost-effective to have many of these functions out-sourced to civilian contractors. Often the contractor personnel who perform these functions are former military members who held similar positions when they were in uniform. Now, there are a variety of combinations of military and contractor personnel performing development and instructional functions in training squadrons. However, in all cases, uniformed military personnel perform the in-flight instruction duties. Although there have been proposals to use contractors for in-flight instruction, they have thus far been rejected. The primary reason for continuing with uniformed instructors is that military leaders think it is very important to expose trainees to active duty instructors who have recent experience in operational aircraft and settings. Not only can these instructors provide recency-of-operational-experience examples for the trainees, but the trainees are also presented with role models who can demonstrate the proper military attitudes and demeanor.

### Mission Training

Once pilots earn their wings (graduate) and receive training from an FTU in their assigned operational aircraft (e.g., fighter, bomber, tanker, transport), they move on to an operational unit where they receive continuation training that improves their piloting skills and teaches them specific operations and tactics used in real-world missions. This mission training uses a combination of training strategies and media. Flying training is a key element of mission training; each pilot must fly a certain number of prescribed sorties to learn and practice key operational mission skills.

### Distributed Mission Training

Recent advances in simulation and networking technology have now allowed the Air Force to start using a mission training concept called Distributed Mission Training (DMT; Fig. I-3), which is a combination of live flying, virtual simulation, and constructive (computer generated) modeling of friendly and enemy forces to form a powerful training capability. It is now possible to link various live, virtual, and constructive entities into a synthetic battlefield that can be used for training, acquisition, testing and

FIG. I.3. The top portion of the figure shows an artist's conception of the synthetic battlespace that includes weapon systems of all types from the live, virtual, and constructive domains. The bottom left part of the figure shows the mission control station and large screen display in use at the Warfighter Training Research Division in Mesa, Arizona as part of the Distributed Mission Training testbed. The four small pictures on the bottom right of the figure show the testbed's four F-16C cockpit simulators, surrounded by four full-field-of-view simulator visual displays. USAF photo by Brian Watkins, photographer; Curtis Sayers, illustrator.

evaluation, and tactics development. Powerful networking technologies make this linking possible over secure and rapid local and wide area networks. It is now possible to have different units train with and against each other across the country or around the world using DMT.

When pilots conduct mission training in actual aircraft on a range, they try to "train like they intend to fight." That is, they want the training environment to be as realistic as possible so their training will transfer to actual, real-world missions in a robust manner. This was a lesson learned from the Viet Nam conflict. High U.S. losses compared to enemy losses convinced Air Force leaders that the training for Air Force pilots was not realistic enough. From that experience grew training ranges that were instrumented to track all aircraft in mock battles, aggressor squadrons of Air Force pilots that mimicked the doctrine and tactics of a potential enemy, and realistic training exercises, like Red Flag conducted in Nevada. All of these innovations and a variety of others were meant to make mission training as real as possible.

Live flying training, despite its quality, still has limitations and constraints. For example, there are altitude restrictions below which pilots are forbidden to fly for safety reasons; many electronic warfare tactics cannot be used for security reasons; and when pilots are shot in mock battles, they, of course, do not go down in a ball of flame but rather exit the playing area for a period of time and then come back to fight again. These and many other constraints can make training something less than fully realistic.

DMT's use of networked virtual simulators and constructive models allows many of these constraints to be overcome. Now that simulation technology has improved, it is possible to allow pilots to train their skills on a regular basis. These are skills that they either cannot practice on live ranges, due to the constraints mentioned earlier, or cannot practice very often. In synthetic environments, pertinent skills can be trained on a frequent and regular basis and then honed when pilots fly on training ranges.

The Air Force has begun the process of putting DMT strategies and equipment infrastructure improvements in place. In many cases, existing simulators and constructive models that work well in stand-alone modes are being modified so they can be networked to larger DMT federations. Research and development efforts are underway to continue to improve the simulation and constructive elements of DMT to add ever-increasing measures of realism. In addition, training effectiveness research is being conducted to help Air Force aircrews determine how best to train using DMT's live, virtual, and constructive assets. All of these DMT activities will help Air Force aircrews truly train missions as they intend to fight.

## ACKNOWLEDGMENT

The contents of this introduction and overview and the views expressed herein are those of the authors and do not necessarily represent the official views or policies of the Department of the Air Force or the Department of Defense.

## REFERENCES

Air Force Almanac Commentary. (1999, May). *Air Force Magazine*, p. 147.

History and Research Office, Air Education and Training Command, U.S. Department of the Air Force. Available 16 June 1999 (http://www.aetc.randolph.af.mil/ho).

*Part* **I**

# MODELS AND TOOLS FOR TRAINING OF AIRCREWS

*Chapter* **1**

# A Civil Aviation View of Aircrew Training

Henry L. Taylor
Tom W. Emanuel, Jr.
*Institute of Aviation, University of Illinois*

Civil aviation encompasses those aspects of aviation not involving military aviation, including commercial airlines, business (corporate) aviation, and general aviation (private aviation). Civil aviation differs from military aviation in several areas. In civil aviation, aircrews are not trained as a team nor do crews trained as a team operate the plane as is the practice in military aviation. Furthermore, commercial airlines do not typically teach beginning pilots to fly, but military pilots receive all of their flight training in the military. In the commercial airlines' cockpit, the captain is in charge of the aircraft and typically "flys the left seat," whereas in the military, the aircraft commander is in charge. The second in command in civil aviation is the first officer, in the military, the copilot is second in command. Both commercial airlines and military have a flight engineer position, but in civil aviation this person is a pilot, whereas in the military the flight engineer is not trained as a pilot.

In the United States, a number of organizations play an important role in aircrew training for civil aviation. These include the Federal Aviation Administration (FAA), the airlines, the military, universities, fixed base operators, aircraft manufacturers, training device and simulator manufacturers, and organizations that represent aircrews. Fixed base operators provide a variety of services at civilian airports including charter flights, line service (refueling), maintenance, aircraft rental, aircraft storage, and flight training. These organizations and a number of important trends are shaping current civil aircrew training and are expected to have an impact in the future.

Although the diversity of the pilot work force is changing, pilots continue to be predominately White and male. In the past, the U.S. military supplied major air carriers with approximately 75% of their pilots. However, in 1995, the number of military pilot or pilot trainees in their first year of service was less than 40% of the number in 1985, a reduction of more than 60% (National Research Council [NRC], 1997). There is no question that in the future, airlines will depend heavily on nonmilitary trained pilots for their workforce. In addition, there is an increased demand for trained aircrew members. Current FAA regulations require that captains and first officers, but not flight engineers, retire from their positions at age 60 (age 60 rule). About 1,000 airline pilots reached age 60 in 1996, and retirements due to the age 60 rule reached 1,400 in 1998. Retirements are expected to average 1,568 per year from 1999 through 2005 (a range from 1,447 to 1,625 is expected). The average number of age 60 retirements is predicted to increase to 2,002 (range from 1,847 to 2,197) between 2006 and 2010 (Darby, K., personal communication, April 20, 1998).

Civil aviation is regulated by the FAA under the Federal Aviation Regulations (FAR) contained in the Code of Federal Regulations: 14CFR Parts 1-199 (Federal Aviation Regulation). The FAA regulates, through the various parts of the FAR, basic training practices as well as specific training requirements for all aspects of civil aircrew training. Recent changes in the FARs under which civil aircrew training is conducted constitute an important trend.

The increased levels of cockpit automation with the advent of the *glass cockpit* have resulted in substantial changes in the way civilian aircrews are trained. Aircraft with glass cockpits will soon dominate air carrier fleets. Glass cockpits are advanced cockpits consisting of electronic displays rather than conventional instruments. Aircrews behave differently in glass cockpit aircraft than in traditional aircraft, but it is not certain if these differences are related to training or to the characteristics of particular automation philosophies and aircraft design (Helmreich, 1991). The increased number of glass cockpit aircraft and the increase in automation also meant an end to the era of the flight engineer and the three-pilot flight deck. Automation allowed the number of crew members to be reduced to two as both aircraft systems and flight management (navigation) were more automated. Duties previously assigned to the flight engineer are now largely the direct responsibility of the first officer. When glass cockpit aircraft were introduced, air carriers initially experienced higher failure rates in transition training than they had experienced previously. The airline industry was able to bring the training problems under control, and by the middle of the 1980s, failure rates in transition training to glass cockpit aircraft were no greater than those found in conventional aircraft (Wiener, 1993).

Another major trend is the development of crew resource management (CRM) and line-oriented flight training (LOFT). LOFT was introduced by Northwest Airlines in 1975. This training is full mission training and involves conducting real-time, enroute training exercises with aircrews in full mission flight simulators. CRM is defined as ". . . the effective use of all resources available to the flight crew, including equipment, technical/procedural skills, and the contribution of flight crew and others" (Taggart, 1994, p. 309).

The final major trend is the increased use of training devices and flight simulators in all phases of civil aircrew training, particularly in airline aircrew training. Air carriers in the United States currently can perform, in FAA qualified flight simulators, initial qualification and transition, as well as recurrent training and checking. This trend reflects an increase in the realism of flight simulators as well as a progressively changing regulatory environment.

This chapter discusses each of these trends in terms of their impact on civil aircrew training. The focus is on commercial airline aircrew training, but aircrew training for business (corporate) aviation is also discussed.

## SOURCE OF PILOTS FOR U.S. AIRLINES

In the late 1980s, a period of increased growth in commercial aviation, combined with the initial downsizing of the number of pilots in the military, resulted in predictions of a shortage of trained aircrew personnel. A Blue Ribbon Panel concluded that there would be a sufficient number of pilots who hold a commercial pilot's certificate and have 250 to 300 hours of flight time, but there was an impending shortage of aircrew members who met the qualifications necessary to operate technologically complex aircraft in the future airspace system (Federal Aviation Administration, 1993). In addition, the number of military pilots available to the airlines was predicted to fall by 30% by 1998. Military pilots traditionally constituted approximately 75% of the air carrier work force (NRC, 1997), but in 1996 and 1997, more than half of the pilots (58%) hired by U.S. airlines were trained in the civilian sector with no military flight training (Darby, 1998).

Military pilots have been attractive to the airlines because they are a highly select population, they have operated in a team-oriented and disciplined environment, they have operated high technology aircraft, they have approximately 8 years of mission-oriented experience, and they require little additional investment by the air carriers. Hansen and Oster (NRC, 1997) found that when the number of pilots hired by major air carriers is low, the percent of military pilots hired is high, but when the number of pilots hired is high, the percentage of military pilots is lower. They concluded that the airlines prefer military pilots, and when they need a few pilots, they hire military pilots. When large numbers of pilots are needed, the airlines hire a substantial number of pilots without a military background. Military pilot separations from the services were over 3,000 per year from 1990 to 1992. These military pilots provided a substantial resource for the air carriers. Levy (1995) projected that future military pilot separations would decrease to 1,700 annually between 1995 and 1999 and to 1,500 by 2002. Consequently, the number of military pilots available to air carriers after 2002 will be less than half of those available in 1992.

In 1997, a National Research Council study considered the implications of military downsizing on training demand and work force diversity, the capacity for civilian training organizations to meet the demand for pilots and mechanics, and the options for attracting underrepresented groups for aviation training (NRC, 1997). They reported that in 1993, 101,000 pilots were in the work force and a total of 53,341 pilots were employed by commercial airlines. Of this total, 46,000 pilots were employed by major air carriers (NRC, 1997). Since that time, there has been substantial growth of the regional airline industry. Indeed, many regional airlines are currently purchasing regional jets to replace their turboprop fleet of aircraft. Regional jets are smaller airplanes compared to the current fleet of jet aircraft. They typically fly a stage length (from one airport to another) of about 500 to 750 miles and have 35 to 70 passengers. Demand for aircrew members is strongly tied to economic conditions. During the past 10 years, new pilot hiring varied substantially from year to year, ranging from a high of 13,401 in 1989 to a low of 3,256 in 1993 (NRC, 1997). Another study by Darby reported that from 1993 to 1997, the new pilot hires for major air carriers ranged from a low of 463 in 1993 to 3,854 in 1997; major carrier hiring was expected to peak in 1998 at 4,300. In the future, airlines will increasingly depend on civilian trained pilots to meet their aircrew requirements.

Equipment changes by major air carriers are also creating increased demand for pilots. TWA is typical of a major U.S. national and international carrier in that they are attempting to utilize more fuel-efficient aircraft. Some of these aircraft carry fewer passengers than the aircraft that they replace, which requires more crewmembers and aircraft to carry the same number of passengers. For example, TWA is phasing out L-1011s and B747s and replacing them with new B-757s, MD-80s, and A-330s. They are hiring fewer flight engineers and more first officers. TWA has 2,900 pilots, and between 1997 and 2003, 1,200 are scheduled to retire due to the age 60 rule. The TWA pilot recruitment coordinator estimated that TWA needs to recruit about 1,500 pilots during that 6-year period (Treon, 1997). Delta Airlines and TWA use the Universal Pilot Application Services (UPAS), an electronic pilot application data base (Flint, 1997). Other electronic employment services include AIR Inc. and Airline Employment Assistance Corps (AEAC).

Pilot screening at TWA has two phases. Phase 1 requires one day and includes testing, the interview, and the simulator evaluation. The simulator evaluation is conducted in a Frasca 142, a twin-engine simulator. The evaluation consists of basic flying skills, instrument visual scan, command (maintaining control), one emergency situation, and basic climbs and turns. If candidates pass phase 1, a background check is conducted that takes about 30 to 60 days. A background check includes criminal and driving records. If the background check is satisfactory, the candidate is scheduled for a final interview—phase 2. In phase 2, candidates are interviewed by senior flight operations management, and the candidates receive a TWA physical examination. If candidates pass phase 2, they are scheduled for training. TWA initial training consists of equipment ground school (5 weeks), simulator training (2 weeks), and initial operating experience (2 to 3 weeks). When pilots complete this 10 weeks of training, they are fully line-qualified crew members (Treon, 1997).

Pilots can obtain FAA commercial pilot certification under a section of the FAR (Part 61) without attending formal aviation training programs by passing FAA computerized knowledge examinations and receiving flight instruction with an independent FAA certified flight instructor (CFI). The FAA publishes books for private pilot, instrument rating, and commercial pilot computerized knowledge examinations. In addition, private companies publish FAA knowledge exam guides for the same certificates and ratings; these contain questions and answers organized by topic with explanations for each question. Pilots have the choice of more than 1,700 flight schools. Of this number, 600 schools are certified by the FAA under Part 141 (approved school; FAA, 1997). Kitely (1995) reported that more than 200 U.S. colleges offer aviation majors, 93 offer bachelor's degrees, and 112 offer associate (2-year) degrees. All major U.S. airlines require a baccalaureate degree as a condition of employment as a first officer, so the inclusion of flight training programs in college and university curricula is an appropriate and convenient pairing. More than 200 U.S. colleges enroll over 18,000 students pursuing programs in flight education (Crehan, 1995).

Flight schools that meet specified personnel, aircraft, facilities, curriculum, and other operating requirements are certified under Part 141 of the FAR. All schools certified under Part 141 may apply for examining authority and, if granted, can recommend their graduates for private and commercial certificates and instrument ratings without the candidates having to take the FAA flight or knowledge test or both. The Institute of Aviation at the University of Illinois is an example of a certificated Part 141-approved school with examination authority. In addition, since 1986, the Institute has been granted an exemption from the flight hours requirements of Part 141 and has developed and maintained a proficiency-based flight training program. The Institute is one of only a few Part 141 schools having this exemption.

The two principal differences between Part 61 and Part 141 flight training are curriculum approval and the minimum required flight hours to certification. Part 141 schools must submit a proposed training course outline to the regional Flight Standards District Office of the FAA. The training course outline must follow the guidelines for both ground and flight instruction listed in Part 141. The minimum flight times permitted under Part 141 are less than the minimum required of Part 61 training. Part 141 allows the commercial pilot certificate to be issued with a total of 190 flight hours, whereas Part 61 requires 250 hours. The chief pilot and assistant chief pilot(s) for Part 141 schools must meet specific aeronautical and teaching requirements not required under Part 61. Additionally, Part 141, but not Part 61, requires certain physical facility requirements for ground schools and ground trainers (if used).

Many collegiate aviation programs (also called professional flight, professional pilot, or career pilot programs) focus on pilot training. The core curriculum typically provides instruction leading to certification as a private and commercial pilot (single engine and multiengine land) and an instrument rating. Most collegiate aviation programs have options for a certified flight instructor rating and certified flight instructor-instrument rating. A few collegiate programs with FAA-approved visual simulators are offering advanced training beyond the core curriculum (see Civil Simulators later in this chapter). After initial flight training, civilian pilots typically progress through on-the-job training. After graduation from college, pilots without military experience usually work as flight instructors or fly as commercial pilots for cargo or air taxi (charter) operations for their first or second jobs. The career path may include advancing from turboprop aircraft to jet aircraft within an airline or from a regional to a major airline with a job as a flight engineer or first officer. Pilots flying for corporations may receive type-specific training in the corporate fleet aircraft from companies such as Flight Safety International or Simuflite who specialize in both initial and recurrent corporate training. Type-specific training includes from 35 to 80 hours of ground school depending on the complexity of the aircraft, 10 hours of simulator training in the left seat, and an indeterminate number of hours of in-flight training in the actual corporate aircraft. Depending on the level of the fidelity of the simulators used for the training, current regulations may not require flight time in the aircraft prior to initial certification (type rating). Pilots who are employed by regional airlines typically receive aircraft-specific training from the airline. In many instances, the airline subcontracts the training to Flight Safety International, Simuflite, or another organization.

## REGULATION OF CIVIL AVIATION AIRCREW TRAINING

Air carriers that fly aircraft of 30 or more seats with a payload capacity of 7,500 or more pounds are regulated under Part 121 of the Federal Air Regulations. Air carriers that operate aircraft with fewer than 30 seats and a payload of less than 7,500 pounds operate under Part 135 of the FAR. All major, national, and regional airlines must have training programs that are approved by a designated FAA Principal Operations Inspector (Birnbach & Longridge, 1993). The standards for approval are contained in the Air Transportation Operations Inspectors Handbook. The primary areas of review are curriculum, course materials, facilities, and procedures for record keeping. Part 121 specifies training and practice on maneuvers listed in Appendix E and proficiency checks on those listed in Appendix F and Appendix H when conducted in a simulator.

The Advanced Qualification Program (AQP) "is an alternate method of qualifying, training, certifying, and otherwise ensuring the competency of flight crew members, flight attendants, aircraft dispatchers, instructors, evaluators, and other operations personnel subject to the training evaluation requirements of Federal Aviation Regulation (FAR) Parts 121 and 135" (FAA, 1991b). The purpose of the AQP is to encourage innovative use of technology and training methods to provide cost-effective training in order to increase operator proficiency.

Special Federal Air Regulation 58 represents the most significant change in requirements for air carrier training programs and crewmember qualification since 1970. The AQP permits cost-effective use of advanced simulators and computer-based training. The regulation and use of advanced simulators under this program or other air carrier training programs is discussed in detail later. Qualification by the FAA of the AQP is based on individual and team performance expressed as proficiency objectives. Individual and team proficiency, and the program itself, is empirically validated by data collection and analysis, and training is systematically developed with an audit trail for all training and data requirements. The methods used for development, implementation, and maintenance of program operations are continued throughout the life of the program (FAA, 1991b).

The AQP accommodates make, model, and series aircraft. Three curricula are required for every aircrew position: (a) indoctrination for new aircrew, instructors, and evaluators (called initial training), (b) qualification (called transition training), and (c) continuing qualification (called recurrent training). Under the program, the airline must provide data that validates that aircrew proficiency meets or exceeds Part 121 and/or Part 135 standards. AQP training and evaluation is conducted both individually and in a crew or team environment and provides for training and evaluation of crew resource management, for the use of operational simulations for both training and evaluation, and for the integration of appropriate advanced flight training equipment. The air carrier must train and evaluate instructors and evaluators that are used in the program and must provide a management plan that includes a transition plan.

The AQP for each aircraft has five phases of development that require FAA approval. Phase 1—initial application; Phase 2—curriculum development; Phase 3—training system implementation; Phase 4—initial operations; and Phase 5—continuing operations. During the initial application phase, the air carrier must provide the FAA with a transition plan and supporting data that results in Phase I approval. Phase 2, the curriculum development phase, has three steps. In Step 1, proficiency objectives are developed based on a task analysis. The objectives include an entry-level analysis as well as qualification standards. In Step 2, lessons are developed and are divided into modules and segments. In Step 3, an implementation/operations plan is developed. In Phase 3, courseware is developed and tested in the training environment, and instructors and evaluators are qualified. Finally, a formative evaluation is conducted by using small groups to test the training. In Phase 4, the initial operations phase, there is a summative evaluation and aircrew training for all curricula. During Phase 5, continuing operations, all training and evaluation is conducted under the AQP. The air carrier is responsible for quality assurance and program maintenance in this phase.

United Airlines, the first airline to develop and receive approval from the FAA for an AQP, is using the program at their training center in Denver, CO. United has developed or is in the process of developing an AQP for each aircraft in their fleet. United and the FAA participate in each phase of the AQP development before going to the next phase (Edwards, J., personal communication, April 13, 1998). For each of the three curricula, the ground school is conducted using computer-based training. This training is self-paced and each student has an opportunity for discussion with an instructor during each day. Next, procedures training in a fixed-base trainer with no force loading as a result of movement of aircraft controls, no motion, and no visual system is conducted. During this training, the student learns the procedures to operate basic aircraft systems, the auto flight systems, and the flight management system (FMS). Following procedures training, the student receives full mission simulator training. During Phase 4, initial operations, LOFT scenarios are used during the training. In this phase, realistic scenarios are used that permit the use of crew resource management techniques. At the end of the simulator training, the aircrew members receive a type rating. The role of the FAA is to validate what United is accomplishing with their training. The FAA determines the validity and reliability of the training from a quality assurance point of view.

The first AQP developed at the United Training Center was for the Boeing 737 model 300 in 1994 as a conversion training program. The Boeing 737-300, first operational in 1984, is a twin-engine jet transport capable of carrying 189 passengers 1,300 miles at 577 mph. When United accepted the Boeing 777 aircraft, all training was done under AQP. The Boeing 777, first operational in 1995, is a twin-engine, wide-body jet transport capable of carrying 440 passengers 8,200 miles at more than 560 mph. The FAA gave the United Training Center final approval of the Boeing 737-300 and the Boeing 777 Advanced Qualification Program for Phase 5, continuing operations. The way in which United uses a particular aircraft in its operations has a major impact on the development of the AQP for that aircraft. For instance, United uses the B-737-300 for shuttle operations along the West Coast. Aircrew members make six or seven landings per day. The evaluation procedures for the Boeing 737-300 emphasizes maneuvers to a significant degree. The Boeing 747 flying over water on international routes has a four-pilot crew and each pilot may not get enough landings in the airplane to remain current. The Boeing 747-400, first operational in 1990, is a four-engine, wide-body jet transport that typically carries 420 passengers up to 8,000 miles at up to 557 mph. Consequently, members of the Boeing 747-400 crew may have to return to the training center periodically to remain current. The Boeing 747-400 curriculum contains extensive training on international procedures and special airports.

United also has a "single visit" exemption for continuing qualification of pilots. Under the exemption, pilots visit the training center once a year. The flight training, all done in simulators, lasts for 3 days, and all training sessions may be videotaped for subsequent review and debriefing. During the first day, trainers and evaluators have the pilot fly certain fixed maneuvers for a "first look" evaluation, which is graded and compared to last year's training. Next, there is a series of short practice LOFT scenarios. During the second day, the crew flies maneuvers according to qualification standards developed by United and approved by the FAA. The aircrew practices workload management during day 2. United standardizes the evaluators and determines the inter-rater reliability of the evaluations. United uses a four-level grading scale with level 2 deficiencies being corrected by debriefing and level 1 being unsatisfactory. During day 2, the crew also receives training on advanced maneuvers such as unusual attitudes. Day 3 is conducted in the simulator using the scenarios. Annually, United develops a basic scenario with a safety theme, and they also develop back-up scenarios. The scenario design consists of a series of event sets that have a trigger event, distracters, and specific criteria for successful completion. The basic criteria that United uses for successful completion are no significant deviation from standard procedures and a safe landing of the airplane. United typically uses eight event sets in a LOFT scenario. A crew fails the LOFT scenario if it fails the same event set twice or fails three of eight event sets. An aircrew can fail two event sets and pass the check if they reaccomplish the failed event sets satisfactorily. Consequently, different event sets must be available during the retest. Aircrew members who fail are retrained and must reaccomplish the testing. These aircrew members also receive special tracking that consists of a line check in 3 months and additional training at the center after 6 months.

## INCREASED LEVELS OF COCKPIT AUTOMATION

A group of pilot training experts participated in a NASA conference in 1981 to discuss the concept of LOFT (Lauber & Foushee, 1981). This meeting was a milestone in pilot training and occurred a few months after the first McDonnell-Douglas MD-80 became operational. This event is considered by Wiener (1993) to be a turning point in the history of cockpit automation. The MD-80, a twin-engine jet transport, is capable of carrying 172 passengers for 3,300 miles at 576 mph. The Boeing 767, which became operational just a few months later, provided an even higher level of flight-deck automation that began the era of the glass cockpit. The Boeing 767, first operational in 1982, is a twin-engine, wide-body jet transport capable of carrying 290 passengers 7,600 miles at 550 mph. Air carriers have developed CRM and LOFT recurrent training modules on crew coordination and workload management that will enable aircrews to successfully fly glass cockpit aircraft (Wiener, 1993).

Increased levels of cockpit automation have affected not only flight operations but also flight training in air carrier operations. Automation tends to lower the physical workload on the flight crew; however, the flight management system (FMS) is yet another highly complex system that needs to be well understood by the crew. Billings (1997, p. 190–191) pointed out that if ". . . (you) cannot *show* the pilot what he or she needs to know in a given situation, then the pilot needs to *know* what he or she needs to know." Even if automation were fully capable of always showing pilots what they needed to know, the aircrew needs to fully understand all systems of the aircraft. In addition to understanding all "traditional" systems on the aircraft (fuel, electrical, hydraulic, etc.), they also need to understand the FMS.

Over the past three decades, the philosophy of air carrier aircrew training has changed. When the Douglas DC-9 first became operational in the 1960s, there was a shift in behavioral objectives in airline training programs. The McDonnell Douglas DC-9, first operational in 1965, is a twin-engine jet transport capable of carrying 139 passengers 2,000 miles at 510 mph. With the elimination of the flight engineer, the workload increased for the captain and first officer. Rather than teach the two remaining flight crew members everything about the DC-9, the training program instead emphasized specific behavioral objectives (Billings, 1997). The result was to reduce the amount of training time, which was a goal set by the airlines. The specific behavioral objectives approach has been both efficient and effective, but as the level of complexity of the FMS increases, problems can arise. Throughout their careers, pilots traditionally have learned all they can about the aircraft in the event that a problem of unforeseen proportions and complexity may arise. Because virtually all pilots learn to fly in aircraft without advanced automation, their exposure to "traditional" aircraft systems is almost inevitably greater than their exposure to the flight management system. When this is coupled with an approach that emphasizes a specific behavioral objective, it should come as no surprise that many cockpit voice recorders have recorded the phrase, "What's it doing now?" with the "it" being the advanced automation system. As Billings (1997, p. 191) says, "(just knowing) . . . how *to operate* these more complex and less transparent systems (the FMS) as opposed to (knowing) how *they operate*. . . ." may not be sufficient.

The question of both philosophy of training and depth of knowledge has been of concern to many airlines and pilots. Delta Air Lines in 1990 developed a philosophy of automation and implemented a new training course, "Introduction to Aviation Automation," which precedes the ground school and is independent of the model of aircraft flown (Wiener, 1993). Federal Express started a similar course in 1991 prior to operating the McDonnell Douglas MD-11. The MD-11 is a three-engine, wide-body jet transport that was first flown commercially in 1990. It can carry 400 passengers for 8,000 miles at 588 mph. Typically there is a lack of training of the effective monitoring of the FMS. Sarter and Woods (1995) noted that whereas crews on traditional aircraft are taught a highly standardized and efficient flight instrument scan, pilots who transition to an aircraft with an FMS complain that they do not receive guidelines concerning how to keep track of the status and behavior of the automation. Sarter and Wickens (1997) found that pilots cannot and do not learn all that they need to know about the FMS during initial transition training. Consequently, the pilots need to continue to learn about the functional structure of the FMS during line flying. However, current FMS design does not aid in this learning because the operating structure of the FMS is neither visible or intuitive to the pilots. In addition to providing for easier acquisition of more knowledge about the FMS, a more transparent design would ". . . also allow them to identify and correct misconceptions in their mental model, thus leading to more appropriate expectations and monitoring." This acknowledgement of the continuum of learning from initial transition training through day-to-day line flying also supports the concept of line-oriented flight training.

These problems are illustrated in a study by Anca (1995). He reported on the training strategies that were used by Philippine Airlines to transition senior pilots from the Boeing 747-200 to the Boeing 747-400. The transition course included 15 days of academic training, 10 days (20 hours per crew) in a fixed-base simulator, 9 days (18 hours per crew) in a full-flight simulator, and the hours per aircrew for training to fly specific routes were negotiated. The results indicated that the majority of failures occurred in the full-flight simulator phase. The highest incidents of pilot failures were data entry and inflight replanning, mastery of procedures and checklists, and aircraft heading and flight techniques. To have a smooth transition to glass cockpit aircraft, Anca (1995) recommended the following to solve these problems: (a) in addition and prior to type-qualification training, pilots should receive an introductory course in automation; (b) pilots designated for transition to glass cockpit aircraft should be given preparatory information concerning the demands of automated flying prior to transition training; and (c) type-qualification and recurrent training syllabi for automated aircraft must contain modules that focus on cognitive skills.

Glass cockpit technology, which includes FMS, electronics, flight instruments, electronic centralized aircraft monitoring, and push button technologies, has required airlines to reevaluate transition and recurrent training for pilots flying new-technology aircraft. Buehle, the training manager for Airbus for Lufthansa, described an integrated type rating used for transition training for pilots transitioning to glass cockpits (Buehle, 1995). The objectives of the course were to provide more emphasis on procedural knowledge, operational knowledge, and crew resource management. The first step was to develop training requirements. To meet training requirements, the transition training was divided into a ground school for cognitive learning, simulator training for aircraft handling characteristics, and crew resource management. An interactive, computer-based training system was used for procedures and systems learning. A Level D simulator with a high resolution visual system was used. The integrated type rating had five training phases: phase 1—basics; phase 2—normal procedures; phase 3—abnormal procedures; phase 4—line-oriented flight training, abnormal; phase 5—line-oriented flight training, line introduction (Buehle, 1995). No evaluation of the training was provided, but Buehle stated that the integrated type rating for Lufthansa aircrew training "is answering the demands of a new training philosophy in connection with the necessity of training their pilots on high technology aircraft for a safe line operation" (Buehle, 1995).

## CREW RESOURCE MANAGEMENT AND LINE ORIENTED FLIGHT TRAINING

In 1975, the FAA issued Northwest Airlines an exemption from Part 121.409 to permit a test program for what later became known as line-oriented flight training (LOFT). In 1977, Part 121 of the Federal Air

Regulations was amended by the FAA to permit LOFT to be a part of any airline training program. An advisory circular, Line Operational Simulations AC120-35A, published in May, 1978, provided the airlines guidelines for recurrent LOFT. This advisory circular was revised in 1990 (FAA, 1990).

Prior to CRM, superior pilot skills were considered to be the most important factor in determining a pilot's ability to safely perform aircrew duties. An analysis of air carrier safety records for a 20-year period indicated that at least 65% of air carrier accidents and incidents were due to inadequate leadership qualities, communication skills, crew coordination, and decision making. Crew resource management, embedded in LOFT scenarios, represents a major step in aircrew training that addresses these problems. United Airlines was one of the first air carriers to develop LOFT scenarios that involved the cockpit aircrew. United developed courses for its aircrew members in command/leadership/resource management (C/L/R) in the early 1980s, and later the courses were available to all air carriers through seminars (Taggart, 1994). The United Airlines C/L/R training currently includes flight attendants.

The required elements to develop an effective LOFT program include task and goal analysis, baseline CRM training, LOFT evaluation skills for instructors and check pilots, and a method for tracking and validating the training process (Butler, 1993). The entry point of a LOFT scenario can be varied to permit a number of different options for the aircrew, and there should be a wide variety of outcomes that depend on the decisions and actions taken by the aircrew.

Recurrent LOFT enables aircrew members to practice CRM skills and to deal with various normal and emergency situations that will be experienced in line operations. To be effective, a LOFT scenario should include preflight preparation, trip paperwork, the use of manuals, and communications. The scenarios should foster a situation under which each aircrew member provides the necessary information to other crew members at the appropriate time. Recurrent LOFT scenarios should be designed so that they are operationally relevant and realistic but also test the aircrew's teamwork skills (Butler, 1993). The LOFT scenarios should not be designed to continuously increase the workload of the aircrew until the crew is overloaded. The aircrew should be required to continue the LOFT scenarios until the problem is solved or the airplane is on the ground. The qualification LOFT is part of a transition and upgrade training program designed to prepare an aircrew member for line operations. The focus is on operational procedures, and it is the last step before approving the aircrew member for line operations. CRM is a factor in qualification LOFT, but it is not the primary purpose of this training (Butler, 1993). Airlines have designed transition training that tailors LOFT scenarios to the technology of the cockpit. For recurrent training, however, airlines have used generic LOFT scenarios that are type-independent (Wiener, 1993).

Crew resource management has been defined as "the effective use of all resources available to the flight crew, including equipment, technical/procedural skills, and the contribution of flight crew and others" (Taggart, 1994, p. 309). The objective of CRM training is to use all available resources to ensure safe and efficient flight operations, while at the same time providing the technical and pilot skills training needed to maintain proficiency in the most effective way possible (Taggart, 1994). CRM was originally developed for the airline industry, but it has been applied successfully to corporate and general aviation. CRM programs have demonstrated that aircrew decision-making skills, which are required for safe flight, can be developed through systematic training. CRM emphasizes the relationships between aircrew members and the need to successfully complete the operational mission (Maurino, 1993). The optimal method to practice CRM is through a LOFT scenario. High fidelity simulation, which duplicates the operational environment, enables aircrew training on line-oriented flight training. The LOFT scenario should produce aircrew performances and behaviors that would be present during an actual line flight, given the same set of conditions. During a LOFT scenario, aircrew members should have access to all the resources that would be available during a line flight (Maurino, 1993).

Helmreich and Wilhelm (1997) indicated that acceptance of CRM has been imperfect despite its evolutionary changes. Open communication between subordinates and superiors has been difficult to implement in cultures where strong hierarchical power is normative. They propose a "fifth generation" of CRM training with the specific goal of managing human error by developing a safety culture. The strongest action to support the safety culture would be a nonpunitive stance toward error and a policy to encourage aircrew members to report errors and safety problems (Helmreich & Wilhelm, 1997).

The FAA recommended that classroom training be provided to aircrew members to raise awareness of crew coordination issues. In addition, aircrew members should have the opportunity to practice and receive feedback on their performances as crew members, and CRM skills should be reinforced through recurrent training. CRM concepts should also be integrated with aircrew proficiency training and evaluation (Chidester, 1993). Most airlines are attempting to comply with these recommendations using advanced qualification programs.

Chidester (1993) described the American Airlines' CRM program that was designed to comply with FAA recommendations. A career approach to CRM was developed and implemented using classroom-based training. American also developed a LOFT program with a video-based debriefing. The career approach begins with an introductory CRM course for beginning flight engineers (½-day course). During the annual recurrent CRM training, all aircrew members participate in a class focused on decision making, communication in time-critical situations, and working with groups outside the cockpit. Aircrew members upgrading to first officer and captain attend 1-day and 1.5 day CRM programs respectively. The program for captains deals with leadership and crew coordination skills and includes a 2-hour session with dispatchers. American provides special training to *check airmen* to reinforce CRM concepts and behaviors. A check airman is a senior airline captain of either gender selected by the airline and approved by the FAA to examine other company pilots and approve (or disapprove) them for upgrade to captain status or for qualification on a different model of airplane. The check airmen also attend the annual recurrent training with line pilots.

## CIVIL SIMULATORS

All airlines have invested heavily in flight simulators that are used to conduct their initial qualification, transition, upgrade, and recurrent training. Simulators have been used by airlines in aircrew training programs and in some checking programs since the middle 1950s (FAA, 1992). As simulator technology developed, the FAA revised the Federal Air Regulations to permit greater use of simulators in the approved training programs of the airlines. Most of the changes have occurred since 1973 when the FAA permitted additional use of simulators with visual systems. Changes in Part 121.439 allowed simulators approved for the landing maneuver to be used in lieu of the airplanes for recency-of-experience qualification. Amendment 161 to Part 121, Appendix H, in 1980, established the Advanced Simulation Plan (FAA, 1980).

Airlines that have a Phase III Training Program approved under Part 121, Appendix H, the Advanced Simulator Plan, can, with certain exceptions, conduct in a simulator all training and checking required in Part 121 and also the certification check requirements of Part 61.157 and Appendix A of Part 61 of the Federal Air Regulations. These exceptions are the line check, static airplane requirements, and operating experience requirement, which must be performed in the airplane (FAA, 1980). Simulator requirements for Phase III include characteristic buffet motions, aerodynamic modeling of aircraft, realistic amplitude and frequency of cockpit noises and sounds, self-testing for simulator hardware and software, diagnostic analysis printout of simulator malfunctions, and visual requirements. The visual requirements are daylight, dusk and night scenes, representative physical relationships, special weather representations, wet and snow-covered runway representations, realistic color and directionality of airport lighting, and weather radar presentations (FAA,

1980). In order to support the Advanced Simulation Plan, the FAA established the National Simulator Evaluation program in 1980. The National Simulator Program Manager position was established by the FAA to administer the program.

The criteria for simulator qualification by the FAA have also evolved with simulator technology and with the increased use of simulators in airline training and checking. The Advisory Circular, Airplane Simulator Qualification AC120-40B, (FAA, 1991a) with change 1, (1/24/92) and change 2, (6/9/93) provides criteria for evaluation and qualification of simulators used in airline training and checking programs. The advisory circular clearly states the FAA position on the increased use of simulators by airlines: "The availability of advanced technology has permitted greater use of flight simulators for training and checking of flight crewmembers. The complexity, costs, and operating environment of modern aircraft also have encouraged broader use of advanced simulation. Simulators can provide a very high transfer of learning and behavior from the simulator to the airplane. The use of simulators, in lieu of airplanes, results in safer flight training and cost reductions for the operators. It also achieves fuel conservation and reduction in adverse environmental effects" (FAA, 1991a).

Advisory Circular AC120-40B divides simulators into levels that permit increased amounts of training and checking that can be accomplished with increased simulator capability. The levels and the relationship with previous simulator definitions in Part 121, Appendix H, are: Level A—Visual; Level B—Phase I; Level C—Phase II; Level D—Phase III. The minimum simulator requirements for qualifying Level D simulators are found in AC120-40B, Appendix I, Simulator Standards. Each of the four levels is more complex than the preceding level, and each contains the features of the preceding levels plus the features for the designated level (FAA, 1991a).

All major air carriers have acquired FAA Level A–D simulators, which permits them to conduct a substantial amount of their initial, transition, and recurrent training and checking in simulators. The LOFT scenarios and CRM training described in the previous section are also conducted in the simulators. The following major or national airlines have training centers with FAA Level A–D simulators: Alaska (2 simulators); America West (5 simulators); American airlines (21 simulators); Continental (10 simulators); Delta (25 simulators); Northwest (25 simulators); Southwest (5 simulators); TransWorld (7 simulators); United (26 simulators); US Airways (18 simulators) (Flight International, 1997). Many of these simulators have cockpit procedures trainers, flight training devices, and flight management system trainers associated with them. Some of these simulators are quite old. For instance, American Airlines has two B-727-100, FAA

Level A simulators that entered service in 1967 and 1968. American's most recent simulators are two Fokker 100 Level C and D simulators acquired in 1992 and 1991 respectively, a MD-82 Level D simulator acquired in 1991, and a Boeing 757-200 Level C simulator acquired in 1992. United Airlines also has a Boeing 727-100 Level B simulator acquired in 1967 and two Boeing 737-200 Level A and C simulators acquired in 1968. United's most recent simulator purchases include a Boeing 747-400 Level D simulator purchased in 1993 and a Boeing 777-200 acquired in 1994.

The technology found in these simulators varies widely. One criterion that can be used to determine the level of advanced simulation technology available to an airline is to examine the number of Level D simulators at the airline's training center. Use of Level D simulators provides the opportunity for airlines to accomplish the maximum amount of flight training permitted by the FAA. As of 1997, most air carriers have at least one Level D device for each of the major aircraft in their fleet. For instance, American Airlines has the following Level D simulators at their training facility in Dallas/Fort Worth: B-727-200, B-757-200, B-767-200, Fokker 100; MD-11, MD-80 (4 simulators), and MD-82. Continental Airlines has one level D simulator acquired in 1995; Delta Airlines has seven Level D simulators at their training center in Atlanta for four aircraft. Northwest Aerospace Training in Minneapolis-St. Paul has six Level D simulators for five aircraft. United Airlines has the following Level D simulators at their training facility in Denver: A-320-200, B-727-200A (2 simulators), B-737-300 (2 simulators), B-747-400 (2 simulators), B-757-200 (2 simulators), B-767-200. USAirways has two Level D simulators at their three training centers in Pittsburgh, Charlotte, and San Diego. Trans-World Airlines has one Level D simulator at their training center in St. Louis, but Southwest Airlines, Alaska Airlines, and America West Airlines have no Level D simulators. Thus, most airlines accomplish a majority of their flight training in simulators.

Advisory Circular AC120-45A, Airplane Flight Training Device Qualification, provides the criteria to evaluate and qualify training devices in which flight training, qualification, or certification of pilots is accomplished (FAA, 1992). In cooperation with the aviation industry, the FAA defined seven levels of flight training devices (Levels 1 through 7). Each higher level is more complex than the previous level. An FAA Principal Operations Inspector authorizes a flight training device to be used for flight training events or flight checking events based on the qualification level of the device and the approved program.

Flight Safety International has 31 training centers with 120 FAA Level A–D simulators. Flight Safety International developed many of the simulators that are used in the training centers (Flight International, 1997).

Flight Safety International currently operates more than 180 corporate and/or regional aircraft simulators worldwide, of which 29 are Level D, and 92 are Level C. In addition to corporate flight training, Flight Safety International represents 15 regional and national carriers. The cost of initial airplane-specific training typically ranges from $ 7,000 to $11,000. Some regional airlines require the pilot to pay this fee. Other airlines cover the cost of the training whether done in-house or by a subcontractor. Simuflite Training International in Dallas, TX, also offers training in 18 simulators (14 Level C, 3 Level D), largely for the corporate flight market (Flight International, 1997). In addition to lecture and simulator training, which are normally aircraft type-specific, a complete training curriculum should consider other crew-related factors. Other topics normally included in a training course include high altitude operations (decompression, hypoxia, etc.), over-water operations (flight planning, ditching, international operations), scheduling and dispatch, crew resource management, and flight/maintenance coordination. A number of vendors provide such training, including the National Business Aircraft Association and universities.

A small number of universities use FAA Level A–D simulators in their training programs. Embry-Riddle Aeronautical University has FAA approved simulators at its Prescott and Daytona Beach campuses. The Prescott campus has two B-727-100, FAA Level A devices acquired from United Airlines in 1994, and the Daytona Beach campus has a B-737-300 FAA Level D device purchased from Flight Safety International in April, 1997. The Daytona Beach campus also has a Beech 1900D with a Vital VIII Visual system and a 6-degree-of-motion axis also purchased from Flight Safety International. Miami-Dade Community College has a Boeing 727-100/200 FAA Level B device that is marketed by Pan Am Academy. Purdue University has a Boeing 727-100, FAA Level C acquired from American Airlines and a Boeing 727-232 FAA Level C device acquired from NASA Ames (Flight International, 1997). The simulators are used to provide advanced flight training to student pilots who have completed their commercial pilot certification.

## SUMMARY

This chapter provided a civil aviation view of aircrew training. The impact of the decreased numbers of military pilots, airline pilot demographics (retirements due to the age 60 rule), and increased demand for pilots was reviewed in terms of the changing source of pilots for civil aviation. Changes in the regulation of civil aircrew training as well as increased cockpit automation and the use of advanced simulation in aircrew training were discussed. The use of the Advanced Qualification Program to present crew resource management training and line-oriented flight training was presented.

## REFERENCES

Anca, J. M. Jr. (1995). Stream gauge to glass cockpit: training strategies for growth transitioning from B747-200 to B747-400. In N. Johnson, R. Fuller, & N. McDonald (Eds.), *Aviation psychology; Training and selection* (pp. 291–297). Aldershot, England: Avebury Aviation.

Billings, C. E. (1997). *Aviation automation: The search for a human centered approach*. Hillsdale, NJ: Lawrence Erlbaum Associates.

Birnbach, R. A., & Longridge, T. M. (1993). The regulatory perspective. In E. Wiener, B. Kanki, & R. Helmreich (Eds.), *Cockpit resource management* (pp. 263–281). San Diego, CA: Academic Press.

Buehle, E. H. (1995). The integrated type-rating. In N. Johnson, R. Fuller, & N. McDonald (Eds.), *Aviation psychology; Training and selection* (pp. 298–301). Aldershot, England: Avebury Aviation.

Butler, R. E. (1993). LINE-ORIENTED FLIGHT TRAINING: Full-mission simulation as crew resource management training. In E. Wiener, B. Kanki, & R. Helmreich (Eds.), *Cockpit resource management* (pp. 231–259). San Diego, CA: Academic Press.

Chidester, T. R. (1993). Critical issues for CREW RESOURCE MANAGEMENT training and research. In E. Wiener, B. Kanki, & R. Helmreich (Eds.), *Cockpit resource management* (pp. 315–336). San Diego, CA: Academic Press.

Crehan, J. (1995). *Educational opportunities in aviation education*. Opelika, AL: University Aviation Association.

Darby, K. (1998). "On the hiring line, A record setting year." *AOPA Air Safety Foundation Project Pilot Instructor Report, 4*(2), 2. Frederick, MD: AOPA Air Safety Foundation.

Federal Aviation Administration. (1980). *Appendix H, Advanced simulator plan. FAR Part 121-132, Amendment #161*. Washington, DC: U.S. Department of Transportation, Federal Aviation Administration.

Federal Aviation Administration. (1990). *Line operational simulations. Advisory Circular AC 120-35B*. Washington, DC: U.S. Department of Transportation, Federal Aviation Administration.

Federal Aviation Administration. (1991a). *Airplane simulator qualification. Advisory Circular AC120-40B*. Washington, DC: U.S. Department of Transportation, Federal Aviation Administration.

Federal Aviation Administration. (1991b). *Advanced qualification program. Advisory Circular AC 120-54*. Washington, DC: U.S. Department of Transportation, Federal Aviation Administration.

Federal Aviation Administration. (1992). *Airplane flight training device qualification. Advisory Circular AC120-45A*. Washington, DC: U.S. Department of Transportation, Federal Aviation Administration.

Federal Aviation Administration. (1993). *Pilots and aviation maintenance technicians for the twenty-first century: an assessment of availability and quality*. Pilot and Aviation Maintenance Technician Blue Ribbon Panel. Washington, DC: U.S. Department of Transportation, Federal Aviation Administration.

Federal Aviation Administration. (1997). *List of certificated pilot schools. Advisory Circular AC 140-2Z*. Washington, DC: U.S. Department of Transportation, Federal Aviation Administration.

Federal Aviation Regulations. Codes of Federal Regulations: 14 CFR Parts 1–199.

*Flight International.* (1997). Civil Simulators 1997 Census, April 23–29, 1997. London: A Reed Business Publication.

Flint, R. (1997). Online employment databases are the latest aids in the search for the right people. *Air Transport World*, November, pp. 63–66.

Kiteley, G. W. (1995). Collegiate Aviation Programs and Options. *Flight Training*, October, C45–C49.

Helmreich, R. L. (1991). Strategies for the study of aircrew behavior. In R. S. Jensen (Ed.), *Proceedings of the Sixth International Symposium on Aviation Psychology* (pp. 338–349). Columbus, OH: The Ohio State University.

Helmreich, R. L., & Wilhelm, J. A. (1997). Crew Resource Management and Culture: National, Professional, Organizational, Safety. In R. S. Jensen, & L. Rabanan (Eds.), *Proceeding of the Ninth International Symposium on Aviation Psychology*. Columbus, OH: The Ohio State University.

Lauber, J. K., & Foushee, H. C. (1981). Guidelines for line-oriented flight training. Vols. I and II. (NASA Conference Publication 2184). Moffett Field, CA: NASA–Ames Research Center.

Levy, C. M. (1995). *The civilian airline industry's role in military pilot retention: Beggerman or thief?* Santa Monica, CA: RAND Report.

Maurino, D. E. (1993). Human factors training in aviation. In R. A. Teller (Ed.), *Aviation instruction and training* (pp. 95–113). Aldershot, England: Ashgate Publishing Limited.

National Research Council. (1997). *Taking flight. Education and training for aviation careers.* J. S. Hansen, & C. J. Oster, Jr. (Eds.). Washington, DC: National Research Council, National Academy Press.

Sarter, N. B., & Wickens, C. D. (1997). *Pilot-automation communication and coordination: in need of support for knowledge- and data-driven attention allocation* (Technical Report ARL-97-13/BOEING-97-1). Savoy, IL: Aviation Research Laboratory, Institute of Aviation.

Sarter, N. B., & Woods, D. D. (1995). "How in the world did we ever get into that mode?" Mode error and awareness in supervisory control. *Human Factors, 37*(1), 5–19.

Taggart, W. R. (1994). Crew resource management: achieving enhanced flight operations. In N. Johnson, N. McDonald, & R. Fuller (Eds.), *Aviation psychology in practice* (pp. 309–338). Aldershot, England: Avebury Technical.

Treon, L. (1997). On the record with Captain Jim Elban, TWA Pilot Recruitment. *Airline Pilot Careers*, October, pp. 24–26.

Wiener, E. L. (1993). Crew Coordination and Training in the Advanced Technology Cockpit. In E. Wiener, B. Kanki, & R. Helmreich (Eds.), *Cockpit resource management*. San Diego, CA: Academic Press.

*Chapter* **2**

# An Instructional Systems Development Primer for Aircrew Training

Dee H. Andrews
Richard Allen Thurman
*Air Force Research Laboratory, Mesa, AZ*

Aircrew training is a complex process. All learning domains (i.e., intellectual skills, verbal information, cognitive strategies, attitude learning, motor skills; Gagne & Briggs, 1979) are involved in a very integrated fashion. This complexity required instructional developers for aircrew training after World Ware II to develop a systematic process for designing, developing, evaluating, and delivering the requisite training. In their search for better ways to produce aircrew training, these designers and developers were at the forefront of adapting systems engineering principles of problem solving for application to the design and development of training. The resulting process, called Instructional Systems Development (ISD), has now been adopted by the community of instructional design in general and continues to evolve as new learning and instructional principles are produced through research and development. In addition, new training technologies (e.g., computer-based training, virtual reality, aircraft simulation, distance learning) required the ISD process to undergo changes. This chapter provides a brief overview of ISD and discusses its application to aircrew training design and development. Although ISD methods and principles were heavily influenced by the aircrew training development community since the 1940s, the development topics discussed can be, and have been, applied to training of all types.

## WHAT IS ISD?

ISD addresses issues of analyzing, designing, developing, implementing, and evaluating instruction. Although the term ends with the word *development*, practitioners usually use the term ISD to imply that all five functions (analyze, design, develop, implement, and evaluate) are accomplished when ISD is used. Essentially, ISD answers the three major questions every trainer faces: "What should I teach?" "How should I teach it?" and "How do I know students have mastered it?"

According to Spector (1992):

> The real value [of ISD] is that it provides a meaningful organizing framework within which development activities can be described, discussed, and possibly actualized. In other words, [ISD provides] a kind of mental model for instructional development and it should be regarded as such, rather than as a rigid set of task prescriptions. (p. 645)

A study of expert instructional designers inferred that they rarely implement the standard activities of ISD serially (Reiser, 1992). Instead, they attempt to cope adaptively to changing environments and constraints.

## WHY USE AN ISD MODEL?

ISD models help designers to fashion a systematic game plan for developing training and instruction. Instructional designers can determine if they need to use a formal ISD model by simply asking themselves four fundamental questions: Does the training or instruction need to be systematic (as opposed to informal or opportunistic)? Is it possible that the training or instruction will need to undergo cycles of testing and improvement (as opposed to a one-shot development effort)? Will the training or instruction be evaluated (as opposed to just assuming it's a good product)? Is there a need for systematic monitoring and control of the design process? Affirmative answers to any of the previous questions is a positive indication that ISD will provide a useful tool.

If it is determined that ISD will be of help in aircrew training development, instructional designers should then decide how to use the ISD principles. The designers could develop their own ISD model from scratch using instructional and systems principles. Certainly, many individuals and groups have developed their own in the past. That is why there are dozens, and perhaps hundreds, of models in existence today. However, probably the most efficient way to get started is to select a model from among the many existing models of ISD, some of which are referenced later. Our experience is that, particularly for aircrew training development, it is wise to use an existing model if at all possible because aircrew training development is such an intensive process and as many resources as possible should be conserved for the hands-on design and development activity versus developing a new ISD model. Before the designers get to the ISD model development or selection process, however, it may be worthwhile to consider what a model is and why an ISD model is necessary at all.

Hayman (1974) indicated that a model is an abstraction and simplification of a defined referent system, presumably having some noticeable fidelity to the referent system. For instructional applications, the referent system is the synthesis of training system components that will be used to impart new skills, knowledge, and attitudes to the trainees. These may include the instructor, curriculum, media, students, evaluation methods, textbooks, etc. ISD models may also include in the referent system the social and economic context of the training system. For example, a designer might want to refer to the economic issues involved in developing flying training using actual aircraft flying hours versus other instructional media. Fidelity refers to how well the model matches the various characteristics and functions of the system it is trying to describe. Generally, as ISD models become less specific about the actual design and development processes they recommend, their fidelity to the real world diminishes.

Reigeluth (1983) contrasted instructional design models with instructional development models, the major difference being that design models indicate what the instructions should be like, whereas development models indicate how to make it that way. Many ISD models examine both design and development.

Instructional development models describe, prescribe, predict, and/or explain. A model may describe the components or activities of instructional design. Of course, merely describing instructional steps is not what most aircrew training developers are looking for. They want a model that prescribes what they should do. If a model prescribes what to do, then the implicit (and sometimes explicit) assumption is that use of the model allows the user to predict the effectiveness of the resulting instruction. Most model builders use theory or experience, or both, in building their model. This theory and/or experience leads them to predict that if the user does what the model prescribes, effective instruction will be produced.

Quality ISD models help the user to explain what is happening in an instructional setting. When users are able to explain what is happening and why, they are much more likely to have positive control over the instructional system. This control is important if the training system is not as effective as it should be because it allows the user to know which instructional system elements are the likely problems. This is particularly important in aircrew training development because of the expense of each of the major instructional components of the system (e.g. aircraft, simulators, training devices, etc.)

An instructional development model should take into account the limitations the external world imposes on the instructional system. The instructional designer is not free to create without regard to issues such as resources, regulations, laws, time, and student and instructor characteristics. Quality ISD models help the user be as creative as possible within the real-world restraints that surround any development efforts.

## PURPOSES OF ISD MODELS

Andrews and Goodson (1980) described four purposes of ISD models:

- Improving learning and instruction by means of the problem-solving and feedback characteristics of the systematic approach.
- Improving management of instructional design and development by means of systematic monitoring and control.
- Improving evaluation processes by means of the designated components and sequence of events, including the feedback and revision events, inherent in models of systematic instructional design.
- Testing or building learning and instructional theory by means of theory-based design within a model of systematic instructional design.

Instructional designers use models to help them fashion game plans for their development activities. The goal of building and using an instructional game plan is to bring planning and standardization to the instructional system, regardless of the content. Standardization of the development steps is helpful in a number of ways: revisions (via formative evaluations) become more systematic and congruent; the development activities take place in a more procedural fashion; media development is more efficient; a continuous product improvement atmosphere is created. Because aircrew training is so expensive, it is vital that this plan be well developed and based on sound instructional principles, before too many resources have been expended.

In addition, standardization forces the developers to concentrate more fully on specific tasks to be performed. Because the development tasks are clearly defined, the development team members are better able to understand what is required of them. Unfortunately, the ISD process has developed a number of "buzzwords" and jargon that are not necessarily interpreted the same way by all team members. *Needs assessment* is a good example. A quality ISD model's clear definitions of terms and concepts aids the efficiency of developers.

Another advantage of ISD standardization is the capability it provides for bringing to bear project management techniques, such as Total Quality Management's statistical process control, the Program Evaluation and Review Technique, or the Planning, Programming, and Budgeting System. These techniques help to assure that the appropriate project resources are brought to bear at the proper times.

Using a model does not guarantee that any of the four purposes described will actually be met. ISD will always have room for subjective interpretation and judgment. Any model can be misunderstood and misused. Also, Lowe and Schwen (1975) found that the documentation of instructional models often omits detailed accounts of how the development process works in various settings. However, even with these possibilities, we believe the chances of developing effective and efficient instruction are improved when a systems approach to instructional development is used.

## A SAMPLE ISD MODEL

The Air Force has revised its ISD model (Clemons & Bills, 1991; Golas & Shriver, 1991, US AirForce Manual 36-2234; Golas, Shriver, Bills, & Bowden, 1992). We believe this new model (the main steps and substeps are shown in the appendix) has all of the features that we recommend as a good example of a generic ISD model. We present here an overview of the model because it meets our criteria for quality and because its revision is based on years of application; much of this "lesson learning" came about in the aircrew training development domain. Consequently, we feel that many of the problems in the original model have been worked out. According to Golas et al. (1992), the Air Force found that the former ISD model worked fairly well for the development of more traditional means of producing instruction. However, the authors stated that recent innovations required a different approach:

> The currently published curriculum design and development procedures appropriate for producing print, audio-visual, or simple computer-assisted instruction provide little guidance for the course developer who is required to adapt the ISD process for interactive videodisc, Digital Video Interactive, external or embedded interactive computer-based instruction, intelligent tutoring systems, or performance support systems. (p. 655)

Golas et al. (1992) conducted a large, insightful survey of users of the original Air Force ISD model. This survey revealed that:

- The ISD model itself was not a problem. It was flexible and systematic.
- They did not like the various "boxes and arrows" in the top-level ISD models.
- They felt evaluation at all points in the ISD model should be increased.

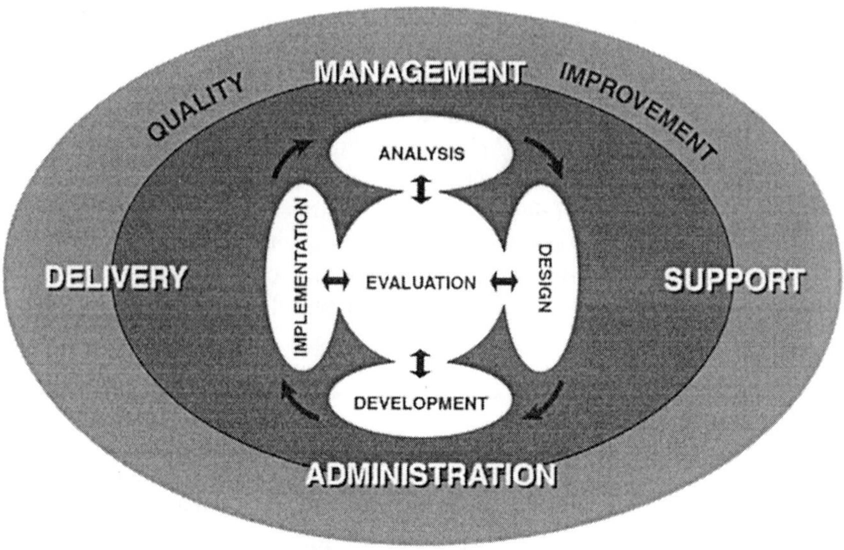


FIG. 2.1. Revised Air Force ISD model.

- The original ISD model was well suited for technical training, but they felt it lacked flexibility for other applications like aircrew training and formal education.
- Emphasis was on following the letter rather than the intent of the procedures.
- There were excessive paperwork, forms, and print report requirements.
- Information as presented in the ISD implementation regulations was too complex. The tables were too bulky, the figures were too complex, the language was not simple, there were not enough examples and samples. This complexity was especially troublesome for inexperienced and novice ISD developers.
- The survey respondents felt the affective and cognitive domains were not well addressed in the ISD model.

Based on the survey findings and a thorough review of recent instructional theories, the following top-level model was produced.

Golas et al. (1992) stated the following about the revised model:

> The revised Air Force ISD model includes the basic phased approach of the ISD process, which is analysis, design, development and implementation. The process is in context of the training or instructional system functions, including management, administration, support, evaluation and quality improve-

> ment. Quality improvement encircles the model, permeating the entire process with a constant quality mindset. Evaluation is internal, providing feedback to the instructional system throughout each phase of the process. (p. 658)

## ISD UTILITY GUIDELINES

Once an ISD model has been selected, there are a number of considerations that are important to ponder as that model is applied to a particular instructional problem. Our experience with the literature of ISD, and with applying ISD models to the development of aircrew instruction, led us to develop a number of utilization guidelines.

1. A major mistake is to follow the ISD model too closely when there may be logical reasons to deviate from the model's prescriptions. There may be development situations where it is justified to skip a model's step or perhaps add a new step. (Although it is possible to take shortcuts, novice developers are urged to do so cautiously. ISD steps can be skipped or rearranged only after considerable aircrew training development experience.)
2. We recommend a time each week where the project's managers, and perhaps the whole ISD team, get together to discuss their experiences with the model that is being used. If changes or deviations are warranted, they should be identified and agreed to by all.
3. The aircrew training developers must be properly trained in how to use the ISD model that has been selected or developed. For example, the military found problems with its early ISD efforts because the manuals for the Interservice Procedures for Instructional System Development were often simply given to service members without proper training in how to implement the model. Many, if not most, ISD models only describe what to do, not necessarily how to do it.
4. Whether you have had to train your ISD developers from scratch or you are fortunate enough to have an experienced team of developers, it is always wise to make sure the entire team understands the model that is to be used. This is especially true for large ISD teams, such as those that work on the development of aircrew training. There are two approaches that you might consider to increase the chances that all team members have a good understanding of the ISD model. The first approach is to have all of the team members work together on producing a small piece of instruction (e.g., training the procedures needed to operate a radar mode on-board the aircraft). All team members would walk through each ISD step. Another approach is to let team members switch positions dur-

ing the course of a project. (Because some team members obviously were hired to perform specific functions for which they have expertise, the managers have to be careful with this approach.)

5. A key feature of any ISD model is the requirement to continually review steps of the model that have already been completed. There is usually a great temptation for an ISD manager and team to ignore the iterative requirement of ISD. Pressure to push on with the project in order to meet deadlines and avoid what is perceived to be possible wasteful resource use can be severe. The revision process is often the first casualty of such pressures. However, the real casualty may be an ineffective instructional product and hence the skill levels of the aircrew trainees who are ultimately shortchanged. For obvious reasons, there is usually great pressure to field the training system before a new aircraft is delivered. We observed a number of instances where this pressure caused unwise shortcuts in the ISD process.

6. The task that is addressed least often by the ISD models is costing. Only 35% of the models Andrews and Goodson (1980) reviewed gave any details about the cost of producing instruction using the model being described. Yet, costing problems can cause more problems in the ISD process than any other ISD task. The up-front costs for most ISD projects are almost always more expensive than any other method for producing instruction. We believe the ultimate product of a systems approach will save money over the life cycle of the instruction that is produced. We urge the reader to make sure costs are realistically estimated before the effort begins. Some models that address costs are: Banathy (1968), Bishop (1976), Briggs & Wager (1982), Friesen (1973), Gagne and Briggs (1979), Kaufman (1988), and Lee (1975).

## CONCLUSION

The ISD approach has served the aircrew training community well. A large percentage of the aircrew training materials used today were designed using an ISD approach. In this chapter, we presented some of the benefits and pitfalls that instructional developers can expect when they adopt a systems approach for aircrew training. There are many quality models of instructional design. We discussed and gave references for those we particularly like. There are also many models that we feel are incomplete in some ways. We urge readers to be cautious when choosing an existing model—caveat emptor!

It is important that ISD users understand that applying a systems model should not limit the teaching strategy choices that can be applied to an instructional setting. Joyce and Weil (1985) and Reigeluth and Stein

(1983) described a number of teaching models. A good ISD model helps the user decide which teaching model best fits the instructional situation. After the "analyze" phase is completed and the "design" and "development" phases are begun, the user is free to consider a whole host of different teaching strategy approaches. The systems approach specifies only a series of logical steps that lead a developer to make a wise strategy choice. It does not require the application of any limited set of strategies for how to conduct instruction. Rather it specifies "what" should be done to get to the "how" decision. If designers want to use teaching strategies that might be considered unusual or nontraditional, ISD does not interfere with that choice. ISD requires the instructional developer to make sure the teaching strategy meets the learner's requirements for gaining skills and knowledge. However, not surprisingly, we prefer the current Air Force instructional design model and recommend it to our readers.

## REFERENCES

Andrews, D. H., & Goodson, L. A. (1980). A comparative analysis of models of instructional design. *Journal of Instructional Development*, Summer, 2–16.

Banathy, B. H. (1968). *Instructional systems*. Belmont, CA: Lear Sieglar, Inc./ Fearon.

Bishop, L. J. (1976). *Staff development and instructional improvement: Plans and procedures*. Boston: Allyn & Bacon.

Briggs, L. J., & Wager, W. (1982). *Handbook of procedures for the design of instruction* (2nd ed.). Englewood Cliffs, NJ: Educational Technology Publications.

Clemons, L., & Bills, C. (1991). Revision of the air force instructional systems development. *Proceedings of the Society for Applied Learning Technology Multimedia Conference*, Orlando, FL. 42–50.

Friesen, P. A. (1973). *Designing instruction: A systematic or "systems" approach using programmed instruction as a model*. Santa Monica, CA: Miller Publishing.

Gagne, R. M., & Briggs, L. J. (1979). *Principles of instructional design*. New York: Holt, Rinehart & Winston.

Golas, K., & Shriver, S. (1991). *Revision of the air force instructional system development process, baseline analysis report*. San Antonio, TX: Southwest Research Institute.

Golas, K. C., Shriver, S., Bills, C. G., & Bowden, P. (1992, November). Update of the U.S. Air Force Instructional Systems Development (ISD) Process. *Proceedings of the 14th Interservice/Industry Training Systems and Education Conference*. American Defense Preparedness Association, San Antonio, TX.

Hayman, J. L. Jr. (1974, May). *The systems approach and education*. The Educational Forum, 491–501.

Joyce, B., & Weil, M. (1985). *Models of Teaching* (3rd ed.). Englewood Cliffs, NJ: Prentice-Hall.

Kaufman, R. (1988). Needs assessment: A menu. *Educational Technology*, July, 21–23.

Lee, B. N. (1975). Instructional system development (ISD)—An Air Force way of life. *Journal of Educational Technology Systems, 4*(1), 33–41.

Lowe, A. J., & Schwen, T. M. (1975). The documentation of instructional development. *Educational Technology*, June, 42–47.

Reigeluth, C. M. (1983). Instructional design: What is it and why is it? In C. M. Reigeluth (Ed.), *Instructional-design theories and models: An overview of their current status* (pp. 3–36). Hillsdale, NJ: Lawrence Erlbaum Associates.

Reigeluth, C. M., & Stein, F. S. (1983). The elaboration theory of instruction. In C. M. Reigeluth (Ed.), *Instructional-design theories and models: An overview of their current status* (pp. 335–381). Hillsdale, NJ: Lawrence Erlbaum Associates.

Reiser, R. G. (1992). What do instructional designers actually do? An initial investigation of expert practice. *Performance Improvement Quarterly, 5*(2), 65–86.

Spector, J. M. (1992, November). Intelligent frameworks for courseware design. *Proceedings of the 14th Interservice/Industry Training Systems and Education Conference*. American Defense Preparedness Association, San Antonio, TX.

U.S. Air Force Manual 36-2234. (1993). *Instructional System Development*. Department of the Air Force, Headquarters U.S. Air Force, Washington, DC 20330-5000.

## SUGGESTED READING LIST

Banathy, B. H. (1991). *Systems design of education: A journey to create the future*. Englewood Cliffs, NJ: Educational Technology Publications.

Banathy, B. H. (1992). *A systems view of education: Concepts and principles for effective practice*. Englewood Cliffs, NJ: Educational Technology Publications.

Branson, R. K. (1978). The interservice procedures for instructional systems development. *Educational Technology*, March, pp. 11–14.

Branson, R. K., & Grow, G. (1987). Instructional systems development. In R. M. Gagne (Ed.), *Instructional technology: Foundations* (pp. 397–428). Hillsdale, NJ: Lawrence Erlbaum Associates.

Case, R., & Bereiter, C. (1984). From behaviorism to cognitive behaviorism to cognitive development: Steps in the evolution of instructional design. *Instructional Design, 13*, 141–158.

Dick, W., & Carey L. (1990). *The systematic design of instruction* (3rd ed.). Glenview, IL: Scott, Foresman.

Forsyth, I., Jolliffe, A., & Stevens, D. (1995). *The complete guide to teaching a course: Practical strategies for teachers, lecturers and trainers*. London: Kogan Page.

Gropper, G. L. (1977). On gaining acceptance for instructional design in a university setting. *Educational Technology*, December, 7–12.

Jones, B. F., Palinscar, A. S., Ogle, D. S., & Carr, E. G. (1987). *Strategic teaching and learning: Cognitive instruction in the content areas*. Alexandria, VA: Association for Supervision and Curriculum Development.

Kemp, J. E., Morrison, G. R., & Ross, S. M. (1994). *Designing effective instruction*. Upper Saddle River, NJ: Merrill.

Mager, R. F., & Pipe, P. (1974). *Criterion-referenced Instruction: Analysis, Design, and Implementation*. Los Altos Hills, CA: Mager Associates.

Merrill, M. D. (1983). Component display theory. In C. M. Reigeluth (Ed.), *Instructional-design theories and models: An overview of their current status* (pp. 279–334). Hillsdale, NJ: Lawrence Erlbaum Associates.

Merrienboer, J. G., Jelsma, O., & Paas, F. G. (1992). Training for reflective expertise: A four-component instructional design model for complex cognitive skills. *Educational Technology Research and Development, 40*(2), 23–43.

Olsen, J. R., & Bass, V. B. (1982). The application of performance technology in the military: 1960–1980. *Performance and Instruction, 21*(6), 32–36.

Reigeluth, C. M., Banathy, B. H., & Olson, J. R. (1993). *Comprehensive systems design: a new educational technology*. New York: Springer.

Richey, R. (1992). *Designing instruction for the adult learner: Systemic training, theory and practice*. London: Kogan Page.

Rothwell, W. J., & Cookson. P. S. (1997). *Beyond instruction: Comprehensive program planning for business and education*. San Francisco, CA: Jossey-Bass.

Saettler, P. (1968). *A history of instructional technology*. New York: McGraw-Hill.

Scanland, W. (1974). *The instructional systems development function—How to get it done*. Paper presented at the National Society for Performance and Instruction National Conference, Miami, FL (ERIC Document Reproduction Service No. ED 095 839).

Tennyson, R. D. (1990). Cognitive learning theory linked to instructional technology. In M. R. Somonson (Ed.), *Proceedings of the 1990 Convention of the Association for Educational Communications and Technology, 6*, 609–619.

Tennyson, R. D. (1994). *Automating instructional design, development, and delivery*. New York: Springer-Verlag.

Tennyson, R. D., & Michaels, M. (1991). *Foundations of educational technology: Past, present and future*. Englewood Cliffs, NJ: Educational Technology Publications.

U.S. Army Combat Arms Training Board. (1975). *Interservice procedures for instructional systems development*. Ft. Benning, GA: Author.

West, C. K., Farmer, J. A., & Wolff, P. M. (1991). *Instructional design: Implications from cognitive science*. Englewood Cliffs: NJ: Prentice-Hall.

## APPENDIX: AIR FORCE'S REVISED ISD MODEL

The following description of each ISD activity is taken from Air Force Manual 36-2234, Instructional System Development.

- Planning
  - Determine Instructional (Need) Requirement. Instructional (need) requirement is the lack of skills, knowledge, and attitudes personnel must have in order to perform an activity adequately.
  - Determine Instructional System Concepts. When you plan the instructional system concept, you make a preliminary estimate of the requirements and constraints to be considered in fielding the total instructional system.
  - Determine Resource Requirements and Constraints. The instructional concept will suggest major considerations from which you can estimate the requirements for equipment, facilities, manpower, and time, which along with funding, should translate into the ideal instruction for meeting the identified need.
  - Determine ISD Management plan. Plan may contain information such as: approval dates/signature, definition of the project, overall project responsibility, individual task responsibilities, milestones, identified constraints, support requirements.
  - Determine ISD Evaluation Plan. The plan may include, but is not limited to, the following information: identification of responsibilities including; taskings, scope, and purpose of the evaluation, how and when the evaluation activities are to be accomplished.

- Analysis
  - Occupational/Educational/Mission Analysis. Occupational (Job) analysis is the process of determining the duties and tasks that make up a job. It starts by breaking the job into the duties which comprise the major segments of the job. It continues by identifying the tasks that make up the duty. Educational analysis is the process of reviewing the educational requirements, developing educational goals, and developing statements of how to achieve the goals. Mission analysis is the process of reviewing mission requirements, developing collective task statements, and arranging the collective tasks in a hierarchical relationship.
  - Task Analysis. The process of breaking a task down to identify the: component steps of a task; sequence of those steps; condition under which the task will be performed such as tools, equipment, and materials required to perform the task.
  - Learning Analysis. The process of analyzing the tasks to be taught to establish learning outcomes in terms of types of learning involved and level of learning desired.
  - Resource Analysis. The resource analysis results in an estimate of resource requirements for the instructional system for both development and operation.
  - Target Audience Analysis. Examples of the data produced are: range of aptitudes, previous background and experiences, previous education, interests, physical characteristics.
  - Update ISD Evaluation Plan
  - Update ISD Management Plan

- Design
  - Develop Objectives. An objective is a precise statement of the performance a student is expected to be able to demonstrate, the condition under which the behavior is to be exhibited, and the minimum standard of acceptable performance.
  - Develop Tests. To ensure that tests adequately measure the objectives they support, the performance required in the test must match the performance required in the objective.
  - Review Existing Materials. . . . one of the first tasks is to determine if materials already exist that will support the objectives. It is possible that some of the material found during the review may not totally satisfy the need. In this case, don't hesitate to modify the materials to "fit" the need.

- Design Instructional Plan.
  - Select Instructional Method. The procedure or process used to attain an objective(s). Examples of instructional methods are: lecture, demonstration, and self-study.
  - Select Media. Media is the means, instrument, or material used to communicate information. In other words, it is a means used to give information to the students.
  - Consider Types of Learning. Examples of types of learning are; forming associates, forming chains, making discriminations, making classifications, using rules, problem solving, performing skilled motor acts, exhibiting desirable attitudes and opinions.
  - Determine Instructional Strategies. Several key learning concepts must be considered when determining instructional strategies. They are: active student participation, student feedback, planned pacing.
  - Design Instructional Activities. Effective and efficient instruction depends on how well the instruction has been sequenced and sized. There are several methods of sequencing instruction. The methods are: job performance order, psychological order and logical order. Job performance order sequences the instruction in the order in which the tasks and subtasks are performed on the job. Psychological order sequences the instructional content based on the ease of learning. Logical order usually sequences instructional activities from simple to complex or from the known-to-unknown.
- Develop Implementation Plan. Plans may include, but not be limited to, information such as: identification information, parameters of instructional system, type of instruction, and instructional content.
- Design Instructional Information Management System. The purpose of an automated instructional information management system is to enable those responsible for managing the various aspects of the instructional system to better manage system information in real time.
- Update ISD Evaluation Plan
- Update ISD Management Plan

■ Development

- Prepare Plan of Instruction/Course Syllabus. The Plan of Instruction or syllabus serves as the overall plan for conducting instruction in a particular course. The Plan or syllabus include such information as:

statement of objectives, instructional hours and approximate allocations of those hours to objectives, equipment utilization, instructor guidance, lesson plans.

- Produce Instructional Materials. Instructional materials refer to printed or other media intended to convey events of instruction or communicate information to the students. Examples are: print-based materials, transparencies, slide/tape, audio/video tapes, computer-based training.
- Install Instructional Management System. During the installation and testing of an instructional information management system, there will likely be documents that guide installation and checklists to test system operation.
- Update ISD Evaluation Plan. To ensure that the plan remains an effective evaluation tool throughout the life cycle of the project, it needs to be updated periodically, especially after each phase of ISD.
- Update Management Plan.
- Validate Instruction. Validation assesses the effectiveness of instruction while it is being developed with the intention of improving it. It is a process of repetitive cycles of development, tryouts, and revisions until evidence shows that the instruction is effective.
  - Develop Validation Plan. Provides instructional designers and instructors with a roadmap for validating the instruction. The plan may include: validation procedures, who may conduct the validation, validation schedules, sources and how the results will be documented.
  - Conduct Internal Reviews. Purpose is to verify the accuracy of the instructional materials as they are developed in order to identify inaccuracies and weaknesses in the materials so they can be corrected. Instructional materials to be reviewed include, but are not limited to: objectives; test items; storyboards/scripts; audiovisual materials such as slides, films, videotapes, transparencies; job aids.
  - Conduct Individual Tryouts. The instruction and materials should be tried out on several students to add validity and reliability to the data collected during the tryouts.
  - Conduct Small-Group Tryouts. The purpose of conducting small-group tryouts is to determine if the instruction and materials work under conditions approximating the actual teaching–learning activity.
  - Conduct Operational (Field) Tryouts. This evaluation activity is conducted under formal operating conditions by an instructor.

Field tryouts of instruction may vary from a single block or module of instruction to an entire course. The instruction to be validated will depend largely on whether it is a new course or a block or two of an existing course that has been revised.

- Finalize Instructional Materials. During finalization, ensure that all of the last minute changes are made to the instructional materials and they are ready for use. Ensure that the materials are; current and accurate, complete, ready to use in the teaching-learning activity.

■ Implementation

- Implement System Functions. System functions provide the structure that supports, operates, and maintains the system.
  - Management Function. The basic activities of management are:
    - planning for the design, development, delivery, support, operation, and maintenance of the instructional system.
    - organizing the resources.
    - coordinating instructional system operation and support.
    - evaluating the effectiveness and efficiency of each element in the instructional system.
    - reporting status and progress of the development of instruction or operation of the instructional system.
  - Administration Function. Provide documents, maintain records, administering student support, scheduling resources, monitoring resources.
  - Delivery Function. The means by which instruction is provided to the students.
  - Support Function. Those long-range, as well as day-to-day tasks performed in order to implement, operate, and maintain an instructional system. Examples are: maintaining equipment and facilities; supplying components for the equipment and materials for the courses; and providing services such as engineering, visuals, and publications.
- Conduct Instruction. At this point in the ISD process, you have made the final check of each component in the instructional system to ensure everything is ready for implementation of the instructional system.
- Operational Evaluation. The continuous process that assesses how well course graduates are meeting the established job performance requirements. Operational evaluation is the process of gath-

ering and analyzing internal and external feedback data to ensure the system continues to effectively and cost-efficiently produce graduates who meet established requirements. It is a quality improvement activity.

  - Internal Evaluation. The acquisition and analysis of internal feedback and management data, such as: test data, student critiques, instructor comments, data correlation from within the instructional system.
  - External Evaluation. The process of gathering and analyzing data from outside the instructional environment in order to determine how well recent graduates are meeting job performance requirements. External evaluations rely on data gathered from such methods as: questionnaires, field visits, and job performance evaluation.

■ Evaluation

- Formative Evaluation. This evaluation is designed to collect data and information that is used to improve the activities and products of the ISD process while the system is still being developed.
  - Process Evaluation. Ensures the quality in the analysis, design, and development activities. It checks each activity against standards, or metrics, established during ISD project planning, to assure process quality, while continually seeking improvements within each activity.
  - Product Evaluation. During product evaluation the focus is again on quality. Products are measured against standards and metrics established in the planning stage of ISD to ensure quality. Two key activities are: validation and quality control.
- Summative Evaluation. Designed to collect data and information during the operational (field) tryouts in order to determine the "summed" effect of the instruction under operational conditions. The only activity within summative evaluation is the operational tryouts. Operational tryouts determine if the instructional system works under operational conditions, provide feedback from a large sample of target population, and identify possible implementation or operational problems.

Chapter 3

# User Involvement in Development of Aircrew Training Systems

Randall Paul Jensen
Garry Boyle
John Fuller
*Air Force Research Laboratory, Mesa, AZ*

Training systems for military aircrews are developed to help the end user (the military aviator) learn quicker and perform better in the aircraft. These training systems are traditionally developed by engineers, training experts, and a customer organization that represents the end user and oversees the development and purchase of the system. End users, on the other hand, are often left out of the development process and do not have the opportunity to participate in the design and development of the systems they are supposed to use. It is our experience that meaningful user involvement is often critical to the successful development of any training system. The most effective structure to support such an effort is an integrated development team in which the user is actively involved with engineers, training experts, and the customer.

The Air Force Research Laboratory was involved with the Air Force Air Combat Command, the Reserve Command, and other users in the development of compact and cost effective F-16 training devices for a number of years (Carroll & Andrews, 1998; Coffman, 1992; Defense Electronics, 1991; Edwards, 1993; Lavitt, 1998; Nordwall, 1993; The Officer, 1993). During this period, the benefits of strong user involvement in a successful training system were evident. This chapter introduces our concepts, recommendations, and principle-centered models for user involvement in developing effective and efficient aircrew training systems.

The complex nature of aircrew tasking requires a sophisticated approach to the design and development of training. The Instructional

Systems Design (ISD) model (discussed by Andrews & Thurman, chapter 2, this volume) provides a good framework for front-end analysis, design, development, implementation, and evaluation. When an ISD approach is used, it typically results in the creation of some type of instructional intervention, such as instructional lessons, as suggested by The Defense Acquisition University (1997), Gagne, Briggs, and Wager (1988), and Dick and Carey (1990), or educational and training programs, such as those discussed by Hannum and Hansen (1989) and Nadler (1982).

Although instructional lessons and programs are important to training systems, they may be unable to accommodate the full range of learning or performance support requirements. This is especially true of complex environments, such as aircrew training, where there is a need for a close association between designers, developers, and end users in order to identify critical areas of need. This close involvement with the end user extends to the development of interventions that are appropriate to meet identified learning and performance support needs. Indeed, this type of involvement with end users was cited as a key component in the success of the development of training systems that helped aircrews develop difficult air intercept skills (Gray & Edwards, 1991; Boyle & Edwards, 1992).

Many lessons were learned from involving end users in the development of interventions that were designed to help in the training of complex aircrew learning tasks. These interventions used emerging technologies to help meet critical training needs. When implemented in the actual user community, each of these interventions represented a significant reengineering of existing training systems (Hammer & Champy, 1993).

## THE TRADITIONAL MODEL FOR DEVELOPING SYSTEMS

The traditional method for procuring and developing a training system focuses on the creation of a rigid set of engineering specifications and requirements that prescribe what the solution should be. In the traditional procurement model, product or system specifications usually include a description of what it will look like, how it will be assembled, how it is supposed to perform, how much it will cost, when it will be delivered, etc. These specifications, however, merely provide a snapshot in time and function as a start point or vector, but do not constitute a vision. As changes occur (i.e., technology, training requirements, resources, priorities, personnel, leadership, etc.), those involved in the project are able to develop a greater vision and understanding of what should be produced in order to most appropriately meet identified needs. Frequently, the developing understanding about what is needed does not match the specifications that were produced to guide the project. This traditional procurement system, with long product development cycles and high cost, is impractical in today's environment. It virtually guarantees that a training system will be technically out of date and not effective in meeting current training needs by the time it is delivered (McClure, 1998).

Project specifications are usually administered through several agencies such as military operational commands or government acquisition agencies. The bureaucratic layers that become involved often result in confusion and misunderstandings that unnecessarily consume a great deal of time and money. The process for administering this traditional model generally follows several major steps that may require years of effort (shown in Fig. 3.1). Initially, the user recognizes that a problem exists in the training system. This problem is eventually communicated to military commanders as a statement of need (Step 1). The military command and government acquisition agency are identified as customers in that they are both involved in the allocation of funds. The military command reviews the statement of need and prioritizes it with other needs that have been

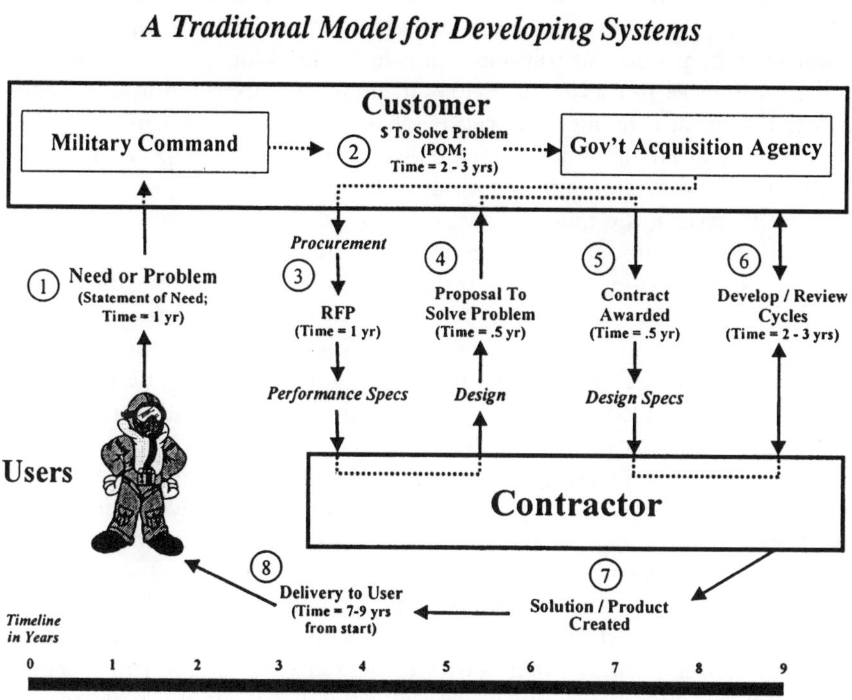


FIG. 3.1. A view of the major steps in a traditional training system development cycle.

identified. If the need is a high enough priority, funds are allocated to the government acquisition agency that will be responsible for meeting the need or solving the problem (Step 2). The government acquisition agency defines a desired procurement and releases a Request For Proposal (RFP) to potential contractors (Step 3). The RFP will also include a description of the desired performance specifications or requirements.

A considerable amount of time has already elapsed since the initial need was recognized by the user. It is very likely that there has already been a great deal of turnover in the user community, and the new users may be largely unaware of the original requirement for the project that is taking place to solve their problem. It is also likely that the nature of the user problem has changed to the extent that the RFP may not reflect the current needs. Nevertheless, the process continues.

When the contractors receive the RFP, they develop a proposed strategy to respond. The contractor's proposal will include a design or description of the product to be developed, a list of the people and skills that will be required for the work group, an estimate of costs (labor, material, travel, etc.), and a timeline estimate (Step 4). The customer (government acquisition agency) examines all of the proposals that are submitted and awards a contract (Step 5). The contract will include a detailed set of design specifications for the product or solution that is to be developed. Work is initiated and continues with a series of reviews to monitor what is being done and ensure that what is being produced is reflective of the specifications in the contract (Step 6). Note that due to the time that may have elapsed, it is very likely that the user community has again changed. This means that those who are to receive the final product or solution are further removed from the circumstances under which the project was initiated.

After being satisfied that the product meets the contract specifications, the finished product or solution is created and signed off by the customer (Step 7). The final step occurs when the completed product is delivered to the user (Step 8). Often, because there has been a disconnect between the user and the production system, the users are surprised and unprepared when the product or solution is delivered.

There are several characteristics of this traditional developmental system that make it difficult to produce a solution that effectively meets the needs of the user. First (as already mentioned), the user is either disconnected or only peripherally involved in the design and development process. Though users may have been involved in the initial identification of a problem or need, once that need becomes a formal procurement, they are often left out as a real player. If users are involved in the process, it is typically only for periodic reviews and not as an integral team member who is involved in day-to-day work and decisions. Second, the documentation and infrastructure that is created to ensure that the needs are met can actually work against success. The big problem is the *time* it takes to respond. Years may elapse between the time at which the initial problem is identified and when a solution is delivered. When the speed of technological change is considered, it is no surprise that the products or solutions that are delivered are usually too late and do not hit the bulls-eye in solving a problem (McClure, 1998).

In summary, a traditional work environment is not able to effectively solve user problems in a timely manner and may work against natural growth processes. Work that cascades linearly from one group to another requires specifications that are too rigid and inflexible. It inevitably results in solutions that are outdated and do not meet current needs by the time they are developed or implemented.

## THE INTEGRATED DEVELOPMENT TEAM

The alternative to the traditional development model is a teamwork approach that we call the Integrated Development Team (IDT). This team consists of the user, the customer, and the contractor (including the engineers and training experts). The IDT is the foundation on which successful new programs will be created and implemented. The IDT is based on the recognition that a fast-changing user environment requires the capability to rapidly assemble the critical players into flexible work groups. This team must be able to quickly accomplish the goal of providing an appropriate solution to meet the need(s) that have been identified. Each member of this team represents an important role in the development of a rational solution that is within budget, on time, performs as required, and meets the needs of the user community. Most importantly, the IDT includes the user as an integral part of the working group. The personnel who are the important members of an IDT, including their roles and responsibilities, are shown in Fig. 3.2.

The *user* must inspire the vision that initially establishes the need for the developmental program. This person represents the community that will utilize and benefit from the training—the people who need the device in order to solve a problem, perform better, or implement new technologies or procedures. The user knows what they want but they may not be able to describe it. Users help to establish the fidelity needed and help to find solutions to problems as they arise during the course of development.

The *customer* is the agency that provides the funding, along with the contracting agency that creates the Request For Proposal (RFP). This includes the military command and the government acquisition agency. The customer reviews and selects the proposals, awards the contract, conducts periodic reviews, and determines the final payment. Ultimately, the

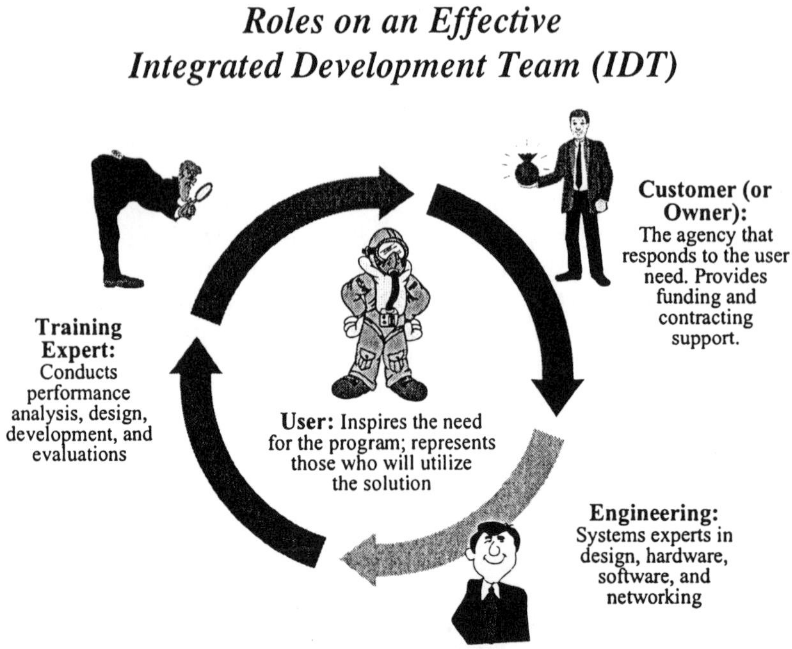


FIG. 3.2. Roles of the members of the IDT and how this fosters an iterative (cyclic or spiral) development process. Arrows represent iterative or "spiral" development cycles. As the team develops, members walk around the problem and view it from different perspectives. They see the problem as other members see it. As they gain understanding of the problem, they begin to understand each other and the contributions each can make.

customer is the "owner" because they are paying for the training system that is being created.

The *engineers* are the systems experts who will ultimately build the training system that is to be developed. Their expertise is in areas such as systems design, hardware, software, and networking. The engineer must figure out how to make the training system work within the technological capabilities and limitations and suggest alternatives based on costs and capabilities. Engineers are not expected to have the expertise to determine priorities for the project, nor do they necessarily have background in instructional or performance objectives.

The *training experts* focus on the performance improvement gains that are to be achieved by the system under development. Their role is to ensure that the solution is effective in helping the user achieve improved performance. The task of the training expert is to capture knowledge from the user and enable the user to put into words what the problem is and how it might be solved. This is accomplished by assessing needs, performing front-end analysis, identifying learning and performance objectives, designing solutions to meet identified needs, assisting in the development of solutions, and performing both formative and summative evaluations. The training expert has background in instructional design, knowledge engineering, and performance improvement technologies.

Past development strategies undervalued the utility of involving the end user as a team member. However, many training system acquisition and development professionals now believe that user involvement is critical to this process. For example, Dobbins and Donnelly (1998) recently performed a study of "critical factors" for program management and found that managers consistently reported "user involvement and support" as one of the most important factors for determining project success. The most important factor was ensuring that the project had a stable and adequate budget (p. 70). Similarly, Delano (1998) surveyed program managers to identify factors that contribute to success and found that "user involvement and input" helped more than any other source (pp. 37–38). He also observed that "Formation of integrated product teams that include the user and contractor serve to foster communication and a joint approach to identifying and solving problems" (p. 45).

All members of the IDT should be able to work in such a way that their natural capabilities, expertise, and leadership can emerge when and where appropriate. There should be no artificial constraints or prescriptions for how the roles are supposed to work at the detail level. Communication flows between all members of the team at all times. The product, or solution to the problem that is being developed, is called the *testbed* and represents an ability to rapidly create prototype solutions. All members of the IDT can work in an environment where a prototype is available and viewable and prompts feedback from all of the members of the team.

## CHARACTERISTICS OF THE INTEGRATED DEVELOPMENT TEAM

It is obvious that rigid engineering specifications are not inherently responsive to rapidly changing environments. Similarly, an effective IDT cannot be created based on a rigid model or set of prescriptions. The high performance IDT consists of users, customers, engineers, and training experts that can bring expertise and interest to the project and be able to respect each other's roles, responsibilities, and capabilities (Delano, 1998; Price & Valentine, 1992). The IDT should be an evolving team, capable of working in a dynamic environment that adopts the appropriate processes that are required for success. The elements that must be present in such a team include vision, adaptability, collaboration, appropriate controls, communication, involvement, synergy, responsiveness, managed expectations, patience, and motivation. When appropriately combined, these elements can forge a special kind of synergistic team in which the whole is truly greater than the sum of its parts. This means that the relationships that are created within the IDT actually become a part of the team itself (Covey, 1989). A team with this type of synergistic interactions will be capable of the innovation and creativity that is needed to solve difficult performance or learning problems.

The difference between the traditional model and the IDT can be understood by a comparison of how teamwork and communication are accomplished in each case. In the traditional model, information is usually provided because it is formally requested or required as part of the process (Fig. 3.1). The process is rigid, linear, predefined, and does not encourage the elements that promote success, such as dyadic interaction, collaboration, two-way communication, responsiveness, vision, adaptability, etc. The IDT, on the other hand, involves participants in a continual dialogue among all team members. Such a dialogue enables the user to clarify the problem and provide a vision for how it might be solved. The user-defined "vision" becomes a common focal point for all team members to refer to as they interpret what is to be done, analyze problems, and discuss ideas that may help solve the problem. The vision does not have an implied solution; rather, it should focus on the overall result. The vision enables the engineers to clarify the capabilities and limitations of technologies that may be used to solve the problem. It also enables the training expert to identify performance problems that may exist and then determine the goals and objectives for improving performance.

### IDT Development Cycle

The IDT development cycle contrasts sharply to the traditional model shown in Fig. 3.1. A vision guides the IDT, whereas engineering specifications guide the traditional development cycle. Specifications represent a single dimension slice in time that may form a good initial vector but does not convey a vision and tends to preordain a type of solution. A vision represents a multi-dimensional dynamic vector. It begins with a current understanding of the situation and the potential solutions and changes position and continues to evolve as members of the team gain a greater understanding of the problem and of the capabilities and limitations of each member of the team. A vision does not have an implied solution, but it does focus on the overall results that need to be achieved.

The vision that guides the IDT must be inspired by the user. In Fig. 3.2, the IDT is portrayed as an iterative model that functions as a spiral. It could also be viewed as a three-element molecule in an upward spiral of quality.

Each player on the team represents a single dimension and only when the major players communicate and understand each other's dimension does the image become three dimensional. The three elements that need to be tightly connected are the user, the engineers, and the training experts. This connection has to be made at the working level. The iterative cycle is a developing spiral where a vision emerges as people "walk around" the emerging solutions and keep talking to each other about what they see in relation to the vision they are trying to work toward.

The differences between the traditional approach and the IDT approach may be represented as similar to the difference between a monocular view that is only capable of portraying a two-dimensional image and a stereoscopic view, which can present images in three dimensions. This 3D view results in a clearer perspective that provides a product that is more in line with what the user wants and needs. Experience has shown that this team approach also shortens the development time by 30% or more, reducing the delivery time shown in Fig. 3.1, from 7–9 years, to 3–4 years.

### Flexibility

There will be no predefinition that will specify the solution, what it will look like, and how it will perform. The solution will literally emerge via the synergy created as the IDT coalesces into a high-performing team. Design ideas are implemented via working prototypes. The value of the prototype is that the members of the IDT are able to touch, view, and walk around their ideas with a tangible model. This has a tendency to jump-start new thoughts and ideas that would not have emerged without something that could be experienced. The danger of this type of iterative development is that ideas can run wild. There must be a means by which this wild flow of new ideas can be managed so that there may be continual progress. The critical factor to all the new ideas is how well they will meet the user needs within the defined framework of costs, time, and capabilities. Because the user is a member of the IDT, its role at this time is to ensure that this is properly managed.

This flexible, synergistic, nonlinear approach to managing the IDT project can be very effective during the initial phases. However, it will change when the project progresses to the production stage. Production requires a much more defined and process-driven approach that moves linearly from point A to point Z.

### Criteria for Success

A successful project enables the user to develop the criteria for success. The user is also the arbiter that balances cost and fidelity. All members of the IDT have absolute accountability at each of their three levels, and each takes ownership for the tradeoffs that must be made in order to ensure success. The engineers answer to the functional fidelity. The training expert answers to whether or not the solution supports improved performance. The user must ultimately be responsible for ensuring that the solution meets the real needs of the user community. The user is not involved unless they are at the working level. They must be "in the ring" and not on the outside looking in.

## AN EXAMPLE OF EFFECTIVE USER INVOLVEMENT IN AN IDT: THE AIR INTERCEPT TRAINER (AIT) PROJECT

An effective IDT was created when users, customers, engineers, production, and training experts were assembled at the Air Force Human Resource Laboratory at Williams Air Force Base in Arizona to work on the problem of improving methods to train F-16 pilots in the air-to-air environment. All members of the team worked together from the beginning to accomplish all the critical steps of the traditional model (shown in Fig. 3.1). This combined team was able to work rapidly in an interactive, spiral way that enabled all the steps to be accomplished concurrently rather than in the traditional serial method. As a member of the team, the customer ensured that the funding and contractual vehicles were in place and kept the users focused on the critical needs (Steps 2 and 5). The users were F-16 pilots who provided the initial identification of the problem and worked as subject-matter experts with the training experts to develop innovative solutions to the problems (Steps 1 and 4). The users and training experts also worked with the engineers and communicated a vision of the desired end results as well as a sense of urgency and need (Steps 3 and 6). Software and hardware engineers created a series of functional prototypes that all team members could try out and provide feedback on as they discussed strengths and weaknesses (Step 6). The group also included production personnel who thus had a window on research and development (Step 7). As a result, the production Air Intercept Trainer (AIT) quickly emerged through this process (Step 8). The AIT was able to meet identified needs and was successfully deployed to a number of sites around the country (Edwards, 1993; Gray & Edwards, 1991; Boyle & Edwards, 1992).

The project was initially funded by the Arizona Air National Guard to assist in their transition from flying the A-7 to the F-16. Most of the A-7 pilots had little or no experience in the air-to-air environment, and the project initially focused on how to train these fighter pilots in the switchology required to use the air-to-air radar. Many of those working on this project had never worked in this type of team environment before. Their initial inclination was to wonder where the specifications were to create the "switchology trainer." But such specifications did not exist. Instead, the user kept talking about the needs to be addressed. The training experts kept asking about the desired skills and performance to be supported and improved. The engineers kept asking for specifications.

As the group held meetings to discuss the problem, each representative tended to reflect his or her "view of the world." At first there was a lot of resistance because each individual tended to hold to the knowledge, processes, methodologies, etc. that he or she knew most about and was comfortable with. The other members of the team tended to use different vocabulary, methods, processes, etc., and these were something of a threat, at least initially. However, as the team gained experience with one another, there was also a growing awareness that each member did have something of value to contribute. What follows is a summary of some of the main lessons that were learned as the team members worked with one another.

### What Users and Training Experts Learned as They Worked Together

The users acted as subject matter experts (SME) for the training experts who spent time with them to analyze the nature of the tasks to be performed. The tasks that were identified for analysis included throttle and side-stick switchology, radar use and interpretation, intercept geometry, and general air-to-air situation awareness. This front-end analysis helped both the users and the training experts to better understand and articulate the primary learning and performance objectives. The users learned that the knowledge, skills, and capabilities that they take for granted are not easy to teach to others, and time must be spent to really focus on where the problems lie. The training experts gained an understanding of complex aircraft systems as well as an appreciation for the number of different tasks that pilots must be proficient in while flying the F-16. This enabled them to gain a much greater understanding of the concept of *task saturation* for fighter pilots.

The collaboration between users and training experts resulted in recommendations for a limited part-task trainer (instead of a full-up simulator), guidelines for training scenarios, user interface to access training scenarios, freeze capability (to analyze the situation while flying an intercept), and an interpretive display that portrayed a side view and a god's-eye view of the intercept (Gray & Edwards, 1991; Edwards, 1993).

### What Users and Engineers Learned as They Worked Together

The users also worked with electrical, hardware, and software engineers to determine what equipment to include and the functional characteristics of both hardware and software. The systems and equipment of interest included the throttle, side-stick, simulated heads-up display (HUD), the radar displays, and simulated flight characteristics. The users quickly realized how difficult it is to specify what one wants to have built in order to solve a problem. The engineers had to understand the constraints imposed on the project, which included the amount of money allocated to the project, timeline to completion, proposed configuration of the training device, and the system requirements for functions and fidelity (amount of realism).

The device that was developed was configured, in part, from the understanding that developed through the interaction between the users and the engineers. Because the users and engineers were also working with the training experts, they were aware of the performance and learning problems that needed to be solved or supported (Boyle & Edwards, 1992).

### What Engineers and Training Experts Learned as They Worked Together

As the training experts worked with the engineers, they learned that technology is not always available the way it is wanted. There are limits to what can be accomplished. There are also limits as to how well humans can make the best use of the available technologies. They learned that an effective use of technology requires a lot of thought and planning so it can provide solutions within the cost and timeline requirements. By working with the training experts, the engineers learned that an effective solution requires more than just a defined set of specifications. They learned that the most important part of any training solution is how usable it is. In other words, can a user approach this device with little training (on how to use it) and use it effectively?

## CONCLUSION

Aircrew training systems produced today continue to have serious problems. They are not delivered concurrently with the aircraft and systems they are supposed to train, development schedules continue to slip, and they do not meet the expectations or needs of the current users. Although the acquisition community is trying to improve development processes, infrastructure and organizational boundaries are tying them to the old linear processes and are preventing them from implementing a system in which the user is appropriately involved.

The U.S. Air Force Research Lab successfully applied these lessons in involving users in the development of our current trainer programs, in particular the F-16C Multi-Task Trainer (MTT) and the C-130H3 Unit-Level Trainer (ULT). Although these were relatively small programs, the trainers were delivered and maintained concurrently with the aircraft and have met with high user acceptance.

## REFERENCES

Affordable experience: MTT steals the show at mid-winter (1993, March). *The Officer, LXIX, Number 3*, 46–47.

AF Human Resources Directorate: Aircrew Training Research Division (1991, November). *Defense Electronics, 23*(11), 38–45.

Boyle, G. H., & Edwards, B. J. (1992, December). Low cost trainers: Lessons for the future. *Proceedings of the Interservice/Industry Training, Simulation and Education Conference*. Orlando, FL.

Coffman, R. (1992). Multi-task trainer: Futuristic in design—operational today. *Citizen Airman, XLIV*(5), 2–3.

Covey, S. R. (1989). *The 7 habits of highly effective people*. New York, N.Y. Simon & Schuster.

Defense Acquisition University (1997, November). *Guide to curriculum development, delivery and evaluation*. Fort Belvoir, VA: Author.

Delano, K. J. (1998). Identifying factors that contribute to program success. *Acquisition Review Quarterly, 5*(1), 35–47.

Dick, W., & Carey, L. (1990). *The systematic design of instruction* (3rd ed.). Tallahassee, FL: HarperCollins.

Dobbins, J. H., & Donelly, R. G. (1998). Summary research report on critical success factors in federal government program management. *Acquisition Review Quarterly, 5*(1), 61–79.

Edwards, B. J. (1993). *Lessons learned from the development and transition of the air intercept trainer (AIT)*. Armstrong Laboratory Human Resources Directorate, Tech. Rep. No. AL/HR-TR-1993-0157. Mesa, AZ.

Gagne, R. M., Briggs, L. J., & Wager, W. W. (1988). *Principles of instructional design* (3rd ed.). New York: Holt, Rinehart & Winston.

Gray, T. H., & Edwards, B. J. (1991). *Aircrew part-task training research and development in the 1980s: Lessons learned*. Armstrong Laboratory Human Resources Directorate, Tech. Rep. No. AL-TR-1991-0005. Williams Air Force Base, AZ.

Hammer, M., & Champy, J. (1993). *Reengineering the corporation: A manifesto for business revolution*. New York: HarperCollins.

Hannum, W., & Hansen, C. (1989). *Instructional systems development in large organizations*. Englewood Cliffs, NJ: Educational Technology Publications.

Lavitt, M. O. (1998). Need for greater value drives innovation—USAF/distributed mission training. *Aviation Week & Space Technology, 148*(16), 72.

McClure, J. T. (1998, December) Working smarter: Training and simulation developed through collaboration [CD-ROM]. *Proceedings of the Interservice/Industry Training, Simulation and Education Conference*. Orlando, FL.

Nadler, L. (1982). *Designing training programs, the critical events model*. Reading, MA: Addison-Wesley.

Nordwall, B. D. (1993). Air Force Lab develops deployable simulator. *Aviation Week & Space Technology, 138*(6), 47.

Price, J. E., & Valentine, M. M. (1992, November). Results-oriented program management as a leadership management model. *Program Manager*, 12–16.

Chapter 4

# Establishing Aircrew Competencies: A Comprehensive Approach for Identifying CRM Training Needs

Clint A. Bowers
Florian Jentsch
Eduardo Salas
*University of Central Florida*

It has been just over a decade since the concept of Cockpit (later: Crew) Resource Management (CRM)[1] was introduced. During this time, CRM has gained almost universal acceptance as a useful addition to aircrew training programs (Wiener, Kanki, & Helmreich, 1993). In fact, CRM training programs are in place in almost every major air operation in the world. As a result, the acceptance of CRM training is, arguably, one the most apparent successes of applied psychology. Yet, we believe that it would be unwise to consider the problem of CRM training as solved. Although the promise of CRM training is well established, the practice of CRM training has recently come under attack. Some reports have suggested that CRM training has done little to impact aviation safety (General Accounting Office, 1997). Other reports have suggested that CRM training can have unintended negative consequences (Helmreich & Wilhelm, 1989). As crew coordination skills are now being evaluated as a qualification and certification requirement for many active pilots (Federal Aviation Administration, 1998; Joint Aviation Authorities, 1998), the pressure to demonstrate the efficacy of these programs is increasing (Mayes, 1998). Thus, optimizing the effectiveness of CRM training has become the next great challenge for aviation psychologists.

[1]The military often refers to this as Aircrew Coordination Training (ACT). For the purposes of this chapter, we use CRM to describe both military and commercial coordination training programs.

In looking at the evolution of CRM training, one apparent weakness has been the lack of a systematic approach to the design and delivery of this instructional strategy. For the most part, CRM training has emphasized the development of positive attitudes towards teamwork. This focus can be effective, but it also has inherent drawbacks (cf. Salas, Cannon-Bowers, Rhodenizer, & Bowers, 1999). In particular, links between attitudes and behaviors are usually not direct. Instead, attitudes work through knowledge and skill to result in change of behaviors.

Consequently, there has been a shift recently to training specific crewmember *skills* as part of CRM training. This is the foundation of the U.S. Navy's Aircrew Coordination Training (ACT) program (Prince & Salas, 1993), and it is emphasized in the FAA's Advanced Qualification Program (FAA, 1998). Although we believe that a skill-based approach offers distinct advantages for CRM training, it, too, imposes some challenges. Most notably, it implies an ability to identify necessary skills. Training designers and developers have hitherto utilized a number of approaches, yielding a variety of skills and skill groupings. For the most part, the resulting answers seem to have been effective in creating skills that guided the development of subsequent training. However, a fragmentation of skill labels between organizations and platforms exists.

Despite this progress, however, there is reason to be concerned about a training approach that is limited solely to skills. In the first place, an exclusive emphasis on skills might cause training developers to overlook the attitude component that has been the foundation of CRM. We do not believe that the recent emphasis on skills was intended to overlook the importance of appropriate attitudes in crew performance. However, this might be an unintentional consequence of a training approach exclusively focused on skills. Secondly, we posit that one might possess a skill without a full understanding of the appropriate use of that skill (Jentsch, Irvin, Bowers, Smith-Jentsch, & Salas, 1997). In fact, pilot assertiveness skill training has come under repeated attack on this account: Practitioners often provide the feedback that junior crewmembers may have learned how to be assertive (i.e., the assertiveness skill component), but lack the judgment about when to apply this skill (Jentsch, Milham, & Bowers, 1997).

Finally, there is a problem regarding the specificity of skills. Some training developers believe that aviation crew skills are universal. Consequently, these trainers say that training programs for different platforms, missions, and environments should all train identical skills. The practice of CRM training, however, has demonstrated clearly that this assumption cannot always be made, as CRM training programs transferred from one organization or platform to another have repeatedly failed. As we describe later, we believe that the causes for such failures can frequently be found in inappropriate transfer, where the assumption of universal applicability of training content could not be made.

Although this situation is clearly problematic, it is equally a truism that overly specific approaches are not necessarily better. Indeed, we have noticed with some chagrin that skill-based training programs for identical platforms flying identical missions (e.g., major airlines) are often vastly different across organizations. In these cases, the differences between programs seem to be the result of philosophical differences among the training developers, rather than the result of the adaptation of training goals, contents, and methods from one organizational setting to another. This is inefficient because it precludes sharing of resources and pooling of training outcome data. Given the concerns described in the preceding paragraphs, there is a need for a technique for identifying training needs that considers and integrates the many separate components that are required for effective crew performance. The purpose of this chapter is to propose one potential technique.

One recent framework that appears to hold some promise in this regard is the construct of *team competencies* (Cannon-Bowers, Tannenbaum, Salas, & Volpe , 1995). According to these authors, team competencies are a constellation of knowledge, skills, and attitudes (KSAs) that foster effective team performance. Furthermore, Cannon-Bowers and her colleagues suggest that there are different types of team competencies. Some competencies, it is suggested, are required in every crew situation. These *team-generic* competencies are required by individual crew members regardless of mission or organization. They include skills such as communication and leadership. Conversely, it is suggested that other competencies are *team-specific*. These skills are only meaningful in specific team situations. These include issues such as knowledge of other team members abilities and team cohesion. They vary from crew to crew.

Cannon-Bowers and her colleagues also suggest that team competencies are influenced by task characteristics. Some competencies are *task-generic*. That is, certain competencies are required in all crew situations. Other competencies are *task-specific*. For example, certain competencies that are required in a search-and-rescue task may not be relevant to transport pilots.

The competency framework described by Cannon-Bowers and her colleagues appears to hold some promise for identifying training needs for aviation crews, allowing the identification of competencies that are generic, and therefore required in every training situation. However, it also allows the flexibility to identify specific skills with which to tailor training to the specific needs of various organizations, platforms, and, indeed, specific teams. Thus far, however, the competency framework has been used only to describe a relatively general universe of skills. To be truly useful, the framework must be utilized in specific training contexts and yield guidance for team training that would not have emerged using traditional approaches. The remainder of this chapter attempts a test of the competency framework for aviation crew skills training.

## CREW COMPETENCIES FOR AVIATION

One might argue that aircrews are simply one type of work team and, consequently, require competencies similar to those articulated for any other team. However, we suggest that there are a number differences between aircrews and other types of work teams. For example, aircrews are predominantly dyads (and less often triads). Thus, they are typically much smaller than many other work collectives (e.g., quality circles). Also, at least the cockpit crew that forms the nucleus of any aircrew is typically made up of similarly trained individuals, rather than of the diverse group of team members that characterizes interdisciplinary teams. As a result, cockpit crews are less subject to some communication problems that relate to diverse task-related vocabularies. Furthermore, aircrews typically have relatively clear role assignments: The captain is legally responsible for the conduct of the flight, and the first officer and (if present) the flight engineer provide support in reaching the objective of a safe and efficient flight. The skills associated with creating role clarity that are central to some teams, such as self-directed work teams, are thus much less relevant for aircrews. Finally, unlike some other teams, aircrews have a relatively confined range of tasks. Many of the necessary procedures are well specified and are documented in the form of handbooks, guidelines, and checklists. Consequently, there is a much smaller need for aircrews to possess the competencies other teams need for creating and organizing their work processes.

By the same token, aviation possesses certain characteristics that might not emerge from an analysis of general work teams. For example, many aircrews are much more fluid in composition than traditional work groups and teams. In commercial aviation in particular, crewmembers might change daily or even several times per day. Thus, there is a need to establish competencies that allow this rapid transition to take place. One must also account for a number of different missions. A framework of aircrew competencies must support such diverse goals as the transportation of passengers and cargo, reconnaissance and surveillance, air combat, search and rescue, and so forth.

Finally, aircrew competencies must consider differences in equipment within the same task environment. For example, some crews perform air transport tasks in highly automated aircraft, whereas others fly traditional aircraft. In fact, research has demonstrated that the level of automation is likely to influence the required competencies (Bowers, Thornton, Braun, &

Salas, 1998). For example, crewmember attitudes toward social interaction can interact with the layout of automated systems to produce vastly different performance outcomes (Hernandez, Dockery, Bowers, & Jentsch, 1995): Crewmembers with high sociability seem to perform better when using automated systems that free them to communicate directly with the other crewmember. Conversely, those with lower sociability scores thrive when using automated systems that remove the opportunity for direct crewmember interaction by requiring considerable interaction with a machine. In subsequent sections, we examine whether the competency approach allows for these important differences.

## Generic Competencies

For the most part, modern CRM training programs have taken a crew- and task-generic approach. Most training programs suggest that the skills they impart are required in all flights, regardless of where one flies and with whom. Thus, there has been substantially more effort dedicated to this cluster than others. However, not all skills within this cluster might have been explored.

***Generic Knowledge Competencies.*** Although contemporary CRM training programs do, of course, impart knowledge, this has not always been an articulated goal of the training. Rather, this knowledge has been communicated largely as a foundation for the targeted skills and attitudes. Also, knowledge of CRM principles is typically not evaluated at the end of training. Instead, it is assumed that successful use of the skills targeted by the training program is also a valid indication that the underlying knowledge has been learned. For example, crewmembers should be familiar with the terms and principles that compose CRM. Consequently, most training programs include a considerable lecture portion designed to provide this knowledge. It should be noted, however, that this training does not necessarily result in learning. A recent paper by Prince (1998, personal communication), in fact, suggests that a substantial number of CRM graduates could not define concepts such as *situational awareness* or *human factors*.

Although knowledge competencies are often not directly targeted in training and evaluation, there is reason to believe that knowledge competencies are an important component of crew coordination. In fact, in a recent study, we were able to directly link trainee knowledge of specific CRM training components to performance on the related crew coordination skills (Jentsch, 1997): Those trainees who knew more about assertiveness and metacognition were able to perform significantly better in situations that required assertiveness skills and situational judgment, respectively, than those who had less knowledge in these areas. Furthermore, theories of learning, such as those motivated by constructivism, emphasize the importance of specific knowledge competencies for successful performance. Thus, a more focused consideration of necessary knowledge competencies might be beneficial.

One construct that may be particularly useful in extending the concept of generic knowledge competencies for aircrews is that of a *shared mental model.* Cannon-Bowers and her colleagues (Cannon-Bowers, Salas, & Converse, 1993) suggested that crew members require a shared mental model of the environment in which they operate (be it equipment, personnel, the external world, etc.). Although this might seem elementary, the advent of automated systems requires that crewmembers be familiar with a large number of systems that are likely to vary from airplane to airplane. Thus, it is important that teams possess the skill to quickly acquire and communicate their understanding of the equipment with which they will perform their tasks.

Similarly, it has been suggested that crews also require interpositional knowledge (Volpe, Salas, Cannon-Bowers, & Spector, 1996). That is, the crew must possess accurate knowledge about the roles and responsibilities of his or her fellow crew members. In aviation terms, this means that all crew members must understand the general responsibilities of the captain, first officer, flight engineer, crew chief, and so forth.

Given these important knowledge competencies, future CRM training should focus on knowledge development and *evaluation.* Recently, research has made great advances in the evaluation and quantification of mental models and their similarity to referent (i.e., expert) models. We see a need here to transfer what has been learned in research to the field, so that practitioners can develop and use valid and reliable measures of aircrew knowledge competencies (Blickensderfer, Cannon-Bowers, & Salas, 1997a).

***Generic Skill Competencies.*** As described earlier, the emphasis on training generic team skills has many advantages for crew training. However, an unfortunate consequence of this emphasis has been a proliferation of skill taxonomies. Many of these categorizations are redundant, contradictory, or simply confusing. So as not to muddy the water any further, we utilize the summary clusters described by Cannon-Bowers and her colleagues (Cannon-Bowers et al., 1995). They are shown in Table 4.1. As is indicated in the table, the skill clusters range from clearly proceduralized skills such as *communication* to less clearly definable skills such as *shared situational awareness.*

Many of the skill dimensions described in the table are represented in skill-based training programs, albeit often under different labels. For example, skills in the cluster *performance monitoring/feedback* may be labeled in different organizations as *workload management, mission analysis, resource management*, or *compensatory behavior*. Although all of these skill labels may describe similar components of the same skill competency, the subtle dif-

TABLE 4.1
Teamwork Skill Dimensions

| | |
|---|---|
| Adaptability | The process by which a team is able to use information gathered from the task environment to adjust strategies through the use of compensatory behavior and reallocation of intrateam resources. |
| Shared situational awareness | The process by which team members develop compatible mental models of the team's internal and external environment; includes skill at arriving at a common understanding of the situation and applying appropriate task strategies. |
| Performance monitoring and feedback | The ability of team members to give, seek, and receive task-clarifying feedback; includes the ability to accurately monitor the performance of teammates, provide constructive feedback regarding errors, and offer advice for improving performance. |
| Interpersonal relations | The ability to optimize the quality of team members' interactions through resolution of dissent, utilization of cooperative behaviors, or use of motivational reinforcing statements. |
| Coordination | The process by which team resources, activities, and responses are organized to ensure that tasks are integrated, synchronized, and completed within established temporal constraints. |
| Leadership/Team management | The ability to direct and coordinate the activities of other team members, assess their performance, assign tasks, motivate team members, plan and organize, and establish a positive atmosphere. |
| Communication | The process by which information is clearly and accurately exchanged between two or more team members in the prescribed manner and with proper terminology; the ability to clarify or acknowledge the receipt of information. |
| Decision making | The ability to gather and integrate information, use sound judgment, identify alternatives, select the best solution, and evaluate the consequences. |

*Note.* Adapted from Cannon-Bowers, Tannenbaum, Salas, & Volpe (1995).

ferences between the various labels make valid comparisons regarding their training difficult. It is therefore important to identify the overlap between these skill components.

The identification of generic team skills clearly represents an area in which aviation psychology has made good progress. Training for some skill areas, such as assertiveness, is well developed and has been scientifically validated (e.g., Jentsch, 1997; Smith, 1994). In other areas, however, this is not necessarily the case. Specifically for skills involving more cognitive components (e.g., decision making, workload management, self-monitoring), validated training is lacking. Only in the last few years have scientific studies been conducted that showed that these generic skills can indeed be trained effectively (e.g., Cohen, Freeman, & Wolf, 1996; Jentsch, 1997). It is therefore incumbent on the research and practice of CRM to validate the effectiveness of generic team skill competencies across a wider range.

***Generic Attitude Competencies.*** At some level, one might argue that the majority of effort spent thus far in CRM development has been targeted towards attitudes, especially in the early CRM training programs. Indeed, a common target of CRM training has been the individualistic, "macho" captain who attempts to solve all problems by himself. It was this general theme that appears to have resonated in the airlines and contributed to the large acceptance of the CRM principle (and its occasional outright rejection). Yet, despite the many programs designed to improve crew attitudes, there is a surprising lack of scientifically validated evidence regarding what attitudes are associated with improved performance.

Clearly, a positive attitude toward teamwork appears to be an important element in crew performance. The work of Helmreich and his colleagues demonstrated that positive attitudes toward teamwork behaviors such as coordination and communication are significant predictors of crew processes (Gregorich, Helmreich, & Wilhelm, 1990) and can discriminate between effective and ineffective managers (Helmreich, Foushee, Benson, & Russini, 1986). Presumably, the existence of positive attitudes toward teamwork behaviors is a necessary condition for skill-based training to be effective. That is, if crews are not inclined to use a behavior, the existence of that behavior in their repertoire will not be effective.

Recently, however, researchers have discussed other generic attitudes that might also be important determinants of crew behavior. One construct that appears to hold some promise in this regard is *collective orientation* (Driskell & Salas, 1992). This attitude describes crew members' inclination to value the crew's goals above their own personal needs. It also encompasses the ability to consider the actions of one's fellow crew members during task performance. Data have indicated that crews with collectively oriented members might be better able to take advantage of crew resources. Selection and/or training toward collective orientation may thus be a useful approach that could augment the traditional attitude component of CRM training programs.

### Team-Specific Competencies

Up until this point, the reader might be thinking that the competency framework to identify crew coordination skills is not different from other approaches that have traditionally been used. We are not wholly inclined to disagree. To date, CRM programs (and team training programs in general) have focused almost exclusively on the generic skill and attitude competencies just described. This might have been due to constraints related to training costs, but it might also be a reflection of the then state-of-the art in team training research. The result of the work done in this area thus far has been a collection of guidelines and suggestions that training developers, presumably, modify to meet the specific needs of any given trainee (or not).

Clearly, however, the next challenge for team training scientists is to approach the problem of *how* to tailor this guidance to the various, yet generally very specific, and often very different situations faced by the trainees in the real world. We believe that considering this challenge in light of team- and task-specific competencies might offer some advantage. We begin with a brief overview of a number of team- and task-specific team competencies, before introducing a model of how these specific competencies can be used in the next generation of CRM training. As we show, the model posits that the real trick of making CRM training work in the cockpit is the appropriate *application* of generic team knowledges, skills, and attitudes in light of the specific team members and the operational environment one is facing. Thus, this model suggests that each situation and each team member require a little more or a little less in terms of the application of a generic skill competency, such as team leadership.

In describing the differences and commonalties of generic and specific competencies, we depart somewhat from the Cannon-Bowers et al. description of team- and task-specific competencies as additional sets of knowledge, skills, and abilities. Although we believe that there are indeed team-specific knowledges and attitudes (see the following), we believe that the same cannot be said for skill competencies. Rather, we consider team-specific skill competencies to be the result of filtering an individual's total repertoire of generic competencies through team- and task-specific knowledge and attitudes. In other words, we believe that team-specific skills are largely a subset of generic competencies that have been selected for use by the individual crewmember based on what he or she knows about the task and team members. Thus, team-specific competencies include not only appropriate knowledge, skills, and attitudes, but also the ability to select skills based on knowledge. We attempt to describe these components more clearly next.

***Team-Specific Knowledge Competencies.*** We argue that knowledge of one's fellow crewmembers is a critical component in selecting the coordination skills to be used with any given team. This is consistent with the assertion by Cannon-Bowers and her colleagues that an accurate mental model of one's teammates is a precursor to effective team performance (Cannon-Bowers et al., 1995). This knowledge includes information such as strengths and weaknesses, preferences, tendencies, and so forth. Recent research with air traffic controllers showed that team members who had better knowledge of the other team members received higher team performance scores than those who had less of this knowledge (Smith-Jentsch, Kraiger, Cannon-Bowers, & Salas, 1998). As a result, helping team members to learn more about one another, and to learn it more quickly, might be a useful way to improve team coordination.

A second necessary knowledge competency is related to knowledge of team roles and norms. Over and above the generic knowledge competency of interpositional knowledge that we already described, there is a need for team members to have a specific understanding of one another's specific role in the crew. This understanding is specific to the team members one is working with (and to the specific tasks the team sets out to perform). It is essential so that crew members are able to utilize *implicit coordination* (Kleinman & Serfaty, 1989). Implicit coordination is the ability for crewmembers to coordinate their actions in the absence of explicit requests for assistance. It is an important cornerstone of successful CRM training, especially in novel situations where creative problem solving is required by the crew.

***Team-Specific Attitude Competencies.*** Although most of the generic attitudes targeted in contemporary CRM training reside within the individual, there are also distinct "team-level" attitudes that are specific to a given combination of team members. One such attitude is *team cohesion*. Team cohesion can be defined as the degree to which members value being a member of the specific team. Team cohesion has been associated with performance gains in sports teams (Carron, Widmeyer, & Brawley, 1988) and laboratory tasks (Zaccaro, Gualitieri, & Minionis, 1995). However, it should be noted that team cohesion has also been associated with negative performance outcomes (Roby, 1953). It is likely that the effect of cohesion on crew performance is dependent on other task conditions (cf. Mullen, Anthony, Salas, & Driskell, 1994). Given the promise of this variable, there is a need to examine it in the aircrew context to determine whether it can help bring about improved performance.

A second related attitude construct that is specific to a given collection of team members is *collective efficacy*. Whereas team cohesion describes the degree to which team members "stick together," the construct of collective efficacy refers to the shared belief in a team's (or crew's) ability to meet specific task demands (Bandura, 1986). It has been suggested that collective efficacy is positively correlated with team performance (Travillian, Baker, & Cannon-Bowers, 1992). There is a need, however, to establish this relationship in aircrews and to develop techniques to increase crew efficacy.

A third important attitudinal variable is *organizational climate*. This construct refers to the degree to which coordination behaviors are accepted and valued within the organization. Several studies have demonstrated that the perceived climate can impact crewmembers' willingness to utilize coordination behaviors (Smith, 1994). Team leaders especially (i.e., captains) can change a generally positive organizational climate into a negative team climate in as little as a few minutes, for example, during a preflight briefing. In addition to establishing organizational norms that emphasize teamwork and CRM during all operations, it is therefore important to also impart positive attitudes towards teamwork onto team leaders and to give them the skills to establish similarly supportive team climates in each new team. This can be achieved through training team leaders in briefing (Tannenbaum, Smith-Jentsch, & Behson, in press) and debriefing skills (Smith-Jentsch, Zeisig, Acton, & McPherson, 1998).

***Team-Specific Skills.*** We suggest that the general skill dimensions described in the generic competencies section are also relevant here. However, team-generic skills represent a selected subset of one's overall skill repertoire. The challenge for the crewmember is to develop an appropriate knowledge structure to represent his or her coworkers. Once this structure is in place, the crewmember selects coordination skills designed to optimize coordination within that specific crew. For example, the assertiveness skills one might choose to deal with a temperamental captain would be different from those one might use in a more open atmosphere.

In this approach, the development of knowledge structures is of critical importance in determining appropriate behaviors. Surprisingly, however, few programs include much consideration of learning about one's fellow crewmembers. In large part, this knowledge is a result of experience with specific individuals. Although this approach might be acceptable for crews that remain fixed for some time, rapidly changing crews require a technique for quickly learning about one another. Some researchers have attempted to respond to this need by including personal information in briefings. However, other techniques should be explored.

It should be noted that another important skill for crewmembers is to select appropriate skills based on knowledge of one's coworkers. This skill has not been addressed directly in the CRM community. However, one training intervention that might be helpful is team self-correction (Blickensderfer, Cannon-Bowers, & Salas, 1997b). This technique asks crew members to replay their performance in a crew-oriented debriefing session. Members discuss their situation assessment, expectations, and actions. They also provide feedback to one another about how to improve crew processes. In this fashion, crewmembers are not only given feedback about how their actions were perceived, but also how to improve them. Data have indicated that team self-correction can have a positive impact on team performance (Blickensderfer et al., 1997b).

## Task-Specific Competencies

Like team-specific competencies, task-specific competencies allow the crewmembers to select generic skills and apply them to specific situations. Thus, generic competencies can also be thought of as "filtered" through task-specific knowledge and attitudes. We discuss these individual components next.

***Task-Specific Knowledge.*** Cannon-Bowers and her colleagues suggested that a shared knowledge of the mission is essential to the development of shared mental models. Thus, a shared understanding of the mission, goals, and resources is required as a knowledge competency. This knowledge should also include consideration of recent and anticipated equipment problems, enemy activity (when appropriate), and preferred alternates and other contingencies.

Developing a better shared understanding of the task parameters in a specific situation is one component of the Advanced CRM (ACRM) research program sponsored by the FAA (Seamster, Boehm-Davis, Holt, & Schultz, 1998). In ACRM, crewmembers explicitly brief "bottom lines" and alternate plans of action before each major segment of the flight (e.g., takeoff, approach). ACRM thus represents a promising approach to creating a shared understanding of the task.

***Task-Specific Attitudes.*** It is not unlikely that crewmembers possess attitudes about coordination as it relates to a specific task. For example, there are experienced pilots who, although believing in CRM in general, deny its importance for single-seat aircraft (even if they fly in formation). This task-specific attitude might impact performance even more drastically than general attitudes towards teamwork.

A second attitude that might be important is pretask expectations. In some situations, crewmembers might have a well articulated vision of the mission's outcome. This belief, if unchallenged, might impact crew processes. This type of expectancy was implicated in the crew's failure in the USS Vincennes incident where the ship's combat information center crew was faced with multiple threats, both on the surface and in the air. Their expectation of an attack from the air led them to selectively attend to information and resulted in the misidentification of an Iranian civil aircraft as a hostile warplane. In the subsequent shoot-down of the Iranian Airbus, 290 people died. As a result, training researchers need to identify how to identify which task-specific attitudes crews need to develop for each task and how they can be improved.

***Task-Specific Skills.*** As with team-specific skills, we posit that task-specific skills are those selected by the crewmember to optimize performance on a specific task. In other words, team members must select the generic coordination skills previously described and apply them as appropriate to the situation. For example, in some cases it might be helpful for the pilot-not-flying to use his or her communication skills and call out altitudes in 1,000 foot increments. On the other hand, a crewmember that calls out altitudes on the entire flight would simply be annoying.

The challenge, then, is to train crew embers to select coordination behaviors. In essence, crew members must be able to mentally simulate the consequences of various behaviors and select the optimum strategy for choosing generic coordination skills. It is in this assessment process where task novices seem to be at particular risk (Jentsch, Irvin, & Bowers, 1997). We found, for example, that prospective first officers tend to overestimate the time available to find a solution and bring it to the captain's attention. Thus, one intervention might be to provide task-specific knowledge to allow novices to "calibrate" their expectations more accurately.

## EMPLOYING TEAM COMPETENCIES IN CRM: A MODEL

Our goal in starting this chapter was to arrive at an approach that would yield the most effective and efficient collection of knowledge, skills, and attitudes for use in CRM training. A successful approach would allow training developers to make use of the resources that exist in the field while also identifying areas that must be specifically tailored in training.

Our point of departure was the competencies framework described by Cannon-Bowers and her colleagues. This approach appeared to offer the advantage of considering attitudes and knowledge as well as specific skills. Furthermore, the authors of this framework considered both generic and specific behaviors, an approach that seems particularly relevant for aviation.

However, we deviated from the original competencies framework in attempting to apply this work to CRM training. The result of the present effort yielded the model illustrated in Fig. 4.1. As indicated by the model, we hypothesize two distinct sets of teamwork skills. One set of skills is referred to as *generic teamwork skill dimensions*. These dimensions are considered to be task- and team-generic. That is, these skills are required in almost every crew situation, regardless of aircrew and flight task. These include broad classes of team behaviors, namely communication, coordination, interpersonal relations, and leadership–team management.

We also suggest a second class of skills that we refer to as *application skills*. These are skill dimensions that allow crews to tailor their responses to be appropriate to their crew and the situation. In other words, these skills allow crews to use their knowledge of the situation and of their crewmates in order to select behaviors from the generic dimensions for use. This clus-

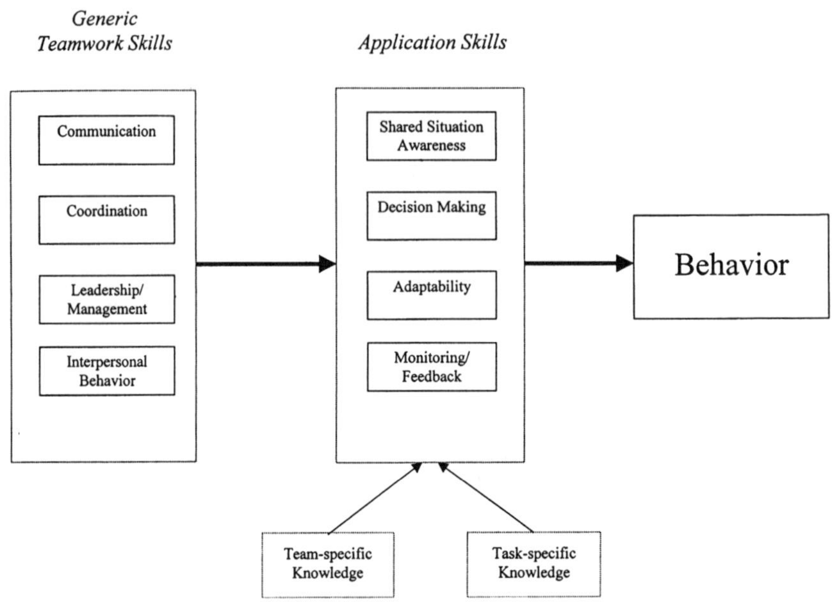


FIG. 4.1. A framework of aircrew competencies.

ter can be likened to a "valve" that allows only specific generic behaviors, in a specific amount, at a specific time, to be selected and applied.

## IMPLICATIONS FOR TEAM TRAINING

The model we described poses several implications for training. First, we suggest that generic teamwork skills can be developed and trained across a fairly broad spectrum. That is, identical training can be provided to a fairly wide grouping of airframes and missions. Indeed, given greater cooperation among training organizations, training of these generic skills could be applied at a very early point in pilot training and should be applicable in most environments the trainee may subsequently face.

At the same time, the training of these application skills poses new challenges for training developers. First, these skills are dependent on thorough and accurate knowledge of the task. Thus, training developers must pay special attention to the technical training, the experience, and the briefings that bring this knowledge about. In fact, the Advanced CRM program (Seamster, Boehm-Davis, Holt, & Schultz, 1998) puts special emphasis on including more task-specific information in briefings to improve this kind of knowledge.

Similarly, application skills are also dependent on knowledge of one's fellow crewmembers. Developing this type of knowledge has been largely overlooked by the aviation training community. Indeed, it is generally believed that this kind of knowledge can be built only by experience. There is a need, however, for techniques that allow crews to quickly learn about one another's preferences, especially in those fleets where crews change frequently. Perhaps briefings can be altered to facilitate this type of knowledge. The previously described team self-correction approach might also offer some promise in this regard.

Once the necessary knowledge is accumulated, there is a need to train crew members to use the application skills. These skills are somewhat more difficult to train than the generic skills. For example, there is a need for more practice and feedback for these skills to be acquired. These skills might also require specialized instructional strategies for optimum skill acquisition. Candidate technologies might include metacognitive training (Jentsch, 1997) and cue-based training (Barnett, Jentsch, & Bowers, 1997).

Despite the challenges, however, we believe that this approach offers a method that allows training developers to take advantage of existing material while optimizing training for the specific user. We hope that this chapter thus assists organizations in developing the next generation of CRM training.

## ACKNOWLEDGMENTS

This chapter was in part made possible by support from the NAWCTSD/FAA/UCF Partnership for Aviation Team Training Research. The opinions expressed herein are those of the authors and do not necessarily represent the position of the organizations with whom they are affiliated.

## REFERENCES

Bandura, A. (1986). *Social foundations of thought and action: A social-cognitive theory*. Englewood Cliffs, NJ: Prentice-Hall.

Barnett, J., Jentsch, F., & Bowers, C. (1997). *Team practice in aviation: Guided practice and cueing*. Technical report submitted to the Naval Air Warfare Center Training Systems Division. Orlando: University of Central Florida.

Blickensderfer, E., Cannon-Bowers, J., & Salas, E. (1997a). Theoretical bases for team self-correction: Fostering shared mental models. In M. Beyerlein, D. Johnson, & S. Beyerlein (Eds.), *Advances in interdisciplinary studies in work teams series* (Vol. IV, pp. 249–279). Greenwich, CT: JAI Press.

Blickensderfer, E., Cannon-Bowers, J., & Salas, E. (1997b). *Training teams to self-correct: An empirical investigation*. Paper presented at the 12th Annual Meeting of the Society of Industrial and Organizational Psychology, St. Louis, MO.

Bowers, C. A., Thornton, C., Braun, C., Morgan, B. B. Jr., & Salas, E. (1998). Automation, task difficulty, and aircrew performance. *Military Psychology, 10*(4), 259–274.

Cannon-Bowers, J. A., Salas, E., & Converse, S. A. (1993). Shared mental models in expert team decision making. In N. J. Castellan, Jr. (Ed.), *Current issues in individual and group decision making* (pp. 221–246). Hillsdale, NJ: Lawrence Erlbaum Associates.

Cannon-Bowers, J. A., Tannenbaum, S. I., Salas, E., & Volpe, C. E. (1995). Defining competencies and establishing team training requirements. In R. A. Guzzo & E. Salas (Eds.), *Team effectiveness and decision making in organizations* (pp. 333–380). San Francisco: Jossey-Bass.

Carron, A. V., Widmeyer, W. N., & Brawley, L. R. (1988). Group cohesion and individual adherence to physical activity. *Journal of Sport & Exercise Psychology, 10,* 127–138.

Cohen, M., Freedman, J., & Wolf, S. (1996). Metarecognition in time-stressed decision-making: Recognizing, critiquing, and correcting. *Human Factors, 38*(2), 206–219.

Driskell, J. E., & Salas, E. (1992). Can you study real teams in contrived settings? The value of small-group research to understanding teams. In R. W. Swezey & E. Salas (Eds.), *Teams: Their training and performance* (pp. 101–124). Norwood, NJ: Ablex.

Federal Aviation Administration (FAA). (1998). *1998 FAR Flight Crew: Federal aviation regulations for flight crew (ASA-98-FAR-FC).* Newcastle, WA: Aviation Supplies & Academics.

General Accounting Office (GAO). (1997). *Human Factors—FAA's guidance and oversight of pilot crew resource management training can be improved* (GAO-RCED-98-7). Washington, DC: GAO.

Gregorich, S. E., Helmreich, R. L., & Wilhelm, J. A. (1990). The structure of cockpit management attitudes. *Journal of Applied Psychology, 75,* 682–690.

Helmreich, R. L., Foushee, H. C., Benson, R., & Russini, R. (1986). Cockpit management attitudes: Exploring the attitude-performance linkage. *Aviation, Space, and Environmental Medicine, 57,* 1198–1200.

Helmreich, R. L., & Wilhelm, J. A. (1989). When training boomerangs: Negative consequences associated with cockpit resource management training. In R. S. Jensen (Ed.), *Proceedings of the Fifth Symposium on Aviation Psychology* (pp. 692–697). Columbus, OH: Ohio State University Press.

Hernandez, E., Dockery, K., Bowers, C. A, & Jentsch, F. (1995, August). *Pilot personality, automation, and performance.* Paper presented at the 103rd Annual Meeting of the American Psychological Association, New York.

Jentsch, F. (1997). *Metacognitive training for junior team members: Solving the "copilot's catch-22."* Unpublished doctoral dissertation, University of Central Florida, Orlando, FL.

Jentsch, F., Irvin, J., & Bowers, C. (1997). *Development and testing of a diagnostic and training tool for pilot-not-flying backup behaviors.* Technical report submitted to the Naval Air Warfare Center Training Systems Division and the Office of the Chief Advisor for Human Factors at the Federal Aviation Administration. Orlando, FL: University of Central Florida.

Jentsch, F., Irvin, J., Bowers, C., Smith-Jentsch, K., & Salas, E. (1997, April). *What determines whether observers recognize targeted behaviors in videotaped modeling displays?* Paper presented at the annual meeting of the Society for Industrial/Organizational Psychology, St. Louis, Missouri.

Jentsch, F., Milham, L., & Bowers, C. (1997). *Identifying critical training needs for junior first officers.* Technical report submitted to the Naval Air Warfare Center Training Systems Division and the Office of the Chief Advisor for Human Factors at the Federal Aviation Administration. Orlando, FL: University of Central Florida.

Joint Aviation Authorities (JAA). (1998). *Flight Crew Licensing (FCL).* Brussels, Belgium: JAA.

Kleinman, D. L., & Serfaty, D. (1989, April). *Team performance assessment in distributed decision making.* Paper presented at the Simulation and Training Research Symposium on Interactive Networked Simulation for Training, Orlando, FL.

Mayes, B. (1998, May). *ALPA AQP chair comments.* Presented at the Air Transport Association (ATA) Advanced Qualification Program (AQP) Conference, Houston, Texas.

Mullen, B., Anthony, T., Salas, E., & Driskell, J. E. (1994). Group cohesiveness and quality of decision making: An integration of tests of the groupthink hypothesis. *Small Group Research, 25,* 189–204.

Prince, C., & Salas, E. (1993). Training and research for teamwork in the military aircrew. In E. Wiener, B. Kanki, & R. Helmreich (Eds.), *Cockpit resource management* (pp. 337–366). San Diego, CA: Academic Press.

Roby, T. B. (1953). *Problems of rational group assembly exemplified in the medium bomber crew* (Technical Report No. 53-18). San Antonio, TX: Lackland Air Force Base, Human Resource Research Center.

Salas, E., Cannon-Bowers, J. A., Rhodenizer, L., & Bowers, C. A. (1999). Training in organizations: Myths, misconceptions, and mistaken assumptions. In G. Ferris (Ed.), *Research in personnel and human resource management* (Vol. 17, pp. 123–161). Stamford, CT: JAI Press.

Seamster, T., Boehm-Davis, D., Holt, B., & Schultz, K. (1998). *Advanced crew resource management (ACRM): A training handbook.* Washington, DC: Federal Aviation Administration.

Smith, K. (1994) *Narrowing the gap between performance and potential: The effects of team climate on the transfer of assertiveness training.* Unpublished doctoral dissertation, University of South Florida, Tampa, FL.

Smith-Jentsch, K., Kraiger, K., Cannon-Bowers, J., & Salas, E. (1998). *A data-driven model of precursors to teamwork: Implications for training.* Paper presented at the 13th Annual Conference of the Society of Industrial and Organizational Psychology, Dallas, TX.

Smith-Jentsch, K., Zeisig, R., Acton, B., & McPherson, J. (1998). Team dimensional training. In J. A. Cannon-Bowers & E. Salas (Eds.), *Making decisions under stress: Implications for individual and team training* (pp. 271–297). Washington, DC: American Psychological Association.

Tannenbaum, S., Smith-Jentsch, K., & Behson, S. (1998). Training team leaders to facilitate team learning and performance. In J.A. Cannon-Bowers & E. Salas (Eds.), *Making decisions under stress: Implications for indiviudal and team training* (pp. 247–270). Washington, DC: American Psychological Association.

Travillian, K., Baker, C. V., & Cannon-Bowers, J. A. (1992, March). *Correlates of self and collective efficacy with team functioning.* Paper presented at the annual meeting of the Southeastern Psychological Association, Knoxville, TN.

Volpe, C. E., Cannon-Bowers, J. A., Salas, E., & Spector, P. E. (1996). The impact of cross-training on team functioning: An empirical investigation. *Human Factors, 38*(1), 87–100.

Wiener, E., Kanki, B., & Helmreich, R. (Eds.) (1993). *Cockpit Resource Management.* San Diego, CA: Academic Press.

Zaccaro, S. J., Gualtieri, J., & Minionis, D. (1995). Task cohesion as a facilitator of team decision making under temporal urgency. *Military Psychology,* 7, 77–93.

Chapter 5

# Using Virtual Reality for Training

Richard Allen Thurman
*Air Force Research Lab, Mesa, AZ*

Trish Russo
*Raytheon Training and Services*

Virtual reality (VR) is the popular term for an interactive experience in which an individual perceives and engages a synthetic (i.e., simulated via computer) environment by means of a particular set of multisensory human–computer interface devices and interacts with synthetic objects in that environment as if they were real (U.S. Congress, Office of Technology Assessment, 1994). The concept of VR and the wave of development accompanying it are creating new forms of simulation that may lead to fundamental improvements in aircrew simulation-based training (Thurman & Mattoon, 1994). Research in VR can lead to an increased understanding of modeling and simulation, as well as multisensory human–computer interface design and development (Burg et al., 1991).

There are a great number of outstanding research issues that need to be addressed before we will have true confidence in the use of VR in training. Although it is possible to utilize previous research (for an overview see Durlach & Mavor, 1995), particularly training simulator studies (Orlansky, Dahlman, Hammon, Metzko, & Taylor, 1994; Thurman, 1993), as a basis for designing and developing VR-based training, our overall understanding of the requirements of VR-based training systems is still limited. It would be wonderful if this chapter contained a strong review of VR aircrew training research, but unfortunately, there are very few training studies that utilize this medium (for exceptions see Dede, Salzman, Loftin, & Sprague, in press; Winn et al., 1997; O'Neil et al., chapter 6, this volume). Instead, we must be content with a review of VR-based training research issues that still need to be addressed.

Most published VR research and development articles deal only with the technological aspects of the field. This is a necessary part of the process of developing a training media, but it is not sufficient. It is also necessary to examine the interaction of the medium with the user and the methods one can use to assess training performance within a VR system. In this chapter, we discuss these aspects of VR-based training research. First we review those human factors issues that should be addressed to achieve effective and efficient training. Then we discuss issues of enhancing VR-based training through training effectiveness research.

## HUMAN FACTORS AND VR RESEARCH

Although most VR technologies emerged from simulation (and particularly flight simulator) research and development, the field is still relatively young, and many of the human factors aspects have yet to be worked out. As Stone (1993) pointed out, the early history behind the emergence of VR is short, incredibly intense, and still in its infancy. Human factors issues need to be addressed particularly because there is a real need to focus the development of the "technologies of immersion" (Stonum, 1998) or how the user's personal viewpoint is surrounded or immersed inside a virtual world. Some of these issues include visual, audio, and haptic interfaces. A *haptic interface* is a force feedback device that allows the user to touch, feel, or manipulate a simulated three-dimensional object in a virtual world.

The term *immersion* refers to how the user's personal viewpoint is surrounded or immersed inside a virtual or synthetic world. An immersive virtual reality system is equipped with a virtual reality display system such as a head-mounted display, a Boom or multiple large projection display such as a Cave Automatic Virtual Environment (CAVE; DeFanti, Sandin, & Cruz-Neira, 1993), which projects the visual display. An immersive virtual reality system may also have some type of interactive touch–feedback system that is often referred to as *haptic* or *tactile* and is a relatively new research area. Few training studies have been conducted on haptics and kinesthetic senses.

### Visual Interface

The human visual system contains substantial capabilities for processing and resolving information such as scene detail, color, texture, and depth (Gibson, 1989; Hendrix & Barfield, 1995). Complex networks of neurons in both the eye and the brain are devoted to visual processing, and in humans, vision is often considered to be the most dominant sense (James

& Caird, 1995; Proctor & Van Zandt, 1994). There is even some evidence suggesting that human cognition is often centered around vision, with people using visual imagery as mediating representations for thought (Kosslyn, 1994). Listed next are several of the most salient human factors issues concerning VR and vision.

***Necessity of Immersion via Head-Mounted Displays.*** Is an immersive head-mounted display (HMD) necessary, or are any of the other VR delivery systems usable as training media? In most cases, the use of an HMD-based system is justified on the grounds that a strong sense of immersion is needed. HMDs supply this sense because of the tight coupling between stereoscopic vision (the ability to see objects in three dimensions) and the capability of changing the visual image in response to head movement (in the case of head-tracked HMDs). In some instances, the capability to head track may be more critical for 3D vision than for stereoscopic vision. For example, Ware, Arthur, and Booth (1993) showed that subjects performed best when using a head-tracked, stereoscopic HMD in a task requiring 3D immersion, but when they uncoupled the two, a head-tracked monoscopic display was superior to a stereoscopic display that did not use head tracking.

***Fidelity of Visual Representation.*** Some training applications may require very accurate, photographic quality, visual displays, others may get by with simple, cartoon-like images. Just where do the trade-offs lie when trying to determine the amount of visual fidelity needed for a given training task? The simulation literature is rife with recommendations to create the highest fidelity system possible (cf. Ray, 1996). However, the justification for doing so is seldom based on empirical data. More research is needed in this area. (See also Alessi, chapter 10, this volume, for an in-depth discussion of the fidelity issue.)

***Visual Acuity Versus Image Resolution.*** Some training tasks are highly dependent on a trainee's *visual acuity* (the ability to see two stimuli separated in space). Others are less so. Research must determine just how sharp or how crisp the visual image needs to be for a given task. For example, in the world of fighter pilot training, students will probably need display systems that supply them with the equivalent of a 20/20 visual display. However, many commercial HMDs cannot supply a visual resolution under 20/70. What are the training impacts when display resolution does not coincide with human visual acuity? (McKenna & Zeltser, 1992)

***Field of View.*** The angle that an eye, or pair of eyes, can see either horizontally vertically is called *field of view*. The total horizontal field of view in humans is about 270° (with eye movement). Vertical field of view is typically more than 120°. Although it is not necessary to have a completely unobstructed field of view to achieve a sense of immersion within the virtual environment, there is some evidence that the trainee's field of view must be more than 90° (Youngblut, Johnston, Nash, Wienclaw, & Will, 1996). Research should continue to determine what field of view is needed to perform a given task effectively.

***Binocular Vision.*** To achieve stereoscopic displays, a degree of binocular overlap is required. In HMDs, partial overlapping of each eye's field of view may be used to produce a wide field of view. However, there is very little information on how much overlap is needed. It does appear that binocular overlap is very important and must be tuned to suit the training application. There is a real need for perceptual and human performance studies to determine if partial overlap solutions are appropriate (Kalawsky, 1993).

***Update Rate Versus Image Fusion.*** Displays that work by presenting a rapid series of images (motion pictures, television, or computer displays) need to introduce successive frames at or above a certain rate. This update rate is critical for image fusion to take place (Davson, 1989). Below this rate, images are perceived as a series of flickering still pictures and do not fuse. However, the update rate at which fusion occurs can vary according to several parameters, including image luminance and size. Although image flicker is undesirable (it makes perception more difficult, is annoying, and disturbs the sense of immersion), there is typically a trade-off between scene complexity and update rate—the more complex the visual scene, the slower the image generator is able to update it. In some training applications it may be preferable to tolerate flicker at least some of the time to gain increased scene complexity (Youngblut et al., 1996), but generally this will not be the case. Research needs to be performed to determine, among other things, what display update rate is acceptable for a given training task.

***System Lag.*** In a VR system, the amount of time between a change in head and/or pointer position and the corresponding change in imagery or response is known as the delay time or system lag. A number of factors contribute to the overall delay as observed by the VR user including the time required to measure a change in position, computational time, image generation time, and system update rate. In general, the overall delay can be considered the additive sum of the individual component delays of the VR system. The amount of delay can have a significant effect on the fidelity of immersive impact of the VR experience. Depending on the VR activity and the experience of the user, delays on the order of 100 msec or more can lead to spatial disorientation, sensory mismatch, and nausea. The effect of excessive delay can be minimized through the use of predictive algorithms and schemes that attempt to estimate future image content based on changes in head and/or pointer position. With prediction algorithms, delays of 50 msec are not unusual for helmet-mounted display systems.

## Audio Interface

Historically, there has been little or no emphasis placed on auditory requirements for VR-based training. Early work in VR tended to focus almost exclusively on visual interfaces. However, the human auditory system provides important sensory information and cannot be overlooked when developing VR-based training applications. Auditory stimulation does much more than simply increase the sense of realism of a simulation; in many instances, audio cues are essential for accurate task performance (Youngblut et al., 1996). Recently, there has been a relatively stronger shift in auditory perception research and concomitant development of devices for presenting 3D audio. For example, Begault (1994) found that pilots had difficulty determining when they had successfully depressed a "virtual" button unless the simulation included a simple auditory cue. Massimino and Sheridan (1993) found that they could substitute auditory feedback for force feedback in several manipulation tasks, and along the same lines Apostolos, Zak, Das, and Schenker (1992) found that auditory feedback decreased the time it took to complete a manipulation task. Hendrix (1995) showed that adding 3D spatialized sound significantly increased the reported sense of immersion in a VR-based simulation. (Curiously, though, 3D audio did not seem to increase the reporting of realism of the VR.)

However, despite these early findings, a considerable amount of research in this area is still needed to focus on salient human factors issues. Major human factors of audio interface research areas include the following.

***3D Auralization.*** The area receiving the most intense VR research in human audition (and indeed much more information is needed) is auralization—the 3D simulation of a complex acoustic field (Youngblut et al., 1996). Although the generation of 3D sound has successfully been achieved with high performance digital signal processors, individual differences in the human auditory system make it extremely difficult to create a general system that can present sound as if it is coming from a specific location.

***Improve Head-Related Transfer Functions.*** Part of the 3D auralization difficulty is due to the interaction between the way sound travels and the way each individual's outer ear is shaped. Because each ear is shaped differently, sound entering the inner ear is significantly different for each individual. Sounds delivered via headphones or speakers, therefore, must take into account and mimic the same transformations that sounds undergo as they travel to the inner ear. These head-related transfer functions are unique for each individual, and to achieve an acceptable degree of spatial auralization, it is necessary to determine each individual user's head-related transfer function and use this in the audio generation system. This is a very involved process and consists, in part, of making acoustical measurements using small microphones placed in the ear canals of individual subjects. Research is needed to understand the complexities of head-related transfer functions and to eventually determine a more generalizable way of presenting 3D audio.

***Externalization.*** Related to those two issues, many users of spatial sound systems complain that the sound appears to be localized within the head. In other words, auditory externalization does not occur (Kalawsky, 1993). Research needs to determine how to overcome this phenomenon.

***Sonification.*** This refers to the transformation and auditory display of information that did not start out as such. There is very little research on the training value of sonification in VR environments; however, what little there is looks promising. For example, Tecumeh and Kramer (1994) found that subjects respond more rapidly and more accurately to simulated operating room emergencies when using an auditory display than when observing a standard visual display. They speculate that users of a system where large numbers of variables are causally and temporally interrelated would profit more from listening to auditory displays as opposed to observing traditional Graphics User Interface displays. This area of study will most likely become increasingly important as the amount (not to mention complexity) of information that needs to be presented to trainees continues to grow.

***Synesthesia.*** In a review of virtual interface devices, Youngblut et al. (1996) reported that some users of 3D audio devices may experience *synesthesia*, the sensation of a separate sensory organ being stimulated by 3D audio input. For example, they report that users of one high-end 3D audio system believe that they are receiving tactile stimulation as well. They report that when users hear a "virtual" soda can being opened near their ear, the users believe they feel the soda bubbles popping on their skin. Research is needed to confirm such findings and to determine how it can be used to advantage in VR-based training.

### Tactile and Kinesthetic Interfaces

In comparison to visual and auditory R&D for VR-based training, research on tactile and kinesthetic interfaces is still in its infancy. Yet tactile and kinesthetic interfaces will provide very important sensory input and will be a prerequisite for many types of VR-based training. For example, a training application probably will not be able to provide a compelling sense of immersion unless simulated contact with objects can somehow be achieved. Training applications that require complex or precise interactions with objects within a "virtual" environment, or between individual participants who are not physically present in the same location, will be severely limited without tactile and force feedback (Youngblut et al., 1996).

Nearly a century of basic psychophysical research shows that the human tactile and kinesthetic senses are more complex than one may think. Fortunately there are many research studies that can support tactile interface technologies R&D. What is needed is applied research on the properties, practicalities, and pitfalls of using certain technologies to give tactile information.

Whereas several different approaches are being undertaken to stimulate tactile and kinesthetic senses, haptic interface technology can support only very limited types of tactile and force feedback. Currently, most interface technologies are confined to force reflecting joysticks, hand–arm exoskeletons, and tactile feedback gloves. Although these are practical starts, they are not yet mature technologies, and they leave much to be desired by way of understanding how they can be used effectively and efficiently in training settings. Listed next are some of the human factors areas and issues concerning VR-based training and haptics.

***Necessity of Providing Tactile Stimulation and Feedback.*** Is it necessary to provide tactile stimulation, or can some other form of human–computer interaction compensate for its loss? Or, for that matter, can another kind of tactile interface be substituted? For example, Ino et al. (1993; as reported in Youngblut et al., 1996) found that control of fingertip temperature can be used to help subjects discriminate object materials—a sense that usually involves determining the shape and texture of objects. Youngblut et al. also cited the findings of Morgan (1965) who varied skin temperature to create the sensation of pressure or object contact.

These questions can be answered only after performing a complete review of a given training task, the training needs of the potential users, and a costs versus benefits analysis of any particular interface device. The worry is that as these technologies mature and more interfaces are fielded, it will become increasingly easy to argue that the use of tactile feedback is justified on the grounds that a strong sense of immersion is needed. However, as with the argument for stereoscopic visual displays (see the Visual Interface section), one should be careful not to fall into an "as high fidelity as possible" trap. That is, it is easy to get caught up in the need to make things seem as real as possible. This need is based on a sense of aesthetics, rather than on reason. A well done cost–benefit analysis will help determine when the high cost of using these devices is justified as opposed to simply satisfying a need for "realism" (Schlager, Bowman, & Piantanida, 1993).

***Necessity of Providing External Force Reflection and Feedback.*** *Kinesthesia* is an awareness of the position and movement of body parts (whether self- or externally generated). Basic psychophysical research shows that this awareness is largely a function of sensors found in the joints, tendons, and muscles (Paillard, 1991; Loeb & Levine, 1990). These sensors provide information about joint angles, muscle length and tension, and their rates of change. So far, there have been no attempts at providing a sense of kinesthesia within a VR by direct stimulation of those sensors. This would involve electrotactile and neuromuscular stimulation—fairly invasive medical procedures at this time. Rather, VR-based applications largely consist of using some form of exoskeleton—a mechanical apparatus that slips over a hand or arm and moves the appendage for you (see Youngblut et al., 1996 for an indepth review of various force reflective devices). To a lesser extent, other applications use prototype devices that one can grasp (such as knobs, joysticks, or pen-like objects), but that still move about in an attempt to give a sense of body-part motion and position (Mark, Randolph, Finch, Van Verth, & Taylor, 1996; Stonum, 1998).

When are the uses of these interfaces justified? Is an exoskeleton device necessary, or can body position and motion be conveyed more conveniently via some other virtual interface? Unfortunately, at this time hardware limitations of these devices are constraining the fidelity with which real world interactions can be simulated. The problem of safety must also be addressed because forces of the order of 10 Newtons will be applied to arms, hands, and legs (Kalawsky, 1993). This is especially a concern when the need arises for the kinesthetic interface to exert forces to oppose a trainee's volitional movements. Fortunately, the field of telerobotic research has attempted to address these, and similar questions and will probably provide a good resource for training designers (Tan, Srinvasan, Eberman, & Cheng, 1994; Stonum, 1998).

***Biomechanics.*** As already mentioned, haptic information is conveyed by sensors found in skin tissue, joints, and muscles and is more complex than one may realize. Haptic simulations will need to take into account the physical properties of surface skin and subsurface tissue. It seems that a main ingredient for the development of a haptic display system (both tactile and kinesthetic) will be a thorough investigation of the biomechanical characteristics of the skin and underlying tissue.

***Bandwidth and Resolution.*** The human hand has at least four kinds of sensory receptors that provide the sense we call touch. The response frequency for these different receptors can range from 40 to 500 msec (Youngblut et al., 1996). Any tactile display will need to take into account the bandwidth of these receptors and their dynamic response (Burdea & Coiffet, 1994). A tactile display will need to have a bandwidth that exceeds that of the skin's tactile sensors. Likewise, a similar requirement exists for force reflective devices, which are meant to supply a sense of kinesthesia (exoskeletons, joysticks, etc.). In addition, the problems of supporting an exoskeleton or other physical manipulandum must be addressed. That is, the system must not only provide the right level of force feedback, but it must also overcome its own mass, inertia, and friction (Kalawsky, 1993). The bandwidth and frequency resolution of force reflecting systems will need careful quantification by experimentation and field testing.

***Modeling.*** Many (if not most) virtual reality systems rely on computer aided design models for their definition of the virtual objects. Computer aided design models supply information on object characteristics such as volume, dimensions, color, etc. However, they do not yet supply information concerning surface characteristics such as roughness or temperature. Research on how to describe and model tactile features into virtual (computer aided design) objects needs to be conducted. In addition, virtual agents will need to be designed and computationally modeled so that they can convey synthesized haptic responses. This will entail computer modeling of the underlying physical properties of skin and surrounding tissues (Kalawsky, 1993).

## TRAINING EFFECTIVENESS AND VR RESEARCH

We expect VR systems to become relatively commonplace within the next 20 years and believe the training R&D community should prepare to take advantage of these technologies as they begin to mature. As the price of implementing VR technologies decreases, the initial questions of cost effectiveness will begin to give way to more specific questions—questions concerning training efficiency and effectiveness, as well as questions of incorporating certain training techniques within these systems. The research recommendations covered in this section address issues that will help designers to determine the appropriate uses of VR in training and testing in order to develop effective applications.

### Determining the Usefulness of VR Training Techniques

VR-based instruction seems to hold great promise for circumventing the physical, safety, and cost constraints that are limiting current training practices. But what about VR as a more general-purpose training tool? Many feel that VR technologies can create an illusion of presence (the degree to which one feels that one is in a real environment as opposed to the virtual; Zeltzer,

1992) that will lead to richer learning experiences than found in other technologies (Mowafy, Russo, & Miller, 1993). These claims are intuitively appealing, but there is little real data to show what effect these technologies have on the acquisition of perceptual, cognitive, and motor skills, knowledge retention, or transfer of training. The best methods for achieving positive learning and transfer effects within VR-based training must first be discovered and understood. There is a real need for research to determine the most effective and efficient ways to use VR in training.

***Determining the Effectiveness of VR in Facilitating Learning.*** Winn (1993) posited that VR provides three kinds of knowledge construction opportunities that are important for learning—size, transduction, and reification. VR technologies make object size easily manipulatable. For example, trainees could find themselves transported through a fuel system (O'Neil et al., chapter 6, this volume), or their view could be enlarged or expanded so that they could view entire scenes taking place at a global level (Mowafy & Thurman, 1993; Russo, Bonilla, & Thurman, 1996). *Transduction* occurs when VR interface devices present information that is not readily available to the human senses. For example, VR will allow fighter pilot trainees to "see" radar cross sections, allowing them to know when they fly too close to potentially dangerous missile sites (Hughes, 1996). *Reification* refers to the use of VR to portray representations of objects and events that have no physical form—mathematical equations, for example. This way of portraying VR capability has a certain appeal to it; however, there is very little data to show that these dimensions are critical for cognitive training.

For example, there is a need for studies to show if trainees develop a more complete or useful functional understanding of an aircraft fuel system from VR-based instruction as opposed to information displayed in two dimensions on a computer screen or even from two-dimensional schematics. Bowman, Piantanida, and Schlager (1993) posited that the value of portraying a third dimension is an advantage in many training applications, simply because it allows more information to be displayed. However, it is not clear whether this advantage will help students develop a stronger functional understanding or correct mental model of the system under study. A recent study by O'Neil et al. (chapter 6, this volume) attempts to address this issue.

***Determining the Suitability of VR in Facilitating Learning at the Skills Level.*** Bowman et al. (1993) asked a second question regarding the effects of VR-based training, the question of multi-sensory involvement. That is, are there advantages to using more than one sensory mode to facilitate learning? Just as dual processing theory (Mayer, 1997) assumes that encoding information through auditory and visual modes improves learning, is it possible that providing haptic involvement (via kinesthetic and tactile feedback) also increases skilled performance? If the incorporation of haptic information into the virtual world facilitates skill development, then its use would be justified. On the other hand, if VR's real value is in the depiction of 3D graphics, then would it not make more sense to provide interactive 3D graphics via conventional screen displays? The use of HMDs is usually more costly and always lowers video resolution.

***Determining Effective Strategies and Tactics for Training in VR.*** VR allows trainees to see the effect of changing physical laws, observe objects and events at differing size and time scales, visualize abstract concepts, and experience environments that would otherwise be prohibited (Thurman & Mattoon, 1994). These capabilities allow VR to support a wide range of training methods that are otherwise unavailable. Perhaps the greatest contribution VR provides to training is the flexibility it gives to instructional designers who want to develop experiential training environments—training environments where trainees learn by doing (Anderson & Adams, 1992; Kolb, 1984). However, once again there is a dearth of information on the proper way to structure and/or sequence VR-based training so that it is effective and efficient. For example, experience with some VR applications (that were not used for training) led some to suggest that users should be free to explore a virtual world without encumbrance, whereas other training studies using non-VR technologies indicated that guided exploration and structured tutorial are more effective for some training objectives (Bowman et al., 1993; Fletcher, 1990). How and how much should a VR-based training system intervene? When does free exploration without structure begin to lose efficiency? When does too much structure begin to interfere with the experiential nature of this training medium?

Youngblut (1998) asks a series of instructional strategies and tactics questions that only careful research can answer. Although the questions are couched in terms of education rather than training, their import is the same. Some of those questions are: How effective is the use of pre-developed VR-based instruction when compared with traditional instructional practices? How effective is the practice of student development of virtual worlds in comparison with traditional instructional practices? How effective is student use of predeveloped virtual worlds when compared with student development of VR? Which is more effective—immersive or non-immersive VR-based instruction? How well does VR support collaborative learning? Is this collaboration effective? Which is more effective—canned VR-based instruction or open-ended (constructivist) VR packages?

***Integrating VR With the Instructional Systems Development Process.*** Andrews and Thurman (chapter 2, this volume) discuss the Instructional Systems Development (ISD) process as it relates to training. This process is an attempt to answer the three major training questions, namely: "What should I train?" "How should I train it?" and "How do I know traineess have mastered it?" In most ISD projects, training designers define objectives and determine what media are most appropriate for achieving a given training objective. However, VR is a relatively new training media, and there are no clear metrics that show when it is and when it is not an appropriate training choice. Research is needed to determine the appropriateness of VR as a trainer. Following is a list of R&D activities that need to be accomplished before VR can be readily used within the ISD process.

***Develop Media Selection Criteria.*** Does VR provide unique attributes that are not usually available through other training media? Are there classes of tasks that clearly need a VR-based training approach? What training objectives call out for VR as the training method of choice? As discussed earlier, VR-based training can provide a more experiential aspect to learning. Trainees can experience the changing of physical laws, visualize abstract concepts, or act within normally restricted environments. Trainees can also experience multiple representations of the same object or phenomenon, and they can experience this at their own pace and according to their own learning style, rather than that of the trainer. Although these capabilities are also found in other training media, VR technology does seem promote their use more. Such general statements, however, are insufficient for selecting and guiding appropriate uses of the technology. What is needed is an understanding of when these (and any other) training needs clearly call for the use of VR technologies. So far, it seems that the selection of VR as a training medium is more a matter of designer taste and personal preference than a rational decision based on training need. This needs to be rectified by careful structured inquiry. There is very little data as to what characteristics of the technology support particular types of training. For example, Youngblut (1998) reviews a range of educational VR studies and asks:

> . . . why did building a set of basic 3D geometric shapes help students to recognize those shapes but did not help them to visualize and draw the shapes? Or why did students receiving auditory and haptic cues, in addition to visual cues, perform better on questions relating to velocity and acceleration, but worse on predicting system behavior? Answers to these types of questions are needed to determine the strengths and weaknesses of the technology and those situations that are most apt to benefit from its use. (p. 97)

***Identify Training Systems Requirements.*** Once training tasks and objectives are clearly identified, it will become necessary to identify and assess the utility of the many different training system configurations into which VR is capable of being incorporated. According to Bowman et al.

(1993) this will necessitate conducting studies to specify several types of training-related requirements, including simulation functional and physical analysis, student data-management capabilities, instructional capabilities, and user interface (i.e., instructor, authoring, student) design.

***Develop Prototype VR Implementations and Conduct Formative Evaluation.*** Studies should be conducted to determine the usability and, in some cases, the feasibility of VR-based training techniques. This can be accomplished by building functional prototypes (not full-fledged VR training systems) that can showcase the particular VR characteristics previously identified. These prototypes should be evaluated to determine their functional utility and to determine any particular design flaws.

***Identify Synergistic Relationships Between VR and Other Media.*** When and in what ways can other training technologies be integrated with VR? What synergistic effects occur when VR is combined with more traditional methods? How can VR and less immersive training techniques be used in concert to optimize training efficiency and maximize effectiveness? Before VR can be optimally implemented into a training curriculum, a designer will need to understand how and when to employ this medium to achieve a given set of training objectives. It will probably be a rare case where VR suffices as a total training package. Rather, it will become increasingly important to determine where and to what extent VR can be incorporated into existing training regimes. Bowman et al. (1993) suggested that VR can optimally be used early in training to familiarize individuals with new job environments, in hands-on training to develop new skills (cognitive, procedural and perceptual), or for qualification testing and recertification. These suggestions provide a foundation for determining where and how VR should be integrated with existing training. Studies should be conducted within this suggested framework to determine VR's ability to maximize training performance when used in concert with other techniques or technologies.

### Create Real-World VR Training Systems and Conduct Formal Evaluations

Currently, the field of VR-based training is almost exclusively populated with isolated examples of what early pioneers and proponents of the technology can produce (Emmerson & Revere, 1998). Their efforts produced trail-blazing examples of the capabilities of such systems (Wexelblat, 1993; Brusilousky, Kommers, & Streitz, 1994). However, their efforts usually consisted of one-shot demonstrations of VR, and they produced very little data on how trainees respond to the technology over the long term. Instead, because the developers are enthusiastic supporters of the technology, it seems that their evaluation efforts consisted largely of gathering impressions and support from demonstration participants.

This is to be expected, to some extent, when a new technology emerges. (See for example Rogers' [1995] discussion of the process of technology adoption.) However, at some point, a concerted effort must be made to gather useful information on the utility and value of real-world VR-based training systems. As the technologies underlying VR mature, and as the costs of implementing VR-based training systems continue to decrease, the ability to conduct formal evaluations should become easier. Following are some evaluation issues that need to be addressed to assure that each VR-based training system provides an effective and efficient training medium.

***Appraising the Training Effectiveness of the System.*** First and foremost among evaluation issues is the need to certify the effectiveness of the VR-based training system. This quality control procedure is an integral part of the ISD process as discussed by Andrews and Thurman (chapter 2, this volume), and is essential for demonstrating the worth or value of the VR environment for training. Certifying VR-based training is essentially a form of summative evaluation. Issues to address during this evaluation period should include at least the following questions: Can trainees achieve the terminal objectives for which the system was designed? What are the training system's strengths and weaknesses? What components of the training are most and least successful? Are there parts of the system that need to be modified or improved? These and other related questions must be addressed so that the system's overall worth, value, and merit can be appraised. Only then can objective, rational decisions be made about fielding the system, continuing its development, or deciding to discontinue it.

***Identifying Supplemental or Incidental Effects of the System.*** As mentioned previously, at this point in time, most VR-based training applications consist of one-shot demonstrations or prototypes. In reality, there are very few fielded VR training systems. Because of this, there is little information on the effects that VR systems do or will have on individuals. More in-depth studies are called for to determine what effect VR-based systems have on individuals (other than what the training was supposed to accomplish). At least three general classes of supplemental information could be gathered when studying VR-based training. First, data concerning the views and reactions of groups connected with the training system could be gathered. This would provide information about how the training system is being perceived by various stakeholders and would alert evaluators about the need for additional information as to why the training system is being viewed in a certain way. Second, information could be gathered concerning trainee performances not specified in the original training objectives. Third, information concerning side effects of the training could be gathered. These effects can be beneficial, as when Youngblut (1998) reported that VR can incidentally increase problem-solving skills, or they can be negative, as when Stanney and Kennedy (1998) reported on the duration of the aftereffects that sometimes occur from exposure to an immersive VR.

***Determining Where the VR-based Training System Should or Should Not Be Used.*** Along with the previous concern, training designers need to know when and why it is appropriate to use VR-based training. For example, Youngblut (1998) asked, "Why did students receiving auditory and haptic cues, in addition to visual cues, perform better on questions relating to velocity and acceleration, but worse on predicting system behavior? Answers to these types of questions are needed to determine the strengths and weaknesses of the technology and those situations that are most apt to benefit from its use" (p. 97). There is very little data to show what characteristics of the technology support particular types of training.

In addition, it is not enough to be concerned over the issue of VR and training type—one must also be concerned over the issue of VR and *trainee* type. That is, are there interactions between the type of learners people may be and the type of VR-based instruction they receive? Some learners seem to be more visually oriented, for example, whereas others seem to profit more by hearing or reading (Mayer, 1997; Mousavi, Low, & Sweller, 1995; Tarquin & Walker, 1997). Do these same learner preferences transfer to VR-based training? The hypothesis that VR-based training will have a greater impact on those students who are visually based comes as no surprise, but there is no hard evidence to indicate that this is true.

***Determining the Real Costs of the VR Training Environment.*** Calculating training costs can be a very difficult and involved enterprise. Many are inclined to view cost estimation as somewhat fixed information, but most authorities in the field steadfastly emphasize its "slippery character" (Wolf, 1979, p. 97). For this reason, many training system evaluations tend to gloss over the need to determine the real costs of a system. It is expected that VR-based training evaluations will continue this trend. To remedy this shortfall, costs should at least be ascertained on the basis of time (of trainers, system maintainers, system administration, students, etc.), space (specialized classrooms, offices, etc.), equipment (VR hardware, software, and supporting equipment), and supplies. Once a training system's real cost is estimated, it can then be compared with the costs of training via other methods and technologies. Cost comparisons can be painful undertakings on the part of VR system designers and developers. As most early adopters of any given technology find out, the cost of creating a new training medium can be exorbitant. What must be kept in mind, however, is that there are different kinds of costs and that comparisons should only be made with like kinds of costs.

## CONCLUSION

There are still many outstanding research issues that should be addressed before we can have true confidence in the use of VR for training. It is reasonable to use previous research, particularly training simulator studies, as a basis for designing and developing VR-based training. However, as this chapter shows, our vision is limited due to lack of systematic training data. Our understanding of the requirements for VR-based training is still quite restricted. Answering the research questions outlined in this chapter will provide a good starting point for improving our knowledge of how to use this valuable technology for training.

## REFERENCES

Anderson, J., & Adams, M. (1992). Acknowledging the learning styles of diverse student populations: Implications for instructional design. In L. Barder & N. Chism (Eds.), *Teaching for Diversity: Vol. 49. New Directions for Teaching and Learning* (pp. 19–33). San Francisco, CA: Jossey Bass.

Apostolos, M. K., Zak, H., Das, H., & Schenker, P. S. (1992). Multisensory feedback in advanced teleoperations: Benefits of auditory cues. *Proceedings of the SPIE - The International Society for Optical Engineering, Vol. 1828. Sensor Fusion V* (pp. 98–105). Bellingham, WA: SPIE.

Begault, D. R. (1994). *3-D sound for Virtual Reality and Multimedia*. Boston, MA: AP Professional.

Bowman, D. K., Piantanida, T. P., & Schlager, M. (1993). *Virtual environment systems for Maintenance training: Volume 4 - Recommendations for government-sponsored R&D* (Final Report No. 8216). Menlo Park, CA: SRI International.

Brusilousky, P., Kommers, P., & Streitz, N. (1994). Multimedia, hypermedia and virtual reality models, systems and applications. *Proceedings of the first international conference, MHVR '94*. Berlin: Springer-Verlag.

Burdea, G., & Coiffet, P. (1994). *Virtual Reality Technology*. New York: Wiley.

Burg, J., Hughes, C. E., Lisle, C., Moshell, J. M., Carrington, J., & Xin, L. (1991, May). Behavioral representation in virtual reality. In D. E. Mullally, M. Petty, & S. Smith (Eds.), *Second Behavioral Representation and Computer Generated Forces Symposium Proceedings* (pp. 115–121). Orlando, FL: Institute for Simulation and Training.

Dede, C., Salzman, M. C., Loftin, R. B., & Sprague, D. (in press). Multisensory immersion as a modeling environment for learning complex scientific concepts. In N. Roberts, W.

Feurzeig, & B. Hunter (Eds.), *Computer modeling and simulation in science education*. New York: Springer-Verlag.

Davson, H. (1989). *Physiology of the eye* (5th ed.). New York: Pergamon Press.

DeFanti, T. A., Sandin, D. J., & Cruz-Neira, C. (1993, October). A Room with a View. *IEEE Spectrum,* 30–33.

Durlach, N. I., & Mavor, A. S. (1995). *Virtual reality: Scientific and technological challenges*. Washington, DC: National Academy Press.

Emmerson, T., & Revere, D. (1998) *Virtual reality in training and education: Resource guide to citations and online information* [On-line]. Available: http://www.hitl.washington.edu/projects/knowledgebase/edvr/

Fletcher, J. D. (1990). *Effectiveness and cost of interactive videodisc instruction in defense training and education*. Alexandria, VA: Institute for Defense Analysis.

Gibson, J. J. (1989). *The ecological approach to visual perception*. Hillsdale, NJ: Lawrence Erlbaum Associates.

Hendrix, C. M. (1995). *Exploratory Studies on the Sense of Presence in Virtual environments as a Function of Visual and Auditory Display Parameters*. [On-line]. Available: http://www.hitl.washington.edu/projects/afosr/hendrix/

Hendrix, C., & Barfield, W. (1995). Presence in virtual environments as a function of visual and auditory cues. *IEEE, 1*, 74–82.

Hughes, D. (1996, March 18). USAF finds C3I uses for virtual reality. *Aviation Week & Space Technology, 145*, 50–52.

Ino, S., Shimizu, S., Odagawa, T., Sato, M., Takahashi, M., Izumi, T., & Ifukube, T. (1993, November). *A tactile display for presenting quality of materials by changing the temperature of skin surfaces*. Paper presented at the IEEE 2nd International Workshop on Robot and Human Communications, Tokyo, Japan.

James, K. R., & Caird, J. K. (1995). The effects of optic flow, proprioception, and texture on novice locomotion in virtual environments. *Proceedings of the Human Factors and Ergonomics Society, USA, 39* (pp. 1405–1406). Santa Monica, CA: Human Factors and Ergonomics Society.

Kalawsky, R. S. (1993). *The Science of Virtual Reality and Virtual Environments*. Reading, MA: Addison-Wesley.

Kolb, D. A. (1984). *Experiential learning: Experience as the source of learning and development*. Englewood Cliffs: Prentice-Hall.

Kosslyn, S. E. (1994). *Image and brain: The resolution of the imagery debate*. Cambridge, MA: MIT Press.

Loeb, G. E., & Levine, W. S. (1990). Linking musculoskeletal mechanics to sensorimotor neurophysiology. In J. M. Winters & S. L.Woo (Eds.), *Multiple muscle systems: Biomechanics and movement organization* (pp. 165–181). New York: Springer-Verlag.

Mark, W., Randolph, S., Finch, M., Van Verth, J., & Taylor, R. (1996, August). *Adding Force Feedback to Graphics Systems: Issues and Solutions*. Paper presented at the annual meeting of the ACM SIGGRAPH, New Orleans, LA.

Massimino, M., & Sheridan, T. (1993). Sensory substitution for force feedback in teleoperation. *Presence, 2*(4), 344–352.

Mayer, R. E. (1997). Multimedia learning: Are we asking the right questions? *Educational Psychologists, 32*, 1–19.

McKenna, M., & Zeltzer, D. (1992). Three dimensional visual display systems for virtual environments. *Presense, 2*(4), 421–458.

Morgan, C. T. (1965). *Physiological Psychology*. New York: McGraw-Hill.

Mousavi, S. Y., Low, R., & Sweller, J. (1995). Reducing congitive load by mixing auditory and visual presentation modes. *Journal of Educational Psychology, 87*, 319–334.

Mowafy, L., Russo, T., & Miller, L. (1993). *Is "presence" a training issue?* Paper Presented at the IEEE Symposium on Research Frontiers in Virtual Reality, Atlanta, GA.

Mowafy, L., & Thurman, R. (1993). Training pilots to visualize large scale spatial relationships in a stereoscopic 3-D display. In J. O. Merritt & S. S. Fisher (Eds.), *Science & Technology/Stereoscopic Displays and Applications IV, Proceedings of the 1993 International Symposium on Electronic Imaging* (pp. 72–81). Bellingham, WA: SPIE.

Orlansky, J., Dahlman, C., Hammon, C., Metzko, J., & Taylor, H. (1994). *The value of simulation for training*. (Report No. IDA-P-2982). Alexandrea VA: Institute For Defense Analyses.

Paillard, J. (1991). *Brain and space*. London: Oxford University Press.

Proctor, R. W., & Van Zandt, T. (1994). *Sensory input: Human factors in simple and complex systems*. Boston: Allyn & Bacon.

Ray, P. A. (1996). Quality flight simulation cueing: Why? *Proceedings of the AIAA Flight Simulation Technologies Conference* (pp. 138–147). Reston, VA: American Institute of Aeronautics and Astronautics.

Rogers, E. M. (1995). *Diffusion of Innovations* (4th ed.). New York: The Free Press-Macmillan.

Russo, T., Bonilla, V., & Thurman, R. (1996). *The basic fighter maneuver visualization trainer* (Technical Paper AL/HR-TP1996-0049). Brooks AFB, TX: Air Force Armstrong Laboratory.

Schlager, M. S., Bowman, D. K., & Piantanida, T. P. (1993). *Virtual environment systems for maintenance training: Volume 3 - cost-effectiveness analysis* (Final Report No. 8216). Menlo Park, CA: SRI International.

Stanney, K. M., & Kennedy, R. S. (1998). Aftereffects from virtual environment exposure: How long do they last. *Proceedings of the Human Factors and Ergonomics Society, USA, 42,* (pp. 1476–1480). Santa Monica, CA: Human Factors and Ergonomics Society.

Stone, R. J. (1993). Virtual reality: Toys or tools of the trade? *Proceedings of the ASLIB, UK, 45*. 167–181.

Stonum, R. (1998). *Survey of immersive technology for maintenance evaluations*. (Technical Report No. AFRL-HE-WP-TR-1998-0038). Wright-Patterson AFB, OH: Human Effectiveness Directorate, Air Force Research Laboratory.

Tan, H. Z., Srinvasan, B., Eberman, B., & Cheng, B. (1994). Human factors for the design of force-reflecting haptic interfaces. *Dynamic Systems and Control, 55*, 353–359.

Tarquin, P., & Walker, S. (1997). *Creating success in the classroom: Visual organizers and how to use them*. Englewood, CO: Teacher Idea Press.

Tecumeh, W., & Kramer, G. (1994). Sonifying the body electric: Superiority of an auditory over a visual display in a complex, multivariate system. *Proceedings of the Santa Fe Institute Studies in the Science of Complexity, Vol. XVIII* (pp. 307–325). Reading, MA: Addison-Wesley.

Thurman, R. A. (1993). Instructional simulation from a cognitive psychology viewpoint. *Educational Technology Research and Development, 41*(4), 75–89.

Thurman, R. A., & Mattoon, J. S. (1994). Virtual Reality: Toward fundamental improvements in simulation-based training. *Educational Technology, 34*(8), 56–64.

U.S. Congress, Office of Technology Assessment. (1994). *Virtual reality and technologies for combat simulation—Background paper* (OTA Publication No. OTA-BP-ISS-136). Washington, DC: U.S. Government Printing Office.

Ware, C., Arthur, K., & Booth, K. S. (1993). "Fish tank virtual reality." *Interchi '93: Bridges Between Worlds* (pp. 37–42). Reading, MA: Addison-Wesley.

Wexelblat, A. (1993). *Virtual reality applications and explorations.* Cambridge, MA: Academic Press Professional.

Winn, W. (1993). *A conceptual basis for educational applications of virtual reality* (Tech. Rep. No. R-93-9). Seattle, WA: University of Washington, Human Interface Technology Laboratory.

Winn, W., Hoffman, H., Hollander, A., Osberg, K., Rose, H., & Char, P. (1997, March). *The effect of student construction of virtual environments on the performance of high- and low-ability students*. Paper presented at the 1997 annual meeting of the American Educational Research Association, Chicago.

Wolf, R. M. (1979). *Evaluation in Education*. New York: Praeger.

Youngblut, C. (1998). Educational Uses of Virtual Reality Technology. (Institute for Defense Analysis Paper D-2128). Alexandria, VA: Institute For Defense Analysis.

Youngblut, C., Johnston, R., Nash, S., Wienclaw, R., & Will, C. (1996). *Review of virtual environment interface technology* (Institute for Defense Analysis Paper P-3186). Alexandria, VA: Institute For Defense Analysis.

Zeltzer, D. (1992). Autonomy, interaction, and presence. *Presence, 1*(1), 127–132.

*Chapter* **6**

# Instructional Strategies for Virtual Aviation Training Environments

Harold F. O'Neil, Jr.
*University of Southern California*

Richard E. Mayer
*University of California at Santa Barbara*

Howard E. Herl
*University of California, Los Angeles*

Colleen Niemi
*Advance Design Information, Sherman Oaks, CA*

Karen Olin
*ISX Corporation, Westlake Village, CA*

Richard Allen Thurman
*Air Force Research Laboratory, Mesa, AZ*

Training simulations provide a virtual practice setting in which skill and knowledge development can be accomplished (Office of Technology Assessment, 1995; O'Neil & Robertson, 1992). However, it is very difficult to create instructionally effective simulations because the state of the art in instructional technology for simulation is weak. Although there are several theoretical frameworks offered to describe instructional simulations (Alessi, 1988, chapter 10, this volume; Reigeluth & Schwartz, 1989; Thurman, 1990), there is practically no empirical research using these or any other frameworks. Unfortunately, there is little funding in either civilian or military settings for this type of research. Moreover, the empirical evidence that exists for the effectiveness of instructional strategies indicates that *free play* instructional simulations are not instructionally effective (Fletcher, 1990; Kulik, 1994). Free play simulation implements a discovery learning approach in which the participant completely controls the pacing, sequencing, etc. of the simulation. There is very little instructional structure. In general, there is little instructional research using virtual environments (but see exceptions: Dede, Salzman, Loftin, & Sprague, in press; Winn et al., 1997). The limited research base has focused mainly on feasibility studies (e.g., Mowafy & Miller, 1993; Williams et al., 1997), comparisons with other approaches to teach the same instructional goals (e.g., Grant & Magee, 1998; Rothbaum et al., 1995), Virtual reality-specific issues such as telepresence (e.g., Draper, Kaber, & Usher, 1998), and general descriptions of virtual environments (e.g., Durlach & Mavor, 1995). Likewise, there have been few evaluation studies (but see exceptions: Hettinger, Brickman, Roe, Nelson, & Haas, 1996; Jacobs et al., 1994; Johnson & Wightman, 1995; Youngblut, 1998).

This chapter is in four sections. In the first section, an introduction provides background and context for a virtual reality (VR) experimental study that investigated various instructional strategies to improve training. The next section describes the instructional design. The third section specifies the research issues and methodology for our research. The final section contains the results and discussion.

We investigated several Air Force aircrew training domains, including (a) the fuel system of the F-16, (b) the B-2 main landing gear (Pizzini & Munro, 1995), and (c) the jet engine of the F-16. Our approach focused on two critical attributes for domain selection: the availability of sufficiently detailed, written information to describe the domain, and the availability of off-the-shelf software objects to describe the domain. Unfortunately, after extensive searching, we were not able to find any usable software objects. However, we did have acceptable information about the fuel system of the F-16. Thus, the F-16 fuel system was chosen as the training domain in which to investigate the impact of selected instructional strategies for virtual environments.

Furthermore, the rationale for using a virtual environment was to permit the design and development of a research context that allowed individuals to "see" structures (e.g., inside a fuel system) that were not observable in real life. For example, trainees could learn the structure and function of an aircraft fuel system by being placed in a virtual environment via head mounted display, image generator, etc. The instructional system could move the trainee along the various fuel lines, fuel filters, etc. Such technologies of "immersion," employing wide-angle, stereoscopic, head and gaze coupled views, are commonly referred to as synthetic or virtual environments (VE). Virtual environments offer the potential to create a cost-effective visualization training environment. Immersion in this simulated environment could, we hypothesized, increase an individual's knowledge of the structure and function of the fuel system.

Such virtual environment instruction could also improve an individual's mental model of an F-16 fuel system. A mental model can be thought of as a person's conception of a system (e.g., a fuel system). Mental models help people describe the purpose of the system, help them explain the system's function and what it is doing, and help them predict what the system will do in the future (Hong & O'Neil, 1992; Rouse, Cannon-Bowers, & Salas, 1992; Senge, Kleiner, Roberts, Ross, & Smith, 1994).

In summary, we sought to improve knowledge about instructional uses of virtual environments in two ways: by providing an explicit representation of both a model of learning and a model of instruction, and by use of state-of-the-art hardware or software environments for VE simulation.

## A MODEL OF PROBLEM SOLVING

For this virtual environment study, the type of learning we wished to investigate is problem solving. Our definition of problem solving is based on that of Mayer and Wittrock (1996): "Problem solving is cognitive processing directed at achieving a goal when no solution method is obvious to the problem solver" (p. 47). This definition is further analyzed into components suggested by the expertise literature, that is, content understanding or domain knowledge, domain-specific problem-solving strategies, and self-regulation (O'Neil, in press). Our approach to measuring problem solving in this study was to use knowledge maps to assess content understanding (Herl, Baker, & Niemi, 1996), lists to assess domain-dependent problem-solving strategies, and finally a questionnaire to measure self-regulation. These measurement approaches are defined later.

### An Instructional Theory

Our theory is based on two key assumptions: It is mandatory for instructional design to specify the outcome or objective of the learning (e.g., cognitive strategy of problem solving); and various outcomes (e.g., problem solving) have specific learning conditions (e.g., instructional strategies as external conditions of learning and motivation as an internal condition of learning). The second assumption was adopted from the frameworks of Gagné, Briggs, and Wager (1988) and Merrill (1983). Although reflecting a predominantly behavioral–cognitive orientation rather than a cognitive science–social cognitive theory, the work by Gagné et al. and Merrill offers a known, effective technology. Thus, learning conditions are dependent on outcome (e.g., cognitive strategy), but are independent of content area (e.g., fuel systems).

Therefore, these learning conditions and strategies are assumed to be independent of subject matter or application area. Nonetheless, these conditions (or strategies) must be instantiated or contextualized for a particular domain. A critical distinction is that of task strategies (the *what* of learning) versus instructional strategies (*how to teach* task strategies). We believe that although task strategies may differ enormously, task-specific instruction is best conceived of as instantiations of domain-independent instructional strategies rather than as domain-dependent instructional strategies unique to a particular domain. For this VE research, we first created a training environment via a detailed instructional design. Then we manipulated an important instructional variable in an experimental study.

## INSTRUCTIONAL DESIGN FOR A GUIDED EXPLORATION OF AN F-16 FUEL SYSTEM

### Project Goals and Rationale

The goal of this project was to examine training applications of a virtual environment simulation. We chose, as a domain of interest, understanding an F-16 aircraft's fuel system. Although the technology now exists for the creation of virtual environments, there is a need to develop research-based principles for how to design useful training programs within virtual environments. Given the potential of virtual environments for improving training, the goal of our project was to explore instructional design principles for maximizing that potential.

### Simulation Description

The project focused on a virtual environment that allows a user to navigate around and through the internal components of an F-16 fuel system. Examples of components in the system include fuel tanks located in the wings of the aircraft, reservoir tanks, fuel lines connecting the tanks with each other and with the engines, and pumps to move the fuel (e.g., air ejector pump, transfer pump, and fuel flow proportioner). An example of process is fuel transfer, including both primary (siphoning) and secondary (powered transfer) phases.

### Instructional Goals

The participants were instructed that they would be able to examine the F-16 fuel system and how it works. The instructional goals were (a) to help users build a mental map of the system—that is, a representation of the major elements and their connections; (b) to help users build mental models of the components and subsystems—that is, cause-and-effect representations of how they work; and (c) to help users be able to use this knowledge to solve practical problems involving the fuel system—that is, problem solving in a maintenance context. The overarching instructional design issue concerns how to design the training system so that users can build an understanding of the fuel system, so it is important to include dependent measures that broadly assess the user's knowledge. Further, it is important to investigate various ways to teach this understanding.

The effect of the instructional variations were evaluated by using both cognitive and affective measures. Cognitive measures focused on content understanding and problem-solving skills. Affective measures assessed anxiety (i.e., state worry), student effort, and self-efficacy. In addition, the impact of instructional interventions was assessed on participants' mental models via a knowledge (or concept) map test.

## An Instructional Design Issue That Was Examined

Because resources permitted us only a single research study, we fixed most basic instructional design issues and permitted only one to vary. We fixed the variables of degree of scaffolding or amount of guidance, role of metaphors and analogies, and role of playback and route-monitoring features. The scaffolding was provided in the context of a guided discovery mode (Mayer, 1997). In a guided discovery mode, the user is expected to take a specified route through the system and to conduct a specified series of tests while traveling through the system, while the system provides hints and guidance whenever the user strays from the path. Moreover, there were no analogies or metaphors. With respect to the role of playback and route-monitoring mode, we added a navigational aid in which the user is able to see a bird's-eye view of the system.

In contrast, we varied the type of presentation for verbal instruction. One type of presentation was textual display; for the second type of presentation, the same information was provided auditorily. Thus, this instructional variable had two values (text vs. auditory). In this study, we compared the use of auditory narration with on-screen "pop-up" text. We expected auditory narration to benefit meaningful learning. The materials were formatively evaluated.

An important issue in complex instructional systems is the potential for working memory overload. We expected the auditory instruction to work better than the visual feedback, as auditory instruction reduces working memory overload. A virtual instructional environment can provide so much information that the user is unable to focus on the essential components and principles. Given the almost overwhelming visual world that the user explores, it should be useful to provide an auditory overlay to guide the user. Previous desktop multimedia research has demonstrated that users can more effectively process verbal information when it is presented auditorily than when it is presented visually (Mayer, 1997; Mayer & Moreno, 1998). In summary, we compared the use of auditory narration with on-screen pop-up text, and we expected better performance in the virtual environment for the auditory instruction group.

## METHOD

### Participants

The 40 paid participants in this study were persons obtained from a temporary job agency in southern California. Nine participants were dropped for various reasons. For the remaining participants, the average age of males ($N$ = 14) was 29.00 years and the average age of females ($N$ = 17) was 30.27 years. We asked for male and female participants who were over 18 years old and not prone to motion sickness, and we desired participants to have some computer experience. We asked for persons not prone to motion sickness because we were concerned about the possibility of simulation sickness as a side effect of immersion in our virtual environment (Lampton et al., 1994; Singer, Allen, McDonald, & Gildea, 1997).

We estimated that we would need approximately 15 participants in each condition, based on a power analysis (Minium, King, & Bear, 1993, pp. 355–374). We used the average effect size from two studies reported by Mayer and Moreno (1998). Their average effect size for retention was .69; for a matching test, .54; and for a transfer test, 1.3. As our primary interest was transfer, we used their average effect size of 1.3 ($\alpha$ = .05, $\beta$ = 10) to estimate a sample size of approximately 15 in each treatment for a minimum of 30 subjects. Although we actually tested 40 participants, we dropped 9 participants from the study for the following reasons: pilot participants ($N$ = 4); and severe technical problems, including unclear presentation of audio or text ($N$ = 5). Of these 9 participants, 2 also experienced motion sickness. Of the remaining 31 participants, 16 were in the text group and 15 were in the audio group.

### Virtual Environment System

Our virtual environment system, which was government-furnished, was an off-the-shelf hardware system integrated by Division, Inc. The system consisted of the following hardware components: a Division 3D mouse, a Virtual Research Head Mount Display (FS5HMD), a Polhemus Fastrack and Ascension 6D position tracker, and a Crystal River Engineering, Inc. Acoustetron II Aureal 3D sound server (Crystal River Engineering, 1996). We did not use the 3D audio feature. The helmet provided both a visual and auditory display, with headphones specifically chosen for spatial performance. Stereoscopic images were generated by a Silicon Graphics Indigo computer with a 20-inch monitor. Participants controlled their movement through the virtual environment using the 3D mouse. The mouse permitted the following actions: forward, backward, selection of an object via a virtual white hand, and a help function. Activation of this help function brought the participant to a virtual view of the rear of the F-16 aircraft and repeated the last question asked of the participant. The purpose of this function was to provide a navigational aid (a bird's-eye view). The main software components consisted of Division, Inc.'s dVISE product family (e.g., dV/Player, dV/Review, dV/Reality, dV/ITools, dV/Immersion, dV/Developer, and dV/MockUp) for UNIX workstations (v.4.0). This product family allows the creation and use of virtual learning environments (Division, Inc., 1997). Figure 6.1 shows a picture of one of the authors in the virtual environment laboratory.

The virtual learning environment was built with dVS/dVISE Virtual Reality software (Version 3.1) from Division, Inc. An interpretive event-based scripting language was used to program control of actors and appli-

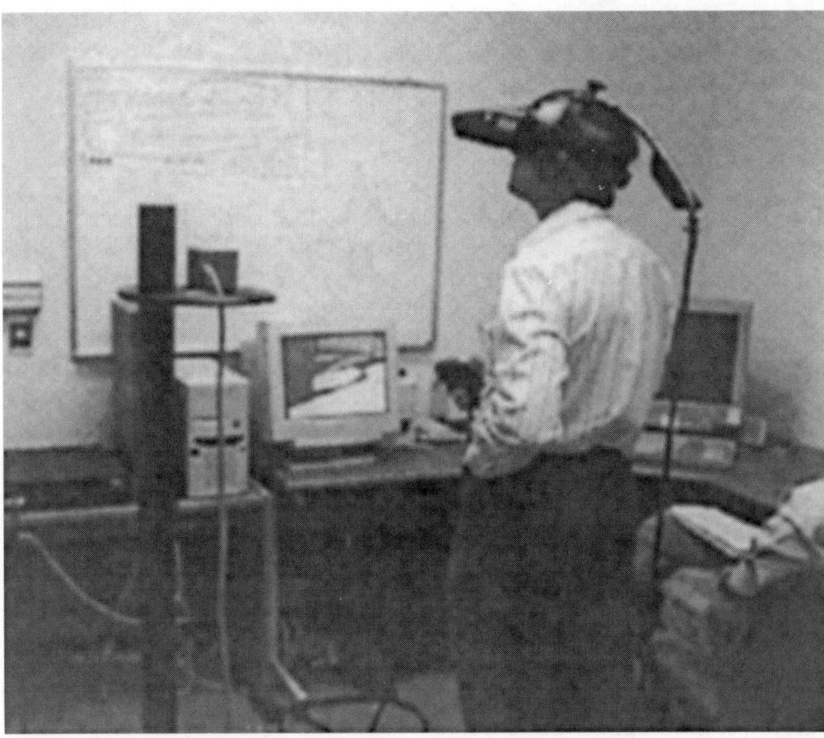

FIG. 6.1. One of the authors in the virtual environment laboratory.

cations through dVISE library functions. External interface functions were provided for MCAD and geometry converters, rendering options for Silicon Graphics, and drivers for a variety of common virtual reality devices.

### Virtual Environment Training Materials

We selected objectives, materials, and test items from Air Force documentation to ensure the accuracy of our training in the virtual environment. The training environment consisted of several components: (a) the Division, Inc. virtual environment development system, and (b) two virtual environments. One virtual environment was a 3D truck engine supplied as a demo by Division, Inc., which we used as training context (e.g., for training in the functions of the 3D mouse and familiarization with the helmet and visual–auditory system). The second virtual environment consisted of our simulation of an F-16 fuel system. The materials that we modified for the virtual environment training for the F-16 fuel system were government-furnished and consisted of the following: F-16 Technical Order (T.O. 1F-16C-1)—in particular, the fuel system (pp. 1-48 to 1-61); the *F-16 Pilot Training Course: Student Study Guide: Fuel System L-140-1* (U.S. Air Force, 1995); the *F-16 Pilot Training Course. Instructor's Manual: Fuel System L-140-1* (U.S. Air Force, no date); and a set of instructor slides (L-140-1, 003-056).

### Design and Measures

The effect of the instructional variation to teach problem solving was evaluated in our work by using both cognitive and affective measures designed or selected to be especially relevant to the interventions. Our approach to measurement of problem solving is shown in Table 6.1. We measured such information before and after interaction with the virtual environment. Cognitive measures focused on content understanding and problem-solving skills. The cognitive measures included a retention test covering basic facts about the system, a knowledge map in which the participant indicated the structure and function of the system (i.e., a mental model), and a transfer test (for example, test items asking how to troubleshoot the system measured problem-solving strategies). Affective measures assessed anxiety (i.e., state worry), effort (state and trait), and self-efficacy (state and trait).

***Assessments: Knowledge Maps.*** Knowledge maps, or concept maps, in cognitive research are a common way to represent knowledge in the form of nodes with links between them, nodes representing concepts or their attributes and links the relations between concepts. Knowledge mapping research has also been investigated as an instructional intervention in multi-

TABLE 6.1
Approach for Measurement of Problem Solving

| *Issues* | *Approach* |
|---|---|
| Definition | "Problem solving is cognitive processing directed at achieving a goal when no solution method is obvious to the problem solver (Mayer & Wittrock, 1996, p. 47) |
| Task area | Fuel system of the F-16 |
| Facets of definition | |
| Content understanding | Measure with knowledge map (Herl, Baker, & Niemi, 1996), multiple-choice knowledge test, and matching test |
| Problem-solving strategies | Measure with transfer questions |
| Self-regulation | Measure with trait metacognition scale (planning, self-monitoring) |
| | Measure with trait motivation scale (effort, self-efficacy) |
| Rubrics | |
| Content understanding | Score knowledge map with existing software—e.g., content score |
| Problem-solving strategies | Use rubric based on Mayer and Moreno's (1998) framework—e.g., compare participant's list of problems/solutions with expert list of problems/solutions |
| Self-regulation | Rubric for scale exists (O'Neil & Herl, 1998) |

ple content domains (Dansereau et al., 1979; Holley, Dansereau, McDonald, Garland, & Collins, 1979; Jonassen, 1996; Novak, Gowin, & Johansen, 1983; Stewart, 1982). In this study, we used knowledge maps as an assessment tool (Herl et al., 1996). Some examples of the links in our knowledge map assessment are (a) is part of, (b) is an example of, (c) transfers fuel to, and (d) is connected to. Some examples of the concepts are (a) storage tanks, (b) reservoir tank, (c) air ejector pump, and (d) transfer pump. The knowledge map task for the F-16 fuel system was developed by project personnel based on Air Force documentation.

We used a knowledge map assessment that was both computer-administered and scored. This software was developed at UCLA/CRESST by a team led by Dr. Howard Herl and Dr. Harry O'Neil. The software was used with CRESST's permission. The computer system consisted of a Macintosh using a hypermedia software (i.e., HyperCard).

## Procedures

The procedures for the study can be found in Table 6.2. These procedures included multiple activities that lasted about 2 hours; the data collection took approximately 3 weeks. Each session consisted of an introduction, pretest, instruction on virtual environment use, virtual environment testing, and posttesting.

***Pretest Activities.*** The pretest activities consisted of four paper-and-pencil activities. A participant biography asked questions on the following topics: (a) gender and highest level of education; (b) activities that reflected mechanical aptitude (e.g., "I own a set of tools including screw drivers, pliers, and wrenches" to "I have an aircraft pilot's license" to "I have been aboard an F-16 military aircraft"); (c) mechanical ability (using a 5-point rating scale ranging from *very little* to *very much* that asked participants to "put a check mark indicating your knowledge of how to fix household appliances and automotive repair"); and (d) knowledge of airplanes, (using a 5-point rating scale ranging from *very little* to *very much* that asked participants to "put a check mark indicating your knowledge of airplanes"). The next activ-

TABLE 6.2
Virtual Environments (VE) Research Sequence

| *Task* | *Design Time* | *Observed Time* |
|---|---|---|
| Introduction | 3 min | 2 min |
| Pretest | | |
| Participant biography | 3 min | 2 min |
| Paper folding test | 4 min | 4 min |
| Card rotations | 4 min | 4 min |
| Trait thinking questionnaire | 6 min | 4 min |
| Instruction on VE use | 15 min | 15 min |
| F-16 poster | | |
| Objectives poster | | |
| 3-D mouse practice | | |
| VE engine room practice | | |
| VE test environment | 15-30 min | 30 min |
| Version 1: audio | | |
| Version 2: test | | |
| Posttest | | |
| State thinking questionnaire —self-efficacy, worry, effort | 6 min | 4 min |
| Knowledge map training | 10 min | 10 min |
| Knowledge map test | 15 min | 15 min |
| Transfer test | 7.5 min | 9 min |
| Retention test | 4 min | 3 min |
| Matching test | 3 min | 3 min |
| Comfort questionnaire | 2 min | 2 min |
| Debrief | 4 min | 3 min |

ity in the set of pretest activities was a paper-folding test and a card rotations test. These two tests measure visualization. We hypothesized that performing in a virtual environment, which is predominantly spatial, may require some sort of visualization ability. Further, performance on one of our major dependent variables—the knowledge map—may also be related to spatial ability. The assumption was that people who do not have good spatial ability might also do poorly on the virtual environment and knowledge mapping task. Both tests were selected from the French Kit (Ekstrom, French, & Harman, 1976).

The final activity was a trait self-regulation questionnaire. This questionnaire (called the Trait Thinking Questionnaire), consists of three subscales: metacognition (i.e., planning and self-checking), self-efficacy, and effort (i.e., motivation). The trait self-regulation questionnaire was developed by O'Neil and Herl (1998). The six items in the metacognition scale (planning and self-checking) and the eight effort items were written by O'Neil and others. The six self-efficacy items were adapted from Pintrich and De Groot's (1990) intrinsic motivation scale (with Dr. Pintrich's permission). Respondents indicate how they generally think or feel using a 4-point Likert scale, ranging from *almost never* (1), to *sometimes* (2), *often* (3), and *almost always* (4). The directions were as follows:

> A number of statements which people have used to describe themselves are given below. Read each statement and indicate how you generally think or feel during learning tasks in general by marking this sheet. There are no right or wrong answers. Do not spend too much time on any one statement. Remember, give the answer which seems to describe how you generally think or feel.

An example of a metacognition item is "I determine how to solve a learning task before I begin." An example of an effort item is "I put forth my best effort on learning tasks." An example of a self-efficacy item is "I expect to do well in this task."

Table 6.3 provides some reliability and validity information from recent studies using previous versions of the trait self-regulation scale. As shown in Table 6.3, the reliability for all scales was acceptable.

***Instruction on Virtual Environment Use and VR Testing.*** Initially, participants were referred to two posters: The first poster provided information about the F-16 aircraft; the second poster provided the objectives that participants were to try to accomplish while in the virtual environment.

Then, a short demonstration (i.e., the truck engine) was given to familiarize the participants with the virtual environment equipment. Participants were then instructed that they would be taken in the virtual environment to

TABLE 6.3
Reliability and Validity Information for Trait Self-Regulation Scales

| *Scales* | *N* | *α* | *Number of Factors* | *Number of Items* | *M(SD)* |
|---|---|---|---|---|---|
| 1. Planning | | | | | |
| O'Neil & Herl, 1998 (pretest)[a] | 181 | .86 | 1 | 8 | 2.75 (.89) |
| O'Neil & Herl, 1998 (posttest)[a] | 93 | .81 | 1 | 8 | 2.98 (.80) |
| Lee, 1998 (pilot)[d] | 302 | .79 | 4 | 8 | 2.88 (.48) |
| 2. Self-checking | | | | | |
| O'Neil & Herl, 1998 (pretest)[a] | 179 | .82 | 1 | 8 | 2.67 (.94) |
| O'Neil & Herl, 1998 (posttest)[a] | 94 | .82 | 1 | 8 | 2.67 (.88) |
| Lee, 1998 (pilot)[d] | 302 | .77 | 4 | 8 | 2.70 (.45) |
| 3. Effort | | | | | |
| O'Neil & Herl, 1998 (pretest)[a] | 173 | .84 | 1 | 8 | 2.88 (.94) |
| O'Neil & Herl, 1998 (posttest)[a] | 90 | .88 | 1 | 8 | 2.86 (.90) |
| Huang, 1996 (pilot)[b] | 276 | .76 | 1 | 8 | 2.66 (.51) |
| Huang, 1996 (main)c | 383 | .74 | 1 | 8 | 2.81 (.35) |
| Lee, 1998 (pilot)[d] | 302 | .77 | 2 | 8 | 2.70 (.47) |
| 4. Self-efficacy | | | | | |
| O'Neil & Herl, 1998 (pretest)[a] | 174 | .89 | 1 | 8 | 2.80 (.93) |
| O'Neil & Herl, 1998 (posttest)[a] | 93 | .91 | 1 | 8 | 2.83 (.93) |
| Huang, 1996 (pilot)[b] | 276 | .90 | 1 | 8 | 2.26 (.71) |
| Huang, 1996 (main)[c] | 383 | .88 | 1 | 8 | 2.54 (.59) |
| Lee, 1998 (pilot)[d] | 302 | .81 | 2 | 8 | 2.89 (.49) |

*Note.* [a]U.S. middle-school students. [b]Taiwanese 10th-grade students. [c]Taiwanese 9th-grade students. [d]Taiwanese 10th-grade students.

the underside of the aircraft and asked to find or select 10 parts over a 30-minute time period. Instructions were given on how to use the 3D mouse and how the virtual environment system communicated with participants. Then the experiment per se began on the fuel system of the F-16. Participants were told that the experiment would take approximately 30 minutes and that an experimenter would always be in the room so that if they became uncomfortable or experienced any motion sickness the experiment would be terminated.

***Posttest Activities.*** The posttest activities consisted of the following: State Thinking Questionnaire, knowledge map, transfer test, retention test, matching test, and comfort questionnaire.

***State Motivation Questionnaire.*** This questionnaire (called the State Thinking Questionnaire) consists of three 6-item motivational scales: self-efficacy, worry, and effort. Participants were instructed to indicate how they thought or felt during the virtual environment task (called VR task or virtual reality task). The state measurement instrument was a modified version of O'Neil and Abedi's (1996) questionnaire, with an added scale for self-efficacy. O'Neil, Sugrue, Abedi, Baker, and Golan (1997) reported acceptable reliability and validity for this scale. O'Neil et al. (1997) used Spielberger's (1975) trait-state anxiety theory to formulate a set of domain-independent state measures for key constructs in this questionnaire. Because, according to O'Neil and Abedi (1996), "states" vary in intensity and fluctuate depending on the situation, the state responses used for this study were rated on an intensity dimension with the following choices: *not at all, somewhat, moderately so*, and *very much so*. The directions were as follows:

> A number of statements which people have used to describe themselves are given below. Read each statement and indicate how you thought or felt during the VR task. Find the word or phrase which best describes how you thought or felt and circle the number for your answer. There are no right or wrong answers. Do not spend too much time on any one statement. Remember, give the answer which seems to describe how you thought or felt during the VR task.

An example of a state effort item is "I worked hard on the VR task." An example of a state self-efficacy item is "I expected to do very well on the VR task." An example of a state worry item is "I was not happy with my performance."

***Transfer Test.*** The four transfer questions were "Suppose you try to transfer fuel from one tank to another but nothing happens. What could be wrong?" "What could be done to make fuel transfer more reliable, that is, to minimize the chances of something going wrong when fuel is transferred from one tank to another?" "What would happen if the air ejector pump stopped working?" and "What would happen if the fuel flow proportioner stopped working?" For each question, participants had 2.5 minutes to write their responses.

***Retention Test.*** Six multiple-choice items with four alternatives each constituted the retention test. An example item is:

Which of the following is the primary method of fuel transfer?

a. Siphoning

b. Powered transfer

c. Use of transfer pumps located in the F-1, F-2, and A-1 tanks

d. Use of boost pumps located in the forward and aft reservoirs

[The correct answer is "a."]

***Matching Test.*** This test consisted of matching 13 terms (e.g., "left internal wing tank" or "transfer pump") to their appropriate place on a graphic of the F-16 fuel system.

***Virtual Environment (virtual reality) System Comfort Questionnaire.*** A list of 15 symptoms (plus space to list others) was provided with the following instructions:

> A number of symptoms which people have used to describe how they felt in the VR environment are given below. Read each symptom and indicate the degree to which you experienced it during the simulation. There are no right or wrong answers. Do not spend too much time on any one symptom. Remember, give the answer which seems to describe how you felt during the VR experience.

Responses were made using a 4-point scale with ratings from *none* to *slight, moderate*, and *severe*. The questionnaire was adapted from Lampton et al. (1994) and Singer et al. (1997).

The second posttest activity—the knowledge map—is discussed in more detail in Herl et al. (1996).

## RESULTS AND DISCUSSION

### Scoring

For each participant, four demographic measures were recorded based on responses to the Participant Questionnaire: sex (male or female), age (in years), educational level (from 1 to 7), and experience level (from 2 to 22). Educational level was determined by giving one to seven points based on which of the seven alternatives the participant marked in response to "Please place a check mark indicating your highest level of education." Participants received 1 point for checking "completed elementary school," 2 points for "some high school," 3 points for "completed high school," 4 points for "some college," 5 points for "completed college degree," 6 points for "some graduate work," and 7 points for "completed graduate degree."

Experience level was determining by tallying the number of items marked on the list following "Please place a check mark next to the things you have done" plus adding 1 to 5 points based on the participant's response to "Please put a check mark indicating your knowledge of how to fix household appliances and automotive repair" and 1 to 5 points based on the participant's response to "Please put a check mark indicating your knowledge of airplanes." Participants received 1 point for checking *very little*, 2 points for checking the line between *very little* and *average*, 3 points for *average*, 4 points for the line between *average* and *very much*, and 5 points for *very much*. These four measures tap basic demographic characteristics that seem most relevant to learning in a virtual environment. No participants indicated any prior knowledge of the F-16 aircraft.

For each participant, four individual attribute measures were recorded based on participants' responses to the paper folding and card rotations tests and the Trait Thinking Questionnaire: spatial ability (from 0 to 20), metacognition (from 6 to 24), effort (from 8 to 32), and self-efficacy (from 6 to 24). The spatial ability score was determining by adding the number of correct answers (out of 10 possible) on the paper folding test and the integer closest to one eighth of the number correct (out of 80 possible) on the card rotations test. The metacognition, effort, and self-efficacy scores of the Trait Thinking Questionnaire were determined by adding the value of the responses (i.e., 1 for *almost never*, 2 for *sometimes*, 3 for *often*, and 4 for *almost always*) for each of six metacognition items, eight effort items, and six self-efficacy items, respectively. These four measures tap cognitive and affective skills that might be relevant for learning in virtual environments.

For each participant, four learning exposure measures were recorded based on participants' performance during learning in the virtual environment: time (in minutes), number of correct answers to the 10 exercises, number of incorrect answers on the 10 exercises, and number of requests for help while working on the 10 exercises. This information was logged by the computer during each participant's virtual reality learning experience. These measures provided information on the degree to which learners were exposed to relevant experiences and engaged in learning while in the virtual environment.

For each participant, we recorded four measures of cognitive learning outcomes: transfer (from 0 to an unlimited number), matching (from 0 to 11), concept mapping (from 0 to 43), and retention (from 0 to 6). The transfer test was scored by tallying the number of acceptable solutions that the participant generated on each of the three transfer test sheets. Examples of acceptable answers for the question about "what could be wrong" included leak or blockage in the fuel line, no pressurization in the tank dispensing fuel, air not expelled from the tank receiving fuel, or the tank receiving fuel is already full. Examples of acceptable answers for the question about making fuel transfer more reliable included the following: add a back-up pumping system, use fewer valves or shorter pipes, filter the fuel to keep it clean, or use pressure sensors or gauges. Examples of acceptable answers to the question about predicting what would happen if the ejector pump stopped working included the following: fuel transfer would stop or slow down, or powered transfer would be needed. Examples of acceptable answers to the question about predicting what would happen if the fuel flow proportioner stopped working included the following: one side of the aircraft could get heavier than the other, or fuel would flow unevenly from the two tank systems. The matching test was scored by tallying the number of correct answers (out of 11). The knowledge mapping test was scored by tallying the number of propositions (i.e., two nodes connected by a link between them) generated by the participant that were identical to those generated by an expert. The retention test was scored by tallying the number of correct answers (out of 6). These measures provided multiple ways of evaluating what was learned in the virtual environment.

For each participant, we recorded four measures of the participants' self-assessment of the motivational and physical aspects of learning based on the State Thinking Questionnaire and the comfort questionnaire: effort (from 6 to 24) , self-efficacy (from 6 to 24), worry (from 6 to 24), and comfort (from 0 to 39). The effort, self-efficacy, and worry scales of the State Thinking Questionnaire were determined by adding the value of the responses (i.e., 1 for *not at all*, 2 for *somewhat*, 3 for *moderately so*, and 4 for *very much so*) for each of 6 worry items, 6 effort items, and 6 self-efficacy items, respectively. The comfort questionnaire was scored by tallying the number for each response on each of the 15 items, with 0 for *none* or *no*, 1 for *slight*, 2 for *moderate* or *yes*, and 3 for *severe*. These measures allow for self-assessment of the participants' motivational and physical states during learning.

## Are the Treatment Groups Equivalent in Demographic Characteristics, Individual Characteristics, and Learning Exposure?

Before examining differences in the learning outcomes of the text and audio groups, preliminary analyses were conducted to determine whether the two treatment groups were equivalent on basic demographic characteristics, individual characteristics, and exposure to learning. As we randomized participants into treatment conditions, it was expected that there would be no significant differences between groups. Table 6.4 summarizes the mean (and standard deviation) of each group on four basic demographic measures, four individual attribute measures, and four measures of exposure to learning.

The top four rows of Table 6.4 summarize the mean (and standard deviation) for each group on four basic demographic measures. The rightmost column indicates whether or not the means differ significantly. A chi-square test (with alpha less than .05) revealed that the percentage of males (or females) in each group was not significantly different [$\chi^2 = .027$, $p > .10$]. *T*-tests (with alpha less than .05) revealed that the groups did not differ significantly in terms of age [$t(29) = .231, p > .10$], educational level [$t(29) = .616, p > .10$], and technical experience [$t(29) = .923, p > .10$]. In

TABLE 6.4
Mean (and Standard Deviation) for Text and Audio groups on Demographic, Individual Attribute, and Learning Exposure Measures

| | *Text Group* | | *Audio Group* | | |
|---|---|---|---|---|---|
| | *M* | *SD* | *M* | *SD* | *Sig. Diff.* |
| Demographic measures | | | | | |
| Percentage of males | 43.75 | (na) | 46.67 | (na) | NO |
| Age | 29.00 | (14.57) | 30.27 | (15.94) | NO |
| Educational level | 4.06 | (0.68) | 3.93 | (0.46) | NO |
| Experience | 7.69 | (4.59) | 6.27 | (3.94) | NO |
| Individual attribute measures | | | | | |
| Spatial ability | 11.75 | (2.93) | 12.53 | (2.62) | NO |
| Metacognition trait | 17.75 | (3.22) | 17.87 | (3.62) | NO |
| Effort trait | 25.38 | (4.15) | 25.87 | (4.16) | NO |
| Self-efficacy trait | 18.25 | (2.77) | 19.67 | (3.11) | NO |
| Learning exposure measures | | | | | |
| Learning time (min) | 25.88 | (3.57) | 27.47 | (5.38) | NO |
| Correct answers | 9.81 | (0.54) | 9.47 | (1.36) | NO |
| Incorrect answers | 3.50 | (2.10) | 4.07 | (3.53) | NO |
| Requests for help | 2.56 | (3.16) | 2.73 | (3.71) | NO |

*Note.* There is no *SD* for percentage of males. None of the differences is statistically significant ($p < .05$) or marginally significant ($p < .10$).

summary, the text and audio groups did not differ significantly on any of the basic demographic characteristics that we examined.

The middle four rows of Table 6.4 summarize the mean (and standard deviation) for each group on four individual attribute measures. *T*-tests (with alpha less than .05) revealed that the groups did not differ significantly in terms of spatial ability [$t(29) = .779, p > .10$] and three measures of motivation and cognitive style [metacognition: $t(29) = .095, p > .10$; effort: $t(29) = .330, p > .10$; self-efficacy: $t(29) = 1.342, p > .10$]. In summary, the text and audio groups did not differ significantly on any of the individual differences measures that we examined.

The bottom four rows of Table 6.4 summarize the mean (and standard deviation) for each group on four learning exposure measures. *T*-tests (with alpha less than .05) revealed that the groups did not differ on the time they spent in the virtual environment [$t(29) = .923, p > .10$], the number of correct answers on the 10 fuel system exercises [$t(29) = .943, p > .10$], the number of incorrect answers on the 10 fuel system exercises [$t(29) = .547, p > .10$], and the number of requests for help while working on the 10 fuel system exercises [$t(29) = .138, p > .10$]. These results are consistent with the observation that the groups had equivalent exposure to the instructional materials—that is, the text and audio groups spent approximately 26 to 28 minutes receiving the same verbal and visual information and working on locating the same 10 items in the virtual environment.

Overall, the text and audio groups were equivalent on each of the 12 measures we examined including demographic characteristics, individual characteristics, and exposure to instructional materials. Together with the fact that participants were randomly assigned to treatment groups, this set of analyzes allows us to attribute any differences in measures of learning outcomes to the effects of the treatment.

### Do the Treatment Groups Differ in Learning Outcomes?

The overarching goal of this research was to identify conditions that foster productive learning in a virtual environment. The primary issue addressed in this study concerned whether the way that verbal and pictorial materials are coordinated in a virtual learning environment affects learning outcomes. Based on a cognitive theory of multimedia learning, we expected the audio treatment to be more effective than the text treatment in helping learners integrate corresponding verbal and visual information. In particular, we focused on the hypothesis that participants in the audio group would learn more deeply than those in the text group.

Table 6.5 summarizes the mean (and standard deviation) of the text and audio groups on four measures of cognitive learning outcomes and four questionnaire scales tapping motivational and physical aspects of the learning experience. The rightmost column indicates whether the difference between the means is statistically significant, based on $t$-tests (with alpha less than .05).

***Learning Outcome Tests.*** The first four rows in Table 6.5 show the performance of the two groups on the measures of cognitive learning that most directly test our hypotheses. The single most important measure of deep understanding is the ability to use what has been presented to generate solutions to new problems—a skill that is tapped by the transfer test. The first row of Table 6.5 shows that the audio group generated 60% more creative solutions to transfer problems than did the text group. A $t$-test revealed that this difference is statistically significant [$t(29) = 2.784$, $p < .01$]. The effect size is .97. This result is consistent with previous research in desktop multimedia environments in which students who viewed animation with concurrent audio narration performed better on transfer tests than did students who viewed animation with concurrent on-screen text (Mayer, 1997; Mayer & Moreno, 1998).

The matching test taps the degree to which participants are able to make mental connections between images (e.g., visually presented parts of the fuel tank system) and words (e.g., names of the parts of the fuel tank sys-

TABLE 6.5
Mean (and Standard Deviation) for Text and Audio Groups on Four Learning Outcome Tests and Four Questionnaire Scales

| | *Text Group* | | *Audio Group* | | |
|---|---|---|---|---|---|
| | *M* | *SD* | *M* | *SD* | *Sig. Diff.* |
| Learning outcome tests | | | | | |
| Transfer | 3.94 | (2.46) | 6.33 | (2.32) | YES |
| Matching | 2.81 | (1.60) | 5.47 | (3.34) | YES |
| Knowledge mapping | 3.00 | (2.66) | 5.73 | (5.22) | YES* |
| Retention | 3.43 | (0.82) | 3.00 | (0.76) | NO |
| Questionnaire scales | | | | | |
| Effort | 21.63 | (2.71) | 22.53 | (2.39) | NO |
| Self-efficacy | 13.75 | (3.28) | 15.40 | (3.96) | NO |
| Worry | 12.06 | (2.91) | 10.60 | (3.04) | NO |
| Comfort | 7.94 | (3.61) | 6.60 | (3.87) | NO |

*Note.* YES indicates that the groups differed significantly ($p < .05$) on the matching and transfer tests; YES* indicates that the groups differed marginally ($p < .10$) on the concept mapping test; NO indicates that the groups did not differ ($p > .10$) on any of the other measures.

tem). According to a cognitive theory of multimedia learning, deep understanding occurs when learners are able to make connections between corresponding visual and verbal information. The matching test provides an important measure of the degree to which the virtual environment was successful in integrating the presentation of visual and verbal information and, thus, in fostering the learner's process of building connections between images and words. The second row of Table 6.5 shows that the audio group produced 90% more correct answers on the matching test than did the text group [$t(29) = 2.854, p < .01$]. The effect size is 1.66. This pattern is consistent with previous research in desktop multimedia environments in which students who viewed an animation with concurrent audio narration performed better on a matching test than did students who viewed animation with concurrent on-screen text (Mayer & Moreno, 1998).

The knowledge mapping test provides another way of measuring the participants' quality of knowledge—that is, by tallying the number of propositions that correspond to those created by experts. The third row in Table 6.5 shows that the audio group produced 90% more expert-like propositions in their concept map than did the text group, yielding an effect size of 1.03. Although the scores parallel our other measures of understanding, a *t*-test produced only a marginally significant difference [$t(29) = 1.856, p < .075$]. This suggests that the knowledge mapping test might be a more useful tool in a larger scale study.

The retention test evaluated participants' retention of specific facts, using the same verbatim wording as in the instruction. Although we expected the audio group to outperform the text group on tests of understanding (including retention of important concepts), we did not predict that the groups would differ on retention of specific facts. The fourth row of Table 6.5 shows that the groups did not differ significantly in remembering specific verbatim facts [$t(29) = 1.548, p > .10$]. The finding that the text group performed as well as the audio group in memory for specific facts is evidence for the claim that both groups received equivalent exposure to the same verbal information. In spite of receiving the same verbal information, however, the audio group was better able than the text group to integrate it with visual information (as tapped by the matching test), organize it into a coherent structure (as tapped by the knowledge mapping test), and use it to solve problems (as tapped by the transfer test).

In summary, by using multiple measures of the participants' learning outcomes, we were able to detect important differences in what was learned in the text and audio groups. The prediction that the audio group would learn more deeply than the text group was supported across three different measures of cognitive learning—the transfer test, the matching test, and the knowledge mapping test. Importantly, the effect size for each of these measures was approximately one or higher, indicating that the benefits of audio over text are substantial. Interestingly, if we had focused solely on a traditional retention test—a multiple-choice test of verbatim memory for the presented words—we would have not detected any differences between the text and audio groups. However, when we examined three different measures of learners' understanding of the presented material, we obtained consistent evidence that the audio group outperformed the text group. Overall, these results provide promising evidence that people learn more deeply in a virtual environment when verbal information is presented as concurrent audio rather than as on-screen text.

***Questionnaire Scales.*** The State Thinking Questionnaire was intended to evaluate the degree to which participants felt motivated, confident, and worried while in the virtual environment. As shown in Table 6.5, the text and audio groups did not differ significantly in their self-assessments of effort [$t(29) = .989, p > .10$], self-efficacy [$t(29) = 1.267, p > .10$], or worry [$t(29) = 1.368, p > .10$]. These findings suggest that the groups judged themselves to be equally hard-working, confident, and free from worry while in the virtual environment. On average, participants reported working moderately hard, feeling somewhat self-confident, and experiencing relatively little worry.

The comfort questionnaire was intended to evaluate the degree to which the participants felt physically comfortable in the virtual environment, with higher scores indicating greater discomfort. The bottom row of Table 6.5 shows that the text and audio groups did not differ in their mean scores on the comfort questionnaire [$t(29) = .996, p > .10$]. This finding suggests that the groups felt equally comfortable in the virtual environment, and that, on average, participants in both groups did not experience severe or even moderate physical discomfort. These results are not unexpected; we selected our participants to be free of motion sickness and dropped two participants from the study that exhibited such sickness in the virtual environment. In summary, the text and audio groups reported equivalent levels of motivation and physical comfort while in the virtual environment.

## Challenges of Training in Virtual Environments

In addition to the quantitative data summarized in the preceding section, in this section we provide a brief summary of the qualitative data, including logs produced by experimenters describing problems in each session and informal observations by the research team. An analysis of the qualitative data revealed three classes of problems with training in virtual environments: technical problems, physical problems, and procedural problems.

The major technical problems involved presentation of visual and verbal material and the system's response to pressing buttons on the hand-held control (i.e., the 3D mouse). In some of the pilot sessions, the text blinked on and off, the audio narration was not audible, and some of the exercises for exploring the virtual environment were not presented. These programming problems were corrected before the main study was begun. In some of the pilot sessions, the system did not respond to button presses on the hand-held control, and in some cases, the button pressed enabled the participant to unintentionally move parts of the fuel system. These programming problems were corrected before the main study was begun. However, a drastic technical problem occurred when an experimenter experienced an electrical shock from the headgear in preparation for testing the 41st participant. Fortunately, there were no long-term effects to the experimenter. On advice of the manufacturer, the experiment was terminated at that point, and no more participants were tested.

Physical problems involved interacting with the headgear and hand-held control (i.e., 3D mouse). Some participants complained that using the headgear was cumbersome; adjusting the focus was particularly difficult when the participant wore glasses. Based on pilot sessions, we determined that participants needed breaks from the headgear after approximately 30 minutes in the virtual environment. In spite of training, some participants had difficulty in operating the hand-held control, such as pressing the wrong button, having trouble in moving to the desired location, and being unable to have the on-screen glove make contact with a target. To compensate for these problems, we decided to allow the experimenters to provide guidance to participants who needed help. Two participants experienced severe dizziness, nausea, and frustration; data for these participants were not used in the analysis.

The major procedural problems involved the degree to which participants were allowed to freely explore the environment. Based on preliminary pilot study work, we determined that free exploration of the fuel tank system did not ensure a complete learning experience for participants. For this reason, we designed a guided exploration technique in which participants were asked to find each of 10 parts of the system. Some pilot participants became lost in the virtual environment, so we added a reset button to the hand-held display to provide a bird's-eye view of the plane. Some participants needed help while in the virtual environment, including reminders about how to use the hand-held controls or what the to-be-found part looked like. We decided to allow the experimenters to provide guidance to the participant concerning how to use the controls whenever the participant asked for help. During pilot sessions, some participants also required a great deal of time to accomplish the 10 learning exercises, and in some cases they were unable to find one or more parts. To standardize the learning experiences, we decided to limit each participant to a maximum of 3 minutes on each learning exercise; if a participant had not found the part within 3 minutes, the experimenter pointed out the part and manually moved the participant to the next exercise. The theme in resolution of these procedural problems was the need to provide some guidance and control for the learner rather than allowing completely free exploration in the virtual environment.

In summary, the technical problems involved programming difficulties and hardware malfunctions, both of which are likely to be reduced with the next generation of VR systems. The physical problems involved poor human factors design of the headgear and hand-held control, both of which are likely to be improved with the next generation of VR systems. Dizziness and nausea were not problems for most participants; we expect the problem to be diminished further as technology improves the speed at which the system responds to head movements. The procedural problems are likely to remain regardless of technological advances; therefore, there is a need for further study of the effects of providing various kinds of guided instructional programs or allowing free exploration.

In conclusion, this research provides one of the first methodologically rigorous studies of the conditions that foster productive learning in a virtual environment. The text and audio groups were equivalent in terms of demographic characteristics (e.g., percentage of males, average age, educational level, and mechanical experience), individual attributes (e.g., spatial ability and three thinking traits), and exposure to learning in the virtual environment (e.g., average time in the virtual environment and equally active participation on each of the 10 learning exercises). In addition, both groups reported equivalent levels of effort, self-efficacy, metacognition, and physical comfort in the virtual environment. However, as predicted, the audio group performed better than the text group on three different learning outcome measures—ability to transfer what was presented to new fuel-system problems, ability to give verbal names to visually presented parts of the fuel system, and ability to create more expert-like knowledge maps of the F-16 fuel system. These results are consistent with the conclusion that participants learn more deeply in a virtual environment when verbal information is presented as concurrent audio rather than as on-screen text. Thus, this study provides important new evidence concerning the instructional design of virtual environments for training. It should be noted, however, that overall performance levels were low, not unexpected for participants with no prior knowledge.

## ACKNOWLEDGMENTS

The research reported here was supported by the Defense Advance Research Projects Agency, contracted by the Navy Research and Development Center, and monitored technically by the Air Force Research Laboratory. The research was conducted by Advance Design Information through a subcontract with ISX Corporation. The research was also supported in part by the Office of Educational Research and Improvement, U.S. Department of Education, Educational Research and Development Centers Program, PR/Award Number R305B60002. We thank Mr. Frank Wetch, who assisted in the programming of the environment, and Mr. Jeff Olin, who assisted in conducting the data collection. The findings and opinions expressed in this work do not reflect the position or policies of the Defense Advanced Research Projects Agency, the Navy Research and Development Center, the Air Force Research Laboratory, or the Office of Educational Research and Improvement, U.S. Department of Education.

## REFERENCES

Alessi, S. M. (1988). Fidelity in the design of instructional simulations. *Journal of Computer-Based Instruction, 15*(2), 40–47.

Crystal River Engineering, Inc. (1996). *Acoustetron II: The Aureal 3D™ sound server.* Fremont, CA: Author.

Dansereau, D. F., McDonald, B. A., Collins, K. W., Garland, J. C., Holley, C. D., Diekhoff, G. M., & Evans, S. H. (1979). Evaluation of a learning strategy system. In H. F. O'Neil, Jr. & C. D. Spielberger (Eds.), *Cognitive and affective learning strategies* (pp. 3–43). New York: Academic Press.

Dede, C., Salzman, M. C., Loftin, R. B., & Sprague, D. (in press). Multisensory immersion as a modeling environment for learning complex scientific concepts. In N. Roberts, W.

Feurzeig, & B. Hunter (Eds.), *Computer modeling and simulation in science education.* New York: Springer-Verlag.

Division, Inc. (1997). *dV/Reality for UNIX workstations v.4.0: user guide.* San Diego, CA: Author.

Draper, J. V., Kaber, D. B., & Usher, J. M. (1998). Telepresence. *Human Factors, 40*(3), 354–375.

Durlach, N. I., & Mavor, A. S. (Eds.) (1995). *Virtual reality. Scientific and technological challenges.* Washington, DC: National Academy Press.

Ekstrom, R. B., French, J. W., & Harman, H. H. (1976). *Manual for kit of factor-referenced cognitive tests.* Princeton, NJ: Educational Testing Service.

Fletcher, J. D. (1990). *Effectiveness and cost of interactive videodisc instruction in defense training and education.* Alexandria, VA: Institute for Defense Analyses.

Gagné, R. M., Briggs, L. J., & Wager, W. W. (1988). *Principles of instructional design* (3rd ed.). New York: Holt, Rinehart & Winston.

Grant, S. C., & Magee, L. E. (1998). Contributions of proprioception to navigation in virtual environments. *Human Factors, 40*(3), 489–497.

Herl, H. E., Baker, E. L., & Niemi, D. (1996). Construct validation of an approach to modeling cognitive structure of U.S. history knowledge. *Journal of Educational Research, 89*, 206–219.

Hettinger, L. J., Brickman, B. J., Roe, M. M., Nelson, W. T., & Haas, M. W. (1996). Effects of virtually-augmented fighter cockpit displays on pilot performance, workload, and situation awareness. In *Proceedings of the Human Factors and Ergonomics Society 40th annual meeting* (pp. 30–33). Santa Monica, CA: Human Factors and Ergonomics Society.

Holley, C. D., Dansereau, D. F., McDonald, B. A., Garland, J. C., & Collins, K. W. (1979). Evaluation of a hierarchical mapping technique as an aid to prose processing. *Contemporary Educational Psychology, 4*, 227–237.

Hong, E., & O'Neil, H. F. Jr. (1992). Instructional strategies to help learners build relevant mental models in inferential statistics. *Journal of Educational Psychology, 84*, 150–159.

Huang, D. (1996). *The role of parental expectation, effort, and self-efficacy in the achievement of high and low track high school students in Taiwan.* Unpublished doctoral dissertation, Los Angeles, CA: University of Southern California.

Jacobs, R. S., Crooks, W. H., Crooks, J. R., Colburn, E., Fraser II, R. E., Gorman, P. F., Madden, J. L., Furness III, T. A., & Tice, S. E. (1994). *Training dismounted soldiers in virtual environments: Task and research requirements* (ARI Res. Rep. 1011; DTIC # ADA 286311). Alexandria, VA: U.S. Army Research Institute for the Behavioral and Social Sciences.

Johnson, D. M., & Wightman, D. C. (1995). *Using virtual environments for terrain familiarization: Validation* (ARI Res. Rep. 1686; DTIC # ADA 304416). Alexandria, VA: U.S. Army Research Institute for the Behavioral and Social Sciences.

Jonassen, D. H. (1996). *Computers in the classroom.* Englewood Cliffs, NJ: Prentice Hall.

Kulik, J. (1994). Meta-analytic studies of findings on computer-based instruction. In E. L. Baker & H. F. O'Neil, Jr. (Eds.), *Technology assessment in education and training* (pp. 9–33). Hillsdale, NJ: Lawrence Erlbaum Associates.

Lampton, D. R., Kolasinski, E. M., Knerr, B. W., Bliss, J. P., Bailey, J. H., & Winner, B. G. (1994). Side effects and after effects of immersion in virtual environments. In *Proceedings of the Human Factors and Ergonomics Society 38th annual meeting* (pp. 1154–1157). Santa Monica, CA: Human Factors and Ergonomics Society.

Lee, C.-Y. (1998). *Reliability and validity of problem solving assessments for Taiwanese high school students.* Unpublished doctoral dissertation. Los Angeles, CA: University of Southern California.

Mayer, R. E. (1997). Multimedia learning. Are we asking the right questions. *Educational Psychologist, 31*, 11–19.

Mayer, R. E., & Moreno, R. (1998). A split-attention effect in multimedia learning: Evidence for dual information processing systems in working memory. *Journal of Educational Psychology, 90*, 312–320.

Mayer, R. E., & Wittrock, M. C. (1996). Problem-solving transfer. In D. C. Berliner & R. C. Calfee (Eds.), *Handbook of educational psychology* (pp. 47–62). New York: Macmillan.

Merrill, M. D. (1983). Component display theory. In C. M. Reigeluth (Ed.), *Instructional design theories and models: An overview of their current status* (pp. 279–333). Hillsdale, NJ: Lawrence Erlbaum Associates.

Minium, E. W., King, B. M., & Bear, G. (1993). *Statistical reasoning in psychology and education* (3rd ed.). New York: John Wiley & Sons.

Mowafy, L., & Miller, L. (1993). *Visualizing spatial relationships: Training fighter pilots in a virtual environment debrief interface* (AL/HR-TR-1993-0148; DTIC # ADA 273972). Dayton, OH: University of Dayton Research Institute.

Novak, J. D., Gowin, D. B., & Johansen, G. T. (1983). The use of concept mapping and knowledge vee mapping with junior high school science students. *Science Education, 67*, 625–645.

Office of Technology Assessment. (1995, September). *Learning to work: Making the transition from school to work* (OTA-EHR-637). Washington, DC: U.S. Government Printing Office.

O'Neil, H. F. Jr. (in press). Perspectives on computer-based performance assessment of problem solving: Editor's Introduction. *Computers in Human Behavior* [Special Issue: Computer-Based Performance Assessment of Problem Solving].

O'Neil, H. F. Jr., & Abedi, J. (1996). Reliability and validity of a state metacognitive inventory: Potential for alternative assessment. *Journal of Educational Research, 89*, 234–245.

O'Neil, H. F. Jr., & Herl, H. E. (1998, April). *Reliability and validity of a trait measure of self-regulation.* Presented at the annual meeting of the American Educational Research Association, San Diego, CA.

O'Neil, H. F. Jr., & Robertson, M. (1992). Simulations: Occupationally oriented. In M. Alkin (Ed.), *Encyclopedia of educational research* (6th ed., pp. 1216–1222). New York: Macmillan.

O'Neil, H. F. Jr., Sugrue, B., Abedi, J., Baker, E. L., & Golan, S. (1997). *Final report of experimental studies on motivation and NAEP test performance* (CSE Tech. Rep. No. 427). Los Angeles: University of California, Center for Research on Evaluation, Standards, and Student Testing.

Pintrich, P. R., & De Groot, E. V. (1990). Motivational and self-regulated learning components of classroom academic performance. *Journal of Educational Psychology, 82*(1), 33–40.

Pizzini, Q. A., & Munro, A. (1995). *The B-2 main landing gear tutor: An application of RIDES* (Behavioral Technology Laboratories USC Working Papers WP95-1). Los Angeles: University of Southern California, Behavioral Technology Laboratories.

Reigeluth, C. M., & Schwartz, E. (1989). An instructional theory for the design of computer-based simulations. *Journal of Computer-Based Instruction, 16*, 1–10.

Rothbaum, B. O., Hodges, L. F., Kooper, R., Opdyke, D., Williford, J. S., & North, M. (1995). Effectiveness of computer-generated (virtual reality) graded exposure in the treatment of acrophobia. *American Journal of Psychiatry, 152*, 626–628.

Rouse, W. B., Cannon-Bowers, J. A., & Salas, E. (1992). The role of mental models in team performance in complex systems. *IEEE Transactions on Systems, Man, and Cybernetics, 22*, 1296–1308.

Senge, P. M., Kleiner, A., Roberts, C., Ross, R. B., & Smith, B. J. (1994). *The fifth discipline fieldbook*. New York: Doubleday/Currency.

Singer, M. J., Allen, R. C., McDonald, D. P., & Gildea, J. P. (1997). *Terrain appreciation in virtual environments: Spatial knowledge acquisition* (ARI Res. Rep. 1056). Alexandria, VA: U.S. Army Research Institute for the Behavioral and Social Sciences.

Spielberger, C. D. (1975). Anxiety: State-trait process. In C. D. Spielberger & I. G. Sarason (Eds.), *Stress and anxiety* (Vol. 1, pp. 115–143). Washington, DC: Hemisphere.

Stewart, J. (1982). Two aspects of meaningful problem solving in science. *Science Teaching, 66*, 731–749.

Thurman, R. A. (1990). Instructional simulation from a cognitive psychology viewpoint. *Educational Technologyh Research & Development, 41*(4), 75–89.

U.S. Air Force (1995, October 25). *F-16 pilot training course. Student study guide: Fuel system L-140-1.* San Antonio, TX: U.S. Air Force, Air Education and Training Command.

U.S. Air Force (n.d.). *F-16 pilot training course. Instructor's manual: Fuel system L-140-1.* San Antonio, TX: U.S. Air Force, Air Education and Training Command.

Williams, F. W., Tatem, P. A., Tate, D. L., Sibert, L., King, T., Hewitt, D. H., Siegmann, C. W., Wong, J. T., Toomey, T. A., & Farley, J. P. (1997). *Virtual environment firefighting/ship familiarization feasibility tests* (NRL/FR/6180—97-9861; DTIC # ADA 332383). Washington, DC: Naval Research Laboratory.

Winn, W., Hoffman, H., Hollander, A., Osberg, K., Rose, H., & Char, P. (1997, March). *The effect of student construction of virtual environments on the performance of high- and low-ability students.* Paper presented at the 1997 annual meeting of the American Educational Research Association, Chicago.

Youngblut, C. (1998). *Educational uses of virtual reality technology* (IDA Document D-2128). Alexandria, VA: Institute for Defense Analyses.

Chapter 7

# Simulation-Based Mission Rehearsal and Human Performance

Robert T. Nullmeyer
*Air Force Research Laboratory, Mesa, AZ*

V. Alan Spiker
*Anacapa Sciences, Inc.*

Detailed digital databases that represent real-world areas can now be developed quickly and accurately. With an appropriate flight simulator, aircrews can interact with this representation of the anticipated environment to rehearse an upcoming mission. Historically, mission rehearsal involved media such as scale models of the objective area or live exercises with mission participants operating in a full-scale replication of an area (see, e.g., Kyle, 1990). Simulation now gives warriors three-dimensional imagery based on National Imagery and Mapping Agency data, enhanced with cultural features and other details. Figure 7.1 shows an image taken from a large area, photo-specific, real-time visual database that is also capable of depicting physics-based, correlated sensor displays. Instead of deploying for live exercises (which is rapidly becoming unaffordable given today's budgets), warriors can now rehearse the planned mission at their home unit in high-fidelity, networked simulators that allow interactions with other members of the tactical team.

A scenario used to test an Air Force flight simulator exemplifies a category of missions that may benefit from simulation-based mission rehearsal (MR). The MH-53J is a heavy lift helicopter that is used to support Special Operations missions. An MH-53J Weapon System Trainer (a high fidelity flight simulator) was used by helicopter crews to help prepare for a joint Army and Air Force training exercise that was executed at White Sands Missile Range in New Mexico (Nullmeyer, Spiker, Bruce, & Reed, 1993). This exercise simulated a strike mission to recover critical equip-

FIG. 7.1. A computer-generated, photo-based depiction of a real-world scene from a rehearsal-quality digital database. Photo reproduced with permission of Aechelon Technologies.

ment from a remote mountain top facility. The concept of operation called for the helicopter crews to insert an Army support team near the objective area to control a barracks housing off-duty security personnel to prevent their entry into the objective area. An assault team would then be taken to the objective area, where they would rappel down ropes from three helicopters to search for the equipment. Two other helicopters were to provide suppressive fire as needed. Following the search, the assault and support teams would be extracted along with the recovered equipment and transported to a safe area. Such a mission requires substantial coordination among the players (intelligence, mission planners, military commanders and decision makers, two teams of ground troops, and the crews of multiple helicopters operating in the same small area), along with strict adherence to a well-formulated plan.

Simulation is generating considerable interest in the military services to help tactical teams prepare for missions such as this. Anticipated benefits include validated mission plans using live players, improved tactics, heightened crew confidence and preparation, and a clearer understanding by command and control elements of battlefield dynamics and risks.

The Army aviation community generated many of the early concepts for MR, including this definition by Wiggers, Hiteshew, and Matusof (1989): "Tactical forces/crews conducting trial performances, to which all factors, including an appropriate level of uncertainty, have been realistically applied to a situation with the intent of preparing for a specific mission" (page 571). MR is one of several powerful mission-planning technologies that are emerging to augment traditional resources like maps and photographs. For example, flight-planning software now allows route planning, including time and distance calculations and automated flight chart preparation, on a lap-top computer. Computer-based mission-planning systems add threat and terrain analysis capabilities. Mission preview systems provide views on a computer monitor of particular areas of interest such as targets or drop zones generated from digital databases. MR is at the high end of this continuum. It immerses crews in the mission environment, allowing interaction with all facets of the environment. Wiggers et al. (1989) discussed the need for realistic simulation of both the appearance and the behavior of a combat mission environment. *Appearance* refers to how the simulated environment is sensed—what it looks and sounds like. *Behavior* refers to how significant elements in the mission environment (e.g., ownship, threats, and other aircraft) act and react. The added value of MR will reflect how these capabilities augment the capabilities of other mission preparation tools.

Monette, Knight, and George (1989) discussed the interrelated nature of combat mission training and MR. They concluded that it is counterproductive for the military to consider MR as a stand-alone function because the technologies needed for MR would also substantially enhance combat mission training. Instead, they recommended integrating mission training and rehearsal to realize the full potential of these technologies in both domains. Early MR systems in the Air Force followed this approach. For these pioneering systems, MR was a natural extension of evolving flight simulation capabilities for training. The first such enterprise was the MH-53J Weapon System Trainer/Mission Rehearsal System. The weapon system trainer is a high-fidelity flight simulator that replicates an MH-53J cockpit on a six-degree-of-freedom motion platform with accurate out-the-window views and correlated sensor displays. The mission rehearsal system linked this high-fidelity simulator to a powerful database generation capability that enabled rapid development of detailed and accurate digital databases that represent specific, real-world areas (Nullmeyer et al., 1993).

Some major differences between MR and traditional simulator training warrant further consideration. One is the scope of the people involved. MR requires coordination among geographically and organizationally diverse people (e.g., a tasking agency, military intelligence personnel, mission planners, and database developers) to provide the simulated mission environment that is needed by aircrews and other mission participants (Nullmeyer, Reed, & Spiker, 1995). The diversity of the MR team heightens the need for feedback, information sharing, and collaborative learning among the players. Operational requirements often limit the time available for MR, necessitating a level of efficiency beyond that of typical training. Moreover, the mission being rehearsed is replete with uncertain or probabilistic information. Factors such as enemy location, weather, and availability of support assets can only be *assumed* at the time of rehearsal and may change once the mission starts (Spiker & Nullmeyer, 1995). Such uncertainty influences the degree to which cognitive flexibility must be incorporated into the MR environment. Similarly, the plan that was rehearsed is often not the one that is executed, as the chaotic tactical environment forces many last-minute changes. Collectively, these issues will drive the desired levels of proficiency, taking into account the dangers of overlearning, and using techniques for "just-in-time" rehearsals (Monette et al., 1989).

MR represents an evolutionary change for flight simulation *technology*, but this *application* represents a substantial departure from tradition. Due to the recency of this application, there is a dearth of direct evidence concerning its value and how to best harness the power of simulation for this purpose. Our goal in the remainder of this chapter is to generate an initial set of recommendations for maximizing MR benefits based on related areas in the scientific literature and lessons learned from early MR experiences.

## RESEARCH EVIDENCE TO GUIDE STRATEGIES FOR EFFECTIVE REHEARSAL

Because direct research on MR is lacking, we looked for generalizable principles from related fields such as psychology and instructional design to identify rehearsal processes that would enhance performance. In line with Norman's (1988) conception of rehearsal as a process in which events "come repeatedly to mind" (page 73), three core activities are relevant: stimulus encoding, practice, and feedback (Spiker & Campbell, 1994). For each area, we describe research where rehearsal has been shown to enhance performance.

### Encoding

Rehearsal processes at encoding can operate at several levels to enhance performance. In clinical studies, rehearsal has been found to induce useful affective properties. For example, visuomotor behavior rehearsal is a cognitive strategy that combines relaxation and imagery training. Though its effects are modest, visuomotor behavior rehearsal has been shown to aid karate, free throw, and tennis performance (Seabourne, Weinberg, & Jackson, 1983) and the performance of beginning archers (Zervas & Kakkos, 1995). The responsible mechanism reflect a combination of reduced anxiety, practice, and focused visualization. A related technique, elaborate rehearsal, has been used to reduce fear of fire and other emergencies among latchkey children (Jones, Ollendick, McLaughlin, & Williams, 1989). This procedure stresses relaxation and provides a coherent rationale as to *why* specific strategies work, by describing how and why each emergency response will be effective.

Cognitive or mental rehearsal has also been found to facilitate performance. Anecdotal evidence comes from surveys of master track and field athletes, where a majority (70%) report using mental practice before their event (Ungerleider, Golding, Porter, & Foster, 1989). More generally, Driskell, Cooper, and Moran (1994) reviewed 35 studies of mental practice and concluded that: (a) mental practice is an effective means for enhancing performance, but less so than overt, physical practice; (b) mental practice is especially effective for cognitive tasks; (c) positive effects of mental practice on performance fade over time; and (d) experienced subjects benefited from mental practice regardless of task type, whereas novices benefited more with cognitive tasks. Guidelines have been developed that encourage the mental practice of tasks (Smith & Beringer, 1987), in which such practice is defined as "doing a task in one's head by rehearsing the steps." Empirical evidence comes from studies of rotary-pursuit tracking (Sterner & Carpp, 1977), where subjects who received instructions to mentally rehearse the movements required to track the target had 33% higher-time-on target scores than a control group that only saw slides. Improved performance from mental practice has also been shown for a variety of other tasks, including typing and knot tying (Annett, 1989), guitar and vocal performance (Theiler & Lippman, 1995), free-throw shooting (Grouios, Mousikou, Hatzinikolaou, Semoglou, & Kabitsis, 1997), and figure skating (Garza & Feltz, 1998). Several explanatory mechanisms have been proposed. Some liken mental rehearsal effects to those found in verbal learning, where rehearsal gives additional cue-storage time to enhance sensory and memory performance (Klatzky, 1980). Alternatively, rehearsal has been postulated to serve a motivating function, by sustaining mental activity during memory transfer (Dwyer & Dwyer, 1991). Though some consider the effects of mental rehearsal to be significant, mental practice should be considered at most an adjunct to, not a substitute for, physical practice (Smith & Beringer, 1987).

Encoding of spatial knowledge is a key component of MR. Thorndyke and Hayes-Roth (1982) compared the spatial knowledge of people who studied a map depicting the floor plan of an office building to that of people who had worked in the building 1 to 2, 6 to 12, or 12 to 24 months. People who studied the map judged distances between two points "as the crow flies" more accurately than people with low experience in the building. Conversely, even the people who worked in the building 1 or 2 months more accurately judged route distance and relative bearing. People with 12 to 24 months' experience in the building performed as well as or better than any other group on all tasks. These results suggest that at least at low experience levels, map study and personal experience produce different types of spatial knowledge.

Seeking to aid the transition from basic instrument flying to contact flying (using information from outside the cockpit, supported with flight instruments, to maintain aircraft attitude), Crosby (1977) investigated the use of cognitive pretraining. This entailed presenting a mix of graphics, vocabulary, rules, and prototypes prior to flying in a simulator. Relative to control groups, Crosby (1977) showed that undergraduate pilots who received cognitive pretraining exhibited superior bank angle and pitch discrimination accuracy in a high fidelity simulator. Prather (1973) described a similar use of mental rehearsal as a rudimentary flight simulator. In either case, it is presumed that cognitive pretraining provides and clarifies conceptual references that tie into the physical features of the cockpit, such as instilling a view-from-the-cockpit schema.

Several encoding principles may help structure an effective MR program. Tulving and Thomson's (1973) *encoding specificity* principle states that recognition memory is better if cues used in the original stimulus presentation are also in the testing-retrieval environment. This implies that the problem during rehearsal is not getting the information into memory (i.e., encoded), but making sure that it will be found later during retrieval. For MR and training, a more comprehensive principle has been proposed—procedural reinstatement—that states that retention occurs to the extent that the specific procedures acquired during learning can be reinstated at the time of recall (Kolers & Roediger, 1984). For MR, the goal is to ensure that the behaviors and cognitions acquired during rehearsal can be executed during the mission. A second encoding mechanism is *metacognition*, in which knowledge of one's own cognitive processes and products improves performance (Flavell, 1977). Metacognitive skills encompass a variety of higher order processes such as predicting outcomes, apportioning time across tasks, monitoring one's own workload, and engaging in long-term planning (Means, Salas, Crandall, & Jacobs, 1993).

## Practice

Through repetition of a behavioral or cognitive response over time, MR affords participants an opportunity to hone or "tune" (Rummelhart & Norman, 1978) the skills and behaviors needed in the criterion environment. In assessing its relation to MR, we must consider the amount, quality, and type of practice. Quantitatively, the relationship is characterized as the log–log linear (or power) law of practice (Crossman, 1959). To the extent that rehearsal duration equates to amount of practice, there is solid evidence for practice effects within the sparse rehearsal literature. For example, visuomotor behavior rehearsal was more effective when practiced over 16 weeks rather than 8 (Seabourne et al., 1983). Longer rehearsal periods are also associated with better performance in domains such as long-term retention of word pairs (Carroll, Nelson, & Kirwan, 1997), theater (Welker, 1971), and chess (Chase & Simon, 1973). Presumbly, the longer an individual can mentally rehearse the content material, the greater the probability that the material will be encoded, learned, and organized in long-term memory (Klatzy, 1980). Also, practice is assumed to select out inefficient actions and support chunking of subelements (Arnett, 1989). Driskell, Willis, and Cooper (1992) reviewed the available literature concerning overlearning. They concluded that, although few studies have been reported in the literature, the existing evidence supports overlearning as an effective means of enhancing subsequent performance, but only if the task that is overlearned is the task called for in the criterion performance setting. Excessive practice can create stereotypic responding that poses problems when varying conditions are encountered in the criterion environment (Dwyer & Dwyer, 1971). Extended mental practice may also lead to a loss of concentration (Driskell et al., 1994). MR must therefore strike a balance between promoting sufficient practice to achieve skilled performance and avoiding rigid, maladaptive stereotyping or loss of concentration. Achieving this balance can be particularly challenging when the response being rehearsed is not clear (Spiker & Campbell, 1994).

Although positive, quantitative practice effects are undeniable, the real story lies in the qualitative effects of practice, as represented by the resulting mental representations of concepts and problems, pattern recognition, task sequencing actions, depth of rules deployed, and procedural automaticity (Law, 1995). Expert–novice differences constitute one of the most robust variables in human performance (Ericsson & Lehmann, 1996), yet experts differ from novices on many factors, including age, quality of learning experiences, motivation, and learning style.

An emerging area that directly addresses both quantity and quality of practice is *deliberate practice*. Ericsson, Krampe, and Tesch-Romer (1993) postulate that special conditions beyond mere experience are needed to achieve the highest performance levels in a host of domains, such as chess, sports, and music. Termed deliberate practice, these conditions entail individualized training activities involving a coach or teacher who organizes task sequences that are specifically designed to improve aspects of an individual's performance through repetition and successive refinement (Ericsson & Lehman, 1996). To receive maximal benefit from feedback, participants must monitor their own progress with effortful concentration, which limits the amount of deliberate practice per day to about 4 hours (Ericsson & Charness, 1994). Elite performers in both laboratory and natural environments exhibit a number of valuable techiques including explanating patterns in deviations while auditing fraud (Ericsson, 1996), anticipating perceptual cues in sports (Starkes, Deakin, Allard, Hodges, & Hayes, 1996), using cues memory retrieval in chess (Charness, Krampe, & Mayr, 1996), and exhibiting overlapping finger movements in typing (Salthouse, 1984). Requirements for deliberate practice tell us much about effective MR strategies, such as having a dedicated instructor, practicing inherently unenjoyable tasks, scheduling uninterrupted practice sessions, providing highly directive feedback, and continual performance tracking. Focused concentration is a key element.

Type of practice also has a major impact on performance. When multiple tasks are to be performed, as in a typical MR context, distributed practice tends to be superior to massed practice (Schmidt & Bjork, 1992) and self-paced practice is more effective than fixed-paced (Swezey & Llaneras, 1997). Some instructional design theorists discount traditional notions of practice, asserting that metacognitive skills are essential for effective functioning in real-world situations where learners must assume the responsibility for establishing and monitoring their own goals and strategies (Law, 1995). The collaborative learning paradigm specifies that learners conceptualize a problem from different perspectives, where effective learning will "emerge" depending on the intentions, elaborations, and internal constructions of intrinsically motivated individual learners. Some even suggest that explicit instruction is unnecessary and should be replaced with resources to help make apprentices become masters (Lave, 1990), encourage strategic exploration of errors, and form coaching partnerships (Greeno, 1991). Adherence to such principles has implications for structuring an effective MR program.

## Feedback

Feedback is an inherent byproduct of practice, and can be either external or internal. External knowledge of results (KR) has been shown to improve performance in virtually every domain studied (Salvendy, 1997). KR is more effective when it provides directional information (e.g., fast/slow) than purely evaluative (e.g., good/bad; Annett, 1989). For individuals, feedback is clearly one of the most powerful performance-improving methods, where internal or kinesthetic feedback improves the acquisition of motor skills beyond that offered by KR (Swezey & Llaneras, 1997). Glaser (1976) cited a study that showed that requiring subjects to verbalize their goals and strategies in each phase greatly enhanced likelihood of problem solution. Campbell (1988) noted that to be effective, feedback must be accurate, timely, credible, and constructive. Indeed, inconsistent, erroneous, and delayed feedback have been shown to dramatically decrease response accuracy and increase response time (Swezey & Llaneras, 1997). Although providing augmented feedback helps initial learning, it can degrade subsequent performance if the conditions between learning and criterion performance are too discrepant (Schmidt & Bjork, 1992), or if the feedback is used as a crutch to suppress initial learning (Annett, 1989).

In military applications, rehearsal is a public rather than private enterprise, in which information sharing and joint decision making are vital. A major goal is to improve individual or group performance through external feedback. Pearson (1991) showed that group feedback was effective in improving the productivity of railway crews. The deliberate practice literature has demonstrated that one main function of a coach or instructor is to provide timely, corrective feedback to shape and reward desired behaviors (Ericsson & Lehmann, 1996). Making feedback external in an MR environment creates the possibility for "shared mental models" (Glaser & Chi, 1988). The provision of collective feedback lets all participants share a common internal knowledge base, increasing the likelihood that team members can anticipate problems, optimize communication flow, and effectively distribute workload (Orasanu & Salas, 1993).

The likely mechanisms underlying feedback have important implications for designing effective MR programs. Because the extensive deliberate practice required of elite performers is effortful and not very enjoyable (Ericsson & Lehman, 1996), it is assumed that feedback provides some type of self-reinforcement. However, there is also a social effect within a shared team environment, with participants observing others being rewarded and punished, that may also have reinforcing effects (Law, 1995). Bandura's (1977) theory posited that the observation of a referent model can improve subsequent performance of nonperforming observers, particularly as it relates to inference of intentions. Yet, there is some question as to the extent of performance improvement that can come from passive observation. Field studies of joint athlete–coach reviews of videotaped performances met with mixed results, as the coach must identify the component action that affected outcome, whereas the trainee must relate that to his memory for the action at the time (Annett, 1989). Zimmerman and Rosenthal (1972) found that observation of the model accompanied by verbal presentation of rules produced the highest level of transfer.

## EARLY EXPERIENCES WITH SIMULATION-BASED MISSION REHEARSAL

Several studies addressed MR directly. Evidence from these early evaluations of this role for simulation provide some interesting suggestions concerning methods for maximizing the value of MR. Nullmeyer, Bruce, Conquest, and Reed (1992) reported an evaluation of the first Air Force MR capability, the MH-53J Weapon System Trainer/Mission Rehearsal System. Aircrews used the system to prepare for a joint Air Force and Army training exercise that was conducted at the White Sands Missile Range in New Mexico. The mission scenario was described at the beginning of this chapter. Reactions to MR were uniformly positive. Following rehearsal, most pilots reported a better understanding of the plan and concluded that rehearsal would likely improve their performance during the training exercise. Following mission execution, all participants reported that MR had helped their actual performance. Perhaps the most compelling data point regarding the value of MR was the decision by military leaders to eliminate a real-world rehearsal for the exercise.

Williams, Hutchinson, and Wickens (1996) compared MR and map study/mental rehearsal for high altitude route study. In one study, an active rehearsal group was instructed to fly the route in a lower capability simulator. A passive rehearsal group observed a digital recording of the out-the-window display from another pilot rehearsing. A map study group was instructed to spend an equivalent amount of time studying the chart and mentally rehearsing the mission. Following this, all pilots flew an evaluation flight in a higher fidelity simulator. Active rehearsal pilots showed less lateral deviation than did members of the other groups. Map study pilots recalled more landmarks during a subsequent reconstruction of the route from memory. All other performance measures showed no reliable difference among groups (e.g., vertical deviations and ability to find the initial point at the completion of the route). In a second study, rehearsal flights were flown at a 40% greater speed, and rehearsal pilots were required to orient their map with north always being up or current heading always up. In this case the map study group showed the least lateral deviation. The authors noted the general robustness of map study as a preparation activity, and cautioned that the value of using simulation to develop geographic knowledge should not be oversold.

Spiker and Nullmeyer (1995) asked MR experts to provide judgments concerning the impact of advanced simulation technologies on 35 selected mission preparation activities, products generated, and impacts on follow-on mission execution. Using a criterion of consensus among six or more of the seven raters, the majority of items (21 of 35) were rated positively. The consensus among raters was neutral for high altitude route study, the task studied by Williams et al. (1996). In fact, high altitude route study was the lowest rated of the 35 items. Raters typically responded that, although this activity could be done in a simulator, traditional map study combined with automated route planning tools provided a faster and easier alternative that is not dependent on a scarce and expensive simulation resource. MR effectiveness clearly needs to be assessed in some of the more probable areas of effectiveness for military crews.

## FOUR STRATEGIES FOR EFFECTIVE SIMULATION-BASED MISSION REHEARSAL

Using the simulator to rehearse a mission as one would a physical mock-up wastes the enormous capabilities offered by an electronic, synthetic environment. Although combat mission training and MR have a number of differences (Spiker & Campbell, 1994), there is no question that MR can benefit from reliance on established principles of instructional systems development (ISD), particularly those grounded in theories of learning and cognition. (See chapter 2 in this book). Next we offer four strategies for conducting effective MR that capitalize on the computational and visual rendering features of high-end simulation and are supported by relevant human performance research. They are also based on more than 5 years experience observing MR exercises.

### Establish Tightly Focused MR Sessions

A major lesson learned in early MRs is that making high-fidelity simulators available does not mean that they will be used effectively. In the absence of some overall structure, crews will "languish in the box," losing valuable opportunities to practice. Because every minute counts in a Special Operations Forces tactical environment, MRs need to be structured around tightly focused, short duration sessions that are tied to specific objectives. This proposal runs counter to the traditional practice of real-time rehearsal of the entire mission as an indivisible unit; however, it takes advantage of some underlying principles of human performance.

We recommend treating MR as a part-task or part-mission rehearsal approach in which the critical mission elements are identified and then rehearsed. By *mission elements*, we mean a set of tactically relevant task-functions that are important to mission success and that could be rehearsed in a 30 to 60 min. block of time. Previously, we identified several component mission functions as good candidates for such specialized rehearsal: waypoint identification, target familiarization and identification, target run-in, engine-out procedures, and objective area tactics (Spiker & Campbell, 1994).

Having identified some suitable candidate elements, a program of utilization could then be constructed that sequences the elements into short duration blocks (say 10 to 15 min.). For example, a program of utilization might schedule elements so that the easier functions (e.g., target area and waypoint familiarization) occur earlier in the session, the difficult ones (e.g., terminal area tactics) have more repetitions, practice is distributed for individual functions across the entire rehearsal session, and there is sufficient relearning of functions whose underlying skills are perishable (e.g., engine-out procedures). If more time becomes available for MR, the program of utilization could be expanded to include additional mission elements, more blocks of practice trials, and longer (e.g., more distributed) periods for relearning critical skills.

### Incorporate Specialized Information Into the MR

Simulation can incorporate mission-specific information into MR that is specifically designed to facilitate the cognitive processing of its participants. There are four high-payoff areas where MR is particularly well-suited to supply such value-added information: self-awareness, active learning, problem solving, and planning.

Military historians have repeatedly noted that planners should "get outside their own heads" and consider the enemy perspective when analyzing a tactical situation. A recent analysis of the Pearl Harbor bombing showed that the commanding officer at Hawaii was "unable to put himself in his opponent's place and see the situation from that perspective," thereby failing to "get a comprehensive assessment of Japanese capabilities from his intelligence staff" (Hone, 1993, p. 60). A corresponding technique for MR entails depicting alternative ingress routes into the target area from vantage points other than the aircraft (Spiker & Campbell, 1994). Other strategies to make crewmembers more aware of their own cognitive processing limitations include cueing them concerning mission segments where workload will be high and prompting them to not spend excessive time on lower priority tasks (weight and balance calculations) at the expense of tactically more pressing ones (selecting a landing zone).

MR can be viewed as a problem-solving exercise in which problems such as identifying the optimum ingress route, determining acceptable gross weight to achieve a given climb altitude, and selecting a suitable landing site must often still be worked out. The computational and visual capabilities of advanced simulators can be configured to support such problem solving by incorporating formal hypothesis testing strategies into MR materials for "just-in training" (Spiker & Campbell, 1994).

As noted previously, MR can be considered an integral part of the mission preparation process. To that end, there are a number of principles of good mission planning that are routinely used by tactical experts. These principles have stood the test of time and could be incorporated into the simulation database and rendered in hardcopy form to promote effective MR (Spiker, Nullmeyer, Tourville, & Silverman, 1997). Table 7.1 lists 10 of these principles, along with a practical application from a Combat Talon I mission (Spiker & Campbell, 1994).

TABLE 7.1
Mission Planning principles and particle Applications

| *Planning Principle* | *Practical Application* |
|---|---|
| 1. Start from the target area and work backwards. | Become familiar with the terrain around the target area and the visibility conditions that will exist at the time-on-target (TOT) to determine constraints on possible tactics. |
| 2. Work from general to specific | Consider possible aircraft platforms (e.g., MC-130E vs. MC-130H) an general tactics (e.g., high altitude drop) before specific avionics configurations and specific tactics (e.g., drop at 5000'). |
| 3. Work from high to low. | Determine timing, communications, and fuel requirements for the high altitude part of the mission first, since low-level route construction will be constrained by these factors. |
| 4. Work from big to small. | Use a small-scale map to visualize the enroute return portion of the mission. Use a large-scale map to visualize the area of operations and target areas. |
| 5. Work from qualitative to quantitative. | Use qualitative rules of thumb during the general planning phase, when mission feasibility and tactics are being established, before working with precise numbers. |
| 6. Build as much slack as possible into the calculations. | Allow a 10% "fudge factor" in aircraft fuel flow calculations. Assume the onboard load will be 10% higher to accommodate unexpected passengers and/or cargo. |
| 7. Always plan for the worst case. | There should be no aspect of the plan that depends on good luck or optimal conditions for its success. |
| 8. Fly as high a the terrain, as low as the threat dictates. | Make sure the planned vertical clearances are not overly low, since crew workload will go up. Make sure that set clearances are not overly high, since large changes in altitude above ground level will be required to avoid threats, which also increases crew workload. |
| 9. Keep the plan as simple as possible. | If the basics of the plan can't be briefed in 10 minutes, then it is too complicated. |
| 10. Remember that time is the true enemy. | Do not use up all the allowable time during general planning. Be sure to save a substantial portion of the time to complete time-consuming, detailed planning (e.g., precise fuel flow calculations, specific waypoint determinations, etc.). |

### Structure MR to Promote Group Decision Making and Shared Understanding

The conduct of any tactical mission is clearly a team effort, and the MR setting should incorporate techniques that enhance group performance. One area where the simulator can foster improvement is in human decision making. Because mission preparation involves making a lengthy series of complex decisions, mission outcome is largely determined by the quality of these decisions (Spiker & Campbell, 1994). In chapter 9, Klein describes how naturalistic decision-making principles can be applied to improve pilot training. There is considerable overlap between the capabilities of simulation-based rehearsal and Klein's recommendations. For example, Klein traces most poor decisions to a lack of experience in anticipating how a course of action could run into difficulties. He also states that if we want to improve the quality of decisions made by aviators, we should focus on the specific types of decisions to be made and the factors that make these decisions problematic. Finally, he reported that better crews mentally rehearsed to prepare themselves for difficult mission segments.

Preflying a mission in a realistic rehearsal environment gives aviators an opportunity to address such issues in the context of the specific mission to be accomplished. MR should go a long way toward helping flight crews identify problematic segments prior to execution. A common database that embodies a common set of assumptions about the mission environment should also facilitate a shared mental model among mission participants. In our early observations of simulation-based rehearsal, we found that even highly experienced crews have difficulty translating maps and photographs into accurate and dynamic mental models of complex environments to which they have never been. Simulation-based rehearsal should help mission participants establish much more accurate expectations concerning the planned course of action, identify decision points that are likely to be encountered, and pull many decisions forward to the more forgiving MR environment. In addition, an MR capability encourages all crews to emulate the rehearsal patterns now exhibited by only the most effective crews.

Although humans are quite proficient at acquiring and synthesizing information in innovative ways, their use of that information to make decisions has been shown to be quite poor, particularly under time-limited and stressful conditions (Klein, Orasanu, Calderwood, & Zsambok, 1993). Some of the most prevalent decision-making biases include overestimating high probability events, underestimating low probability events, overconfidence in stating simple event probabilities, disregarding evidence that conflicts with one's current hypothesis, and using ease of recall to estimate frequencies of classes of events. It is particularly disconcerting that such biases characterize the decision-making processes of experts as well as novices, such that an increase in domain knowledge is not accompanied by corresponding improvement in decision-making quality. One can assume that Special Operations Forces planners, despite their high levels of skill, experience, and motivation, are no less immune to these biases than are their counterparts in other knowledge domains (Spiker & Campbell, 1994).

One strategy for mitigating the effects of these biases would be to reduce participants' reliance on self-generated estimates of frequencies and probabilities by providing them with quantitative data from past missions. In particular, one could develop a formal mission-specific, relational database that could be accessed by the crews as they plan and rehearse their missions. Integration of a computerized system is technically feasible within an MR environment, where such a database could be used to extract more reliable estimates of event frequencies (e.g., likelihood of single engine take-off during missions when density altitude is high), allowing crews to prioritize what-if conditions based on their relative frequency of occurrence in previous missions (Spiker & Walls, 1996). Once developed and refined, such a system could even help restructure the MR environment so that mission crews are rehearsing *decision making* rather than simply performing tasks and functions. To be sure, some decisions would be of the traditional analytic sort, such as route selection, fuel, weight and balance, and so forth. Yet, other to-be-rehearsed decisions would be more intuitive, in keeping with the "fog of war" conditions that inevitably exist. Decisions such as selecting an emergency landing zone, reassigning mission tasks following the loss of an aircraft, and continuing the mission following a failed tanker join-up must often be made on the basis of uncertain and conflicting information. Such decisions can certainly benefit from extensive practice in a controlled rehearsal environment (Spiker & Campbell, 1994).

A second area for improving group performance entails utilizing the multi-media capabilities of the MR environment to promote shared understanding of mission details during rehearsal. One strategy is to use mediated feedback, where one subset of the planning team gives feedback to the other. With this technique, one part of the crew would "go in the box" to rehearse the mission while under the watchful eye of the other part of the crew. Using the audiovisual facilities of the MR setting, the external observers could record thoughts, comments, and critiques during the simulation, and suggest improvements for the next rehearsal "run." Besides the mission crews, this external viewing and mediated feedback could and should be provided by squadron commanders and other key decision-makers.

A less intrusive method for promoting shared understanding during MR is to embed computer-based symbols containing a mix of text, voice, and graphics to record an individual's thoughts during planning (Spiker & Campbell, 1994). These electronic post-its would provide a nonthreatening way to elicit input from individuals who are by nature not very outgoing or verbal. The use of hypermedia programming tools (e.g., Hypercard) makes this capability particularly feasible, and could conceivably be inserted as an add-on module for the computer-based mission planning system of choice in the MR environment.

### Use a Specialized Rehearsal Team to Guide MR

The importance of instilling high levels of user confidence and motivation to ensure device utilization cannot be overstated; these factors are prime determinants of user acceptance of any complex system. A fourth strategy for improving the effectiveness of MR would be to use a specialized rehearsal team to manage and control the conduct of MR sessions (Spiker & Campbell, 1994). In virtually all domains that employ rehearsal as a tool for learning or practice, a team is responsible for conducting the rehearsal sessions. These teams are found in such analogous group domains as the performing arts, psychodrama, and sports (e.g., Welker, 1971). In the beginning, this approach would be implemented on a limited, trial basis to gauge its impact on mission outcome and acceptance by mission participants. However, based on our observations of MRs and comments from numerous participants, the use of a specialist team has the potential to significantly improve the efficiency and effectiveness of MR.

The MR team would direct, block, prompt, and score. These roles could be served by four individuals or one, but all functions are important and should be provided. The director would run the MR and monitor the progress of the rehearsal sessions. This individual would be a highly experienced tactical player who has extensive knowledge of the underlying simulation technology. He would determine optimal stimulus presentation methods, tasks to be rehearsed, and the criterion levels of performance to be achieved.

The prompter would inform crew members when they are to be in the simulator and for how long. He or she plays a key role by ensuring that the overall timetable for mission preparation is adhered to, despite the complexity of scheduling the shorter duration rehearsal "runs" recommended. The blocker is responsible for ensuring the physical location or whereabouts of the crews during rehearsal. Although this role is filled by a separate individual in the performing arts, the duties of prompter and blocker for tactical operations could filled by the same person.

Finally, the scorekeeper is responsible for monitoring and scoring the performance of rehearsal participants during each MR session, and is in charge of ensuring that knowledge of results or feedback reaches each participant in a timely fashion. Importantly, this individual serves as an observer to provide knowledge of results for noninstrumented cognitive and visual tasks during rehearsal. He could also motivate by keeping track of times and accuracies for tasks that might otherwise become tedious. Such scoring would serve to keep motivation high, perhaps by instilling some friendly competition between crews. Tasks for which this gaming function might be helpful include enroute high altitude flying (e.g., who can maintain the truest heading?), aerial refueling (e.g., who can maintain the most accurate orbit?), or terrain familiarization (e.g., who can find the sawed-off cottonwood tree the fastest?).

## FOLLOW-ON RESEARCH NEEDS

Andrews, Nullmeyer, and Fuller (1990) reported a lack of guidelines for effective MR. Similarly, Williams et al. (1996) noted the lack of controlled studies that address the value of simulation-based MR. Little has been accomplished to date to fill these voids.

### MR Strategies

Some theorists have expressed concern that both the military and industry have focused primarily on methods to enhance near-term performance during training at the expense of subsequent, long-term performance in the criterion mission environment (Schmidt & Bjork, 1992). Indeed, mounting evidence suggests many variables that serve to heighten near-term training performance can inhibit transfer across long retention periods (Swezey & Llaneras, 1997). These variables include fixed-block (vs. random) practice trials, massed (vs. spaced) practice (Wickelgren, 1972), concurrent feedback, and continuous (vs. intermittent) feedback (Schmidt & Wulf, 1997).

One critical research goal must be to identify variables with proven ability to facilitate transfer to the criterion mission environment. Although there is little evidence directly relevant to MR in this regard, we can make some reasoned extrapolations from the training literature to identify candidate variables. These include providing MR participants with intermittent feedback that is eventually faded out, varying the conditions of practice as much as possible, as well as inserting longer retention periods between trials to permit some of the original learning to decay (Swezey &

Llaneras, 1997). This latter method capitalizes on Jost's second law, which states that every time "old" material is rehearsed, not only is it relearned more easily than originally (first law) but its rate of subsequent forgetting is slowed down (Annett, 1989). It appears that transfer is enhanced to the extent that the information processes required during acquisition and criterion performance match (Schmidt & Bjork, 1992).

### Establishing the Actual Impacts of MR on Military Applications

MR impacts need to be addressed in the context of military operations. Spiker and Nullmeyer (1995) found high concurrence among Air Force MR experts that planning tasks such as target area familiarization, threat avoidance tactics selection, and low level route development would be greatly enhanced with MR. Experts also felt that MR would improve terrain masking, identifying the target area, and meeting planned time on target during mission execution. Interestingly, they concurred that high altitude route study, the one task that was studied in a laboratory environment, was not a productive task for MR. The experts typically stated that traditional map study provided a faster and easier alternative. A logical next step is to translate the high pay-off areas identified by MR experts into observable events (e.g., the time at which the target is recognized, or the number or severity of surprises during mission execution) and document the impacts of MR. Over time, the demonstrated strengths of MR can be incorporated into more effective rehearsal strategies and specifications for future MR technology acquisitions.

## SUMMARY AND CONCLUSIONS

Increasing simulator capabilities and lower costs, coupled with decreasing military budgets for live exercises, have resulted in a considerable investment by the Department of Defense to develop flight simulators that can support mission rehearsal. Because this application represents a substantial departure from traditional uses of flight simulators, there is little direct evidence concerning its value and how to maximize the effectivenes of simulation in this setting. To fill this void, we identified generalizable principles from psychology and related fields, and generated an initial set of guidelines for maximizing the value of MR.

We found considerable empirical evidence in the literature linking rehearsal to improved subsequent performance over a wide variety of domains including sports, the performing arts, and aviation. Effective rehearsal can assume many forms including cognitive or mental rehearsal. Mental rehearsal was shown to be an effective means of enhancing performance, but less so than overt, physical practice. Experienced performers benefit from mental rehearsal regardless of task type, whereas novices benefit more with cognitive tasks. Overlearning has also been linked with enhanced subsequent performance, but more practice is not always better. Extended practice can lead to loss of concentration and stereotypic responding that may be inappropriate when conditions vary in the criterion environment. Finally, accurate, timely, credible, and constructive feedback can be effective, but feedback can also become a crutch that suppresses learning, and erroneous feedback has been shown to lead to degraded subsequent performance. Effective MR will require the right balance in these areas.

Four strategies for effective MR that capitalize on the capabilities of high-end simulation are supported by relevant human factors research:

1. MRs need to be structured around tightly focused, short duration sessions that are tied to specific objectives.
2. MRs should incorporate mission-specific information that is specifically designed to facilitate the cognitive processing of its participants.
3. MRs need to be structured to promote both group decision making and shared understanding.
4. A specialized rehearsal support team is required for effective MR. Key team functions are directing the MR, controlling the physical location of MR participants, ensuring that the overall timetable for mission preparation is adhered to, and providing knowledge of results and other feedback to the participants.

## REFERENCES

Andrews, D., Nullmeyer, R. T., & Fuller, J. (1990, November). Mission rehearsal behavioral research issues. *Proceedings of the 12th Interservice/Industry Training Systems Conference*, Orlando, FL.

Annett, J. (1989). Training skilled performance. In A. M. Colley & J. R. Beech (Eds.), *Acquisition and performance of cognitive skills* (pp. 61–84). New York: Wiley.

Bandura, A. (1977). *Social learning theory*. Englewood Cliffs, NJ: Prentice-Hall.

Campbell, J. P. (1988). Training design for performance improvement. In J. P. Campbell, R. J. Campbell, & Associates (Eds.), *Productivity in organizations* (pp. 177–216). San Francisco: Jossey-Bass.

Carroll, M., Nelson, T. O., & Kirwan, A. (1997). Tradeoff of semantic relatedness and degree of overlearning differential effects on metamemory and on long-term retention. *Acta Psychologica, 95*, 239–253.

Charness, N., Krampe, R., & Mayr, U. (1996). The role of practice and coaching in entrepreneurial skill domains: An international comparison of life-span chess skill acquisition.

In K. A. Ericsson (Ed.), *The road to excellence: The acquisition of expert performance in the arts and sciences, sports, and games* (pp. 51–80). Mahwah, NJ: Lawrence Erlbaum Associates.

Chase, W. G., & Simon, H. A. (1973). The mind's eye in chess. In W. G. Chase (Ed.), *Visual information processing* (pp. 215–281). New York: Academic Press.

Crosby, J. V. (1977). *Cognitive pretraining: An aid in the transition from instrument to composite flying* (AFHRL-TR-77-62). Williams Air Force Base, AZ: Air Force Human Resources Laboratory, Flying Training Division.

Crossman, E. R. F. W. (1959). A theory of the acquisition of speed skill. *Ergonomics, 2*, 153–156.

Driskell, J. E., Cooper, C., & Moran, A. (1994). Does mental practice enhance performance? *Journal of Applied Psychology, 79*(4), 481–492.

Driskell, J. E., Willis, R. P., & Cooper, C. (1992). Effects of overlearning on retention. *Journal of Applied Psychology,* 77, 615–622.

Dwyer, F. M., & Dwyer, C. A. (1991). *Effects of students' prior knowledge and presentation mode on achievement (visual/verbal testing) of different educational objectives*. Washington, DC: U.S. Department of Education, Office of Education Research and Improvement (ERIC Document Reproduction Service No. ED-323 926).

Ericsson, K. A. (1996). The acquisition of expert performance. In K. A. Ericsson (Ed.), *The road to excellence: The acquisition of expert performance in the arts and sciences, sports, and games* (pp. 1–50). Mahwah, NJ: Lawrence Erlbaum Associates.

Ericsson, K. A., & Charness, N. (1994). Expert performance: Its structure and acquisition. *American Psychologist, 49*(8), 725–747.

Ericsson, K. A., Krampe, R. T., & Tesch-Romer, C. (1993). The role of deliberate practice in the acquisition of expert performance. *Psychological Review, 100*(3),. 363–406.

Ericsson, K. A., & Lehmann, A. C. (1996). Expert and exceptional performance: evidence of maximal adaptation to task constraints. In J. T. Spence, J. M. Darley, & D. J. Foss (Eds.), *Annual Review of Psychology, 47* (pp. 273–305). Palo Alto, CA: Annual Reviews, Inc.

Flavell, J. H. (1977). *Cognitive development*. Englewood Cliffs, NJ: Prenctice-Hall.

Garza, D. L., & Feltz, D. L. (1998). Effects of selected mental practice on performance, self efficacy, and competition confidence of figure skaters. *The Sport Psychologist, 12*, 1–15.

Glaser, R. (1976). Components of a psychology of instruction: Toward a science of design. *Review of Educational Research, 46*, 1–24.

Glaser, R., & Chi, M. T. H. (1988). Overview. In M. T. H. Chi, R. Glaser, & M. J. Farr (Eds.), *The nature of expertise* (pp. xv–xxviii). Hillsdale, NJ: Lawrence Erlbaum Associates.

Greeno, J. G. (1991). Mathematical cognition: Accomplishments and challenges in research. In R. R. Hoffman & D. S. Palermo (Eds.), *Cognition and the symbolic processes: Applied and ecological perspectives* (pp. 255–279). Hillsdale, NJ: Lawrence Erlbaum Associates.

Grouios, G., Mousikou, K., Hatzinikolaou, K., Semoglou, K., & Kabitsis, C. (1997). The effect of a simulated mental practice technique on free throw shooting accuracy of highly skilled basketball players. *Journal of Human Movement Studies, 33*, 119–138.

Hone, T. C. (1993, June). Combine strategy & intelligence. *Navy Proceedings*, 59–60.

Jones, R. T., Ollendick, T. H., McLaughlin, K. J., & Williams, C. E. (1989). Elaborative and behavioral rehearsal in the acquisition of fire emergency skills and the reduction of fear of fire. *Behavior Therapy, 20*, 93–101.

Klatzky, R. L. (1980). *Human memory: Structures and processes* (2nd ed.). San Francisco: Freeman.

Klein, G. A., Orasanu, J., Calderwood, R., & Zsambok, C. E. (Eds.). (1993). *Decision making in action: Models and methods*. Norwood, NJ: Ablex.

Kolers, P. A., & Roediger, H. L. (1984). Procedures of mind. *Journal of Educational Psychology, 53*, 250–253.

Kyle, J. H. (1990). *The guts to try.* New York: Orion.

Lave, J. (1990). Views of classroom: Implications for math and science learning research. In M. Gardner, J. G. Greeno, F. Reif, A. H. Schoenfeld, A. diSessa, & E. Stage (Eds.), *Toward a scientific practice of science education* (pp. 251–263). Hillsdale, NJ: Lawrence Erlbaum Associates.

Law, L. C. (1995). *Constructivist instructional theories and acquisition of expertise* (Research report No. 48). online: http://infix.emp.paed.uni-muenchen.de/lslmandl/.

Means, B., Salas, E., Crandall, B., & Jacobs, T. O. (1993). Training decision makers for the real world. In G. A. Klein, J. Orasanu, R. Calderwood, & C. E. Zsambok (Eds.), *Decision making in action: Models and methods* (pp. 306–345). Norwood, NJ: Ablex.

Monette, R., Knight, S., & George, G. (1989). Achieving mission rehearsal objectives employing mission-oriented simulator development. *Proceedings of the 11th Interservice/Industry Training Systems Conference*, Orlando, FL.

Norman, D. A. (1988). *The psychology of everyday things*. New York: Basic Books.

Nullmeyer, R. T., Bruce, P. D., Conquest, M. T., & Reed, E. T. (1992). The use of the MH-53J weapon systems trainer for mission rehearsal: Initial assessments and lessons learned. *Proceedings of the 14th Interservice/Industry Training Systems Conference*, San Antonio, TX.

Nullmeyer, R. T., Reed, E. T., & Spiker, V. A. (1995, November). Simulation-based mission rehearsal as a human activity system. *Proceedings of the 17th Interservice/Industry Training Systems Conference*, Albuquerque, NM.

Nullmeyer, R. T., Spiker, V. A., Bruce, P. D., & Reed, E. T. (1993). Simulation technology transforms SOF mission preparation. *Armed Forces Journal International, 132*(4), 37–40.

Orasanu, J., & Salas, E. (1993). Team decision making in complex environments. In G. A. Klein, J. Orasanu, R. Calderwood, & C. E. Zsambok (Eds.), *Decision making in action: Models and methods* (pp. 327–345). Norwood, NJ: Ablex.

Pearson, C. (1991). An assessment of extrinsic feedback on participation, role perceptions, motivation, and job satisfaction in a self-managed system for monitoring group achievement. *Human Relations, 44*, 517–537.

Prather, D. C. (1973). Prompted mental practice as a flight simulator. *Journal of Applied Psychology, 57*, 353–355.

Rummelhart, D. E., & Norman, D. A. (1978). Accretion, tuning and restructuring: Three modes of learning. In J. W. Cotton & R. L. Klatzky (Eds.), *Semantic factors in cognition* (pp. 37–54). Hillsdale, NJ: Lawrence Erlbaum Associates.

Salthouse, T. A. (1984). Effects of age and skill in typing. *Journal of Experimental Psychology: General, 113*(3), 345–371.

Salvendy, G. (1997). *Handbook of human factors and ergonomics* (2nd ed.). New York: Wiley.

Schmidt, R. A., & Bjork, R. A. (1992). New conceptualizations of practice: Common principles in three paradigms suggest new concepts for training. *Psychological Science, 3*(4), 207–217.

Schmidt, R. A., & Wulf, G. (1997). Continuous concurrent feedback degrades skill learning: Implications for training and simulation. *Human Factors, 39*(4), 509–525.

Seabourne, T., Weinberg, R., & Jackson, A. (1983). Effect of individualized practice and training of visuo-motor behavior rehearsal in enhancing karate performance. *Journal of Sport Behavior*, 7, 58–67.

Smith, M. J., & Beringer, D. B. (1987). Human factors in occupational injury evaluation and control. In G. Salvendy (Ed.), *Handbook of human factors* (pp. 767–789). New York: Wiley.

Spiker, V. A., & Campbell, J. L. (1994). *Strategies and tactics for effective mission rehearsal: Application of principles from psychology and other relevant domains* (AL/HR-TR-1994-0032). Mesa, AZ: Armstrong Laboratory, Aircrew Training Research Division.

Spiker, V. A., & Nullmeyer, R. T. (1995, April). Benefits and limitations of simulation-based mission planning and rehearsal. *Eighth International Symposium on Aviation Psychology*, Columbus, OH.

Spiker, V. A., Nullmeyer, R. T., Tourville, S. J., & Silverman, D. R. (1997, September). *Effects of mission preparation on crew combat mission training performance*. Paper presented at the 41st Annual Meeting of the Human Factors and Ergonomics Society, Albuquerque, NM.

Spiker, V. A., & Walls, W. F. (1996). *Development of a mission-information collection, analysis, and reporting system (MICARS)* (AL/HRA TR-1996-0043). Santa Barbara, CA: Anacapa Sciences, Inc.

Starkes, J. L., Deakin, J. M., Allard, F., Hodges, N. J., & Hayes, A. (1996). Deliberate practice in sports: What is it anyway? In K. A. Ericsson (Ed.), *The road to excellence: The acquisition of expert performance in the arts and sciences, sports, and games* (pp. 81–106). Mahwah, NJ: Lawrence Erlbaum Associates.

Sterner, R. T., & Carpp, L. (1977). Psychomotor rehearsal: Enhancement of rotary pursuit tracking using a massed-training procedure. *Perceptual and Motor Skills, 44*, 243–248.

Swezey, R. W., & Llaneras, R. E. (1997). Models in training and instruction. In G. Salvendy (Ed.), *Handbook of human factors* (2nd ed., pp. 514–577). New York: John Wiley.

Theiler, A. M., & Lippman, L. G. (1995). Effects of mental practice and modeling on guitar and vocal performance. *The Journal of General Psychology, 122*(4), 329–343.

Thorndyke, P. W., & Hayes-Roth, B. (1982). Differences in spatial knowledge acquired from maps and navigation. *Cognitive Psychology, 14*, 560–589.

Tulving, E., & Thomson, D. M. (1973). Encoding specificity and retrieval processes in episodic memory. *Psychological Review, 3*, 112–129.

Ungerleider, S., Golding, J. M., Porter, K., & Foster, J. (1989). An exploratory examination of cognitive strategies used by masters track and field athletes. *The Sports Psychologist, 3*, 245–253.

Welker, D. (1971). *Theatrical direction: The basic techniques*. Boston: Allyn & Bacon.

Wickelgren, W. A. (1972). Trace resistance and the decay of long-term memory. *Journal of Mathematical Psychology, 9*, 418–455.

Wiggers, R., Hiteshew, L. K., & Matusof, R. (1989) Mission rehearsal: More than just another simulation. *Proceedings of the 11th Interservice/Industry training systems conference*, Ft. Worth, TX.

Williams, H. P., Hutchinson, S., & Wickens, C. D. (1996). A comparison of methods for promoting geographic knowledge in sumulated aircraft navigation. *Human Factors, 38*(1), 50–64.

Zervas, Y., & Kakkos, V. (1995). The effect of visuomotor behavior rehearsal on shooting performance of beginning archers. *International Journal of Sport Psychology, 26*(3), 337–347.

Zimmerman, B. J., & Rosenthal, T. L. (1972). Concept attainment, transfer, and retention through observation and rule-provision. *Journal of Experimental Child Psychology, 14*, 139–150.

*Chapter* 8

# Above Real-Time Training

Peter Crane
*Air Force Research Laboratory, Mesa, AZ*

Dutch Guckenberger
*SDS International, Inc.*

The design of flight training simulators is often based on the concept that "transfer of training is highest when similarity of the training and transfer situations is the highest . . . this is the governing principle for most simulators that are built" (Adams, 1979). Although many research programs have been directed toward identifying the level of fidelity required for effective training (see Adams, 1989, ch. 18), there have been almost no efforts to deliberately reduce simulator fidelity in order to increase training effectiveness. However, in the early 1970s, engineers and test pilots at the National Aeronautics and Space Administration's (NASA) Dryden Flight Research Center attempted to increase the training effectiveness of flight simulators by distorting the apparent rate of time within the simulator (Hoey, 1976; Kolf, 1973). Normally, simulator time—the apparent rate of time within a simulated environment—equals real-time or clock-time, which is the unalterable passage of time in the real world. In above real-time or compressed-time simulation, simulator time passes more quickly than real-time (Crane & Guckenberger, 1977). Time-lapse photography showing a flower's growth for a month in one minute demonstrates compressed-time.

Kolf (1973) and Hoey (1976) documented simulator training interventions that were aimed at improving test pilots' ability to keep up with the pace of events in flight. Kolf noted that "regardless of the type or amount of pre-flight simulator training accomplished by the pilot, the actual flight seems to take place in a much faster time frame than real time" (p. 1).

Hoey (1976) reported that in the X-15 program, pilots typically spent 10 hours in the simulator for each 10 minutes of flight. Even with this preparation, pilots reported that "It sure seems to happen faster in the real airplane," or "I had the feeling that I was 'behind the airplane' " (pp. 2–3).

As an experiment, Kolf (1973) increased the apparent rate of time in the training simulator for the M2-F3 Lifting Body, an experimental aircraft used in development of the space shuttle. Kolf altered the simulator's time integration factor, which is the amount of simulator time that the system assigns to have passed between hardware cycles. In a real-time simulator, the time integration factor equals the interval between cycles. For example, in a system updating at 60 Hz, the integration factor for real-time simulation equals 1/60 second or 166 msec. Changing the time integration factor to 250 msec increases simulator time to 1.5 time real-time. In the modified simulator, a training mission that required 10 minutes in real-time would be completed in only 6 minutes, 40 seconds. Unfortunately, the Lifting Body test program was canceled and there were no opportunities for a pilot to fly a simulator mission in compressed-time before flying the actual mission. However, three M2-F3 pilots flew a familiar mission at 1.5 times real-time and all agreed with "enthusiastic responses" (p. 2) that the modified simulator felt exactly like the aircraft.

The response of the engineering community to Kolf's M2-F3 experiment was that the physical environment of flight is so different from simulation that it was impossible to draw inferences about above real-time simulation and training. Hoey (1976) described an application of compressed-time to a flight test program for a remotely controlled aircraft that a pilot flies from a ground-based console. The training simulator for the remotely controlled aircraft was the same console as for actual flight except that the pilot controlled a virtual rather than an actual aircraft. Nevertheless, pilots of remotely controlled aircraft who used simulation at 1.4 times real-time as final preparation before a flight (pp. 5 and 18) reported being "less rushed and more confident" (p. 18) than when using real-time training exclusively, although no differences in mission performance were reported. A typical practice was to conduct 70% of training at real-time and the last 30% at 1.4 times real-time. Hoey reported, however, that information about these experiments with above real-time simulation was not widely disseminated and no further research was conducted.

## ENGINEERING FOR ABOVE REAL-TIME TRAINING

Modifying real-time simulators for above real-time operations has not proven to be difficult or costly. The major problem has been to overcome misunderstandings regarding the concept of above real-time simulation.

## Effects of Changing Simulator Time

When describing compressed-time applied to an air traffic control task, Vidulich, Yeh, and Schneider (1983) equated simulation at 20 times real-time with "an aircraft flying at 5200 knots" (p. 162). Although this characterization is easy to understand, it leads to misunderstanding. A fighter aircraft simulator operating at 1.5 times real-time might be described as having the aircraft fly at 450 knots rather than at 300 knots. This description is incorrect. An aircraft flying at 450 knots has different handling characteristics, such as turn radius, from the same aircraft at 300 knots. Also, other systems on board the aircraft, such as radar and other entities in the simulation, have not changed their simulated time. More correctly, in a simulation operating at 1.5 times real-time, everything happens 1.5 times faster than in a real-time simulation or the actual aircraft: the simulated aircraft moves over the (virtual) earth faster, the radar antenna sweeps faster, enemy aircraft move faster, and missiles fly faster.

Although pilots in the Kolf (1973) and Hoey (1976) studies described time-compressed simulation as feeling very much like the actual aircraft, some characteristics of a simulated aircraft operating above real-time will be physically incorrect when performance is compared to real-time. If an actual aircraft or a real-time simulation flies at 300 knots with a bank angle of 30°, its turn rate is approximately 126° per minute. When increasing simulated time to 2.0 times real-time, indicated speed and turn rate are unaffected, whereas in the physical world, doubling air speed should decrease turn rate by half for a given bank angle. In above real-time training at 2.0 times real-time, the out-the-window appearance will be that of flying at 600 knots with a turn rate of 252° per minute of clock-time. Such anomalies have the potential for causing pilots to reject the simulation as low-fidelity. However, our observation of pilots flying many different simulators has been that when actively engaged in a task such as intercepting another aircraft, they rarely notice increases in simulator time up to twice real-time. Human time perception is highly plastic, and judgments of absolute time intervals are often biased by concurrent activity (see Crane, Guckenberger, Schreiber, & Robbins, 1997, p. 2). Our experience has been that moderate increases in simulator time ($\leq 2 \times$ real-time) do not affect pilot acceptance of simulation fidelity.

## Algorithms for Effecting Above Real-time Simulation

Two basic types of algorithms changes have been utilized to implement above real-time training. The first method is to change the basic update rate of the system. If a simulator system normally operates at 60 Hz, increasing its update rate to 90 Hz produces above real-time operation at 1.5 times real-time. Conversion of the time base by alteration of the update rate requires no modifications to simulation model software and can be implemented without costly changes. As an illustration, consider the case where a simulator responds to hardware interrupts 30 milliseconds apart. Altering the hardware interrupts to occur every 22.5 milliseconds produces simulator time at 1.5 times real-time. Modification using this technique usually requires a change to the hardware clock and analysis to confirm sufficient spare processing time to satisfy the predicted update rate. This method can be cost effective and requires little change to the software.

The second method is referred to as the *hyper-time algorithm* and has been primarily utilized in converting existing simulation systems (Crane et al., 1997). The hyper-time algorithm is a technique for implementing above real-time simulation by altering the amount of assigned simulated time between frame updates. When the hyper-time algorithm is operating, the simulation model produces the same number of frames updates as during real-time operation, but more or less simulated time has passed between updates. In a system operating at real-time, the time integration factor, delta_t, equals the time between hardware updates. As an example, assume that a system operates at 100 Hz with delta_t equal to 10 msec (.01 sec). By multiplying delta_t by 2, the simulation will calculate the next update as if twice as much time (20 ms) has passed as the real-time simulation (2 × above real-time). If delta_t is multiplied by 0.25, the simulation will calculate the next update as if one fourth as much time (2.5 ms) has passed as the original (0.25 × less than real-time). Update rate or frame time is unaffected. From the pilot's view, aircraft behavior, out-the-window imagery, and instruments are all synchronized to the simulated time. For simulations running at a nondeterministic frame rate, the time constant, delta_t, is calculated between frames by calling an operating system time function each iteration and computing the difference between successive times to determine the current frame time (Guckenberger et al., 1995).

Above real-time training within a network of simulators operating under Distributed Interactive Simulation (DIS) protocols (IEEE Standard 1278.1-1995, Standard for Distributed Interactive Simulation) has been accomplished by implementing the hyper-time algorithm on each simulator (Guckenberger et al, 1995). In a DIS network, each simulator operating independently broadcasts position and other relevant information to all other simulators at a non-deterministic rate several times per second. A demonstration of above real-time operation on a network was conducted at the 1994 Industry/Interservice Training Systems and Education Conference (Guckenberger et al, 1995). Four flight simulators, a computer-generated fighter aircraft, and a viewport not connected to any of the simulated entities (DIS stealth platform) were modified to operate at three times real-time using the hyper-time algorithm with some additional modifications to insure that the DIS dead reckoning algorithms correctly incorporated changes to simulated time. The different simulators used different host computers and operating systems that required unique adaptations of the hyper-time algorithm. However, in all cases the modifications required less than 1 hour. No difficulties with cross-platform compatibility, network stability, or interactions among players such as missile fly-outs were observed.

## TIME AND PERFORMANCE IN TRAINING

### Basic Relationships

The NASA efforts reported by Kolf and by Hoey used simulated time as an independent variable within a training system. More typically, research on training uses the time required to complete a task as a dependent measure of skill acquisition. Fitts and Posner's (1967) model of skill acquisition uses time and errors to define three stages in learning a skill. In the first or cognitive stage, the trainee is given instruction in facts and procedures. Performance is slow and errors are frequent. The trainee must rely on declarative knowledge to perform the task. In the second or associative stage, declarative knowledge is transformed into procedural knowledge and errors in understanding are eliminated. Time required for performance is reduced but performance can be disrupted by unexpected events or distractions. In the third or autonomous stage, performance is most rapid and error-free. In this stage, performance is entirely based on procedural knowledge, and the performer may not be able to verbalize the declarative knowledge being used to solve a particular problem.

Comparison of Fitts and Posner's cognitive and associative phases of skilled performance to the autonomous phase corresponds well with the phenomenon of controlled versus automatic processing (see James, 1890; Schneider, 1985).

> Automatic processing is a fast, parallel, fairly effortless process that is not limited by short-term memory capacity, is not under direct subject control, and performs well-developed, skilled behaviors. Automatic processing typically develops when subjects deal with the stimulus consistently over many trials. Controlled processing is characterized as a slow, generally serial, effortful, capacity-limited, subject-controlled processing mode that must be used to deal with novel or inconsistent information. (Schneider, 1985, pp. 296–297)

Controlled processing is characteristic of a trainee in the cognitive or associative stage, whereas the autonomous stage demonstrates automatic processing. Ericsson and Polsson's (1988) analysis of skilled memory illustrates the differences between controlled and automatic processing in performance of a cognitive skill, and Klein's (1989) model of recognition-primed decisions (see also chapter 9, this book) shows its application to pilot performance.

### Above Real-time Training as an Instructional Technique

NASA's experience with above real-time simulation suggests that compressing time as an independent variable in simulator training may be an effective instructional strategy for developing automatic processing. Hoey (1976) conducted interviews with test pilots who had experienced above real-time training when participating in the remote controlled aircraft program. These individuals were highly experienced pilots with several thousand flight hours. Automatic processing for flight skills was not an issue. However, to maximize data return from a given test flight, mission profiles were designed to be demanding and to increase pilot workload on successive flights. The pilots reported that using above real-time training as the final step in mission preparation helped them feel more confident, less rushed, and better able to handle the workload during the actual mission. They also thought that above real-time training should be applied to training high workload tasks such as instrument approaches or emergency procedures and that it could be used to evaluate task-planning effectiveness and pilot readiness. Overall, the NASA experience suggests that above real-time training provides training benefits that would result from automatic processing: reduced workload and less time required to complete tasks.

## RECENT RESEARCH ON ABOVE REAL-TIME TRAINING

The NASA application of above real-time training was limited to experts preparing for specific missions. Manipulating simulator time has also been evaluated in training research with both novices and experienced individuals as trainees.

### Above Real-time Training Applied to Training Novices

Schneider, Vidulich, and Yeh (1982) and Vidulich, Yeh, and Schneider (1983) used time compression to help train air traffic controllers to perform a turn-point initiation task. The task was for the controller to monitor an aircraft's flight path on a radar display and issue turning instructions so that the aircraft would fly through a specific vector. Actual aircraft would traverse 20 nautical miles and require approximately 5 minutes at 260 knots to complete the turn. These researchers increased the apparent rate of time in the simulator to 20 times real-time so that a turn would be completed in approximately 15 seconds. The primary effect of this high level of time compression was to make it easier for the novice controller to see the aircraft's turn. Secondary effects were that the rapid pace of events focused the trainee controller's attention to the immediate task and allowed more training trials per hour than real-time simulation. Vidulich, Yeh, and Schneider (1983) trained university students over 4 hours to perform the turn-point initiation task. One group of students performed the task in real-time and experienced approximately 32 trials in 4 hours of training. A second group performed the same task at 20 times real-time and received approximately 260 time-compressed trials followed by only three or four real-time trials in 4 hours of training. All trainees were tested at real-time for 2 hours. Students trained using above real-time simulation showed significantly better performance at initiating turns properly. These authors assert that above real-time training encourages automatic processing by allowing many trials, not overloading memory, and training under a mild speed stress.

Guckenberger, Uliano, and Lane (1992) trained novices—university students—in tank gunnery using several above real-time training conditions. In this experiment, students were trained to perform gunnery tasks that required them to detect, identify, and shoot a moving target using a simulator for an M-1 tank. Students received familiarization trials in real-time followed by training trials in real-time or above real-time. Students were then tested in real-time. Subjects trained in above real-time showed better performance on test trials than the students trained in real-time.

Schneider (1989) proposed that the primary effect of time compression is to allow more training trials within a given period of clock-time. In the air traffic control studies, all subjects were given the same amount of training time (clock-time) in the simulator so that the subjects trained in above real-time received more training trials. In contrast, Guckenberger et al. (1992) gave all subjects the same number of training trials so that the above real-time training subjects received less clock-time in training than the students trained in real-time. Because the students trained using above real-time training performed better on real-time test trials than students trained in real-time, Guckenberger et al.'s results indicate that above real-time training has a beneficial effect beyond simply increasing the number of training events. It should be noted, however, that the Guckenberger et al. findings are based on a task familiar to anyone who has played video games (i.e., locate the target, put the cross-hairs on the target, and shoot), so that the subjects were not truly novices.

Crane et al. (1999) trained university students to perform a complex, aerial gunnery task in real-time or at 1.5 times real-time. Students completed a real-time performance test in which test trials were more demanding than training trials, that is, more targets maneuvering more aggressively. The performance test was completed either immediately or after 2 days. Performance scores did not decrease for either group over the 2-day retention interval compared to an immediate test. Further, students trained in above real-time performed on test trials as well as students trained in real-time but with less clock-time spent in the simulator. However, although test performance was equal, performance of students using above real-time simulation during training was depressed compared to students training in real-time, suggesting that initial task training should occur in real-time before implementing compressed time.

### Above Real-time Simulation Applied to Training for Experienced Individuals

Guckenberger, Uliano, Lane, and Stanney (1993) conducted an experiment using 24 experienced F-16C pilots. The pilots used a low-cost, F-16A trainer to fly several different missions. Pilots trained at real-time, 1.5 times real-time, 2.0 times real-time, or with a random mix of apparent times. Pilots then tested at real-time. One task required the pilot to both engage an enemy aircraft and then to perform a complex albeit contrived threat response when a warning was detected. For this dual-threat task, all above real-time training groups showed significantly greater numbers of bandit kills during real-time, test trials. Also, the 2.0 times real-time and mixed above real-time groups showed faster threat response than the group trained in real-time.

Crane and Guckenberger (1997) performed two experiments using above real-time training with experienced and student F-16 pilots. In one experiment, experienced F-16 pilots assigned to operational squadrons conducted single-ship, defensive, counterair missions over a ground target using two scenarios. In one scenario, single-emergency, the pilot's aircraft suffered engine failure. The pilot's task was to restart the engine and then to engage an incoming bandit. In the second scenario, multiple-emergencies, the pilot had to respond to several emergencies and engage two bandits in succession. Pilots received initial training in real-time followed by additional practice in real-time or at 1.5 times real-time. All pilots were then tested in real-time. Pilots in the single-emergency task performed engine restarts and defeated a bandit aircraft as quickly as pilots trained in real-time, but no faster. However, in the more complex multiple-emergency task, pilots who practiced using above real-time training were able to perform two of the three required emergency procedures, and killed both bandit aircraft, significantly faster than pilots who trained in real-time. Above real-time training provided better training than real-time training for these experienced pilots provided that (a) the tasks being trained are highly demanding of a pilot's time and workload management skills, and (b) the pilot has received initial training in real-time.

In a second experiment, student F-16 pilots practiced a radar-skills task as a supplement to their normal training syllabus. Students practiced radar skills in real-time or above real-time after they had successfully completed the air-to-air portion of the F-16 syllabus. In this experiment, rather than receiving an equal number of training trials, pilots using above real-time training received more training trials than pilots trained in real-time using approximately the same amount of clock-time in the simulator. All pilots were then tested on more complex scenarios presented in real-time. Student pilot scores on the test trials were significantly higher for the above real-time group than for the real-time group. Furthermore, the pilots trained in above real-time showed a significant increase in scores from the difficult training trials to the more complex test trial.

## SUMMARY OF RESEARCH ON ABOVE REAL-TIME TRAINING

NASA's experience with above real-time training was that time-compressed simulation can be used to augment conventional simulator training in order to encourage automaticity and reduce pilot workload. Following Fitts and Posner's (1967) three-stage model of skill acquisition, NASA's application of above real-time training suggests that it should be most useful for the second or associative stage of training rather than for the first or cognitive stage. Other recent experimental evidence supports this position. Schneider, Vidulich, and Yeh (1982) and Vidulich, Yeh, and Schneider (1983) used above real-time training with great success with novices as trainees. These studies greatly increased simulated time, 20 times real-time, in order to time-compress events that take too long to see in real-time. The novice subjects in these studies received extensive training on the task, suggesting that for much of their training they were in the associative rather than the cognitive stage. Guckenberger, Uliano, and Lane (1993) also trained novices using above real-time simulation and found that for a tank gunnery task, above real-time training increased performance on a real-time transfer task. Guckenberger, Uliano, Lane, and Stanney (1992) and Crane and Guckenberger (1997) found above real-time training to be beneficial for experienced pilots in a dual or multiple threat task. Crane and Guckenberger (1997) also found above real-time simulation to be beneficial in training student F-16 pilots to perform radar skills tasks that were more demanding than normal, syllabus training. However, Crane et al. (1999) found that performance in training novices can be depressed by above real-time simulation. Overall, research suggests that above real-time simulation should be an effective strategy for training tasks that require significant effort at time and workload management, and trainees who have completed the cognitive portions of skill acquisition. For this combination of task and trainee, above real-time training represents an instructional tool consistent with Schneider's (1985) guidelines for training complex skills:

- Design training to allow many trials of critical skills.
- Maintain active participation by minimizing passive observation of the task.
- Maintain high motivation throughout the training period.
- Train under mild speed stress.
- Train strategies that minimize operator workload.
- Train time-sharing skills for dealing with high-workload environments. (pp. 297–298)

Above real-time training is an instructional strategy that is inexpensive to implement and can increase the effectiveness of simulator-based training. Like freeze-and-reset, replay-and-review, task segmentation, or backward chaining, compressed-time simulation takes advantage of capabilities inherent in flight simulator systems. Flight simulation can both present the student with a high fidelity facsimile of the real world and be configured to provide training opportunities that would be difficult or impossible to create in actual aircraft.

## REFERENCES

Adams, J. A. (1979). On the evaluation of training devices. *Human Factors, 21*, 711–720.

Adams, J. A. (1989). *Human Factors Engineering*. New York: Macmillan.

Crane, P., & Guckenberger, D. (1997). Above real-time training applied to air combat skills. *Proceedings of the 19th Industry/Interservice Training Systems and Education Conference*. Orlando, FL: National Security Industrial Association.

Crane, P., Guckenberger, D., Rossi, M., Ali, A., Archer, M., & Williams, J. (1999, April). *Retention of effects of above real-time training*. Paper presented at International Conference of Aviation Psychology, Columbus, OH.

Crane, P., Guckenberger, D., Schreiber, B., & Robbins, R. (1997, August). *Above Real-Time Training (ARTT) Applied to Air Combat Skills: Final Report* (Tech. Report AFRL-TR-1997-0104). Mesa, AZ: Air Force Research Laboratory.

Ericsson, K. A., & Polson, P. G. (1988). An experimental analysis of the mechanisms of a memory skill. *Journal of Experimental Psychology: Learning, Memory, and Cognition, 14*, 305–316.

Fitts, P. M., & Posner, M. I. (1967). *Human Performance*. Belmont, CA: Brooks-Cole.

Guckenberger, D., Guckenberger, L., Whitney, R., Anschuetz, E., Hall, G., & Mair, D. (1995). DIS above real-time air combat demonstration. *12th Workshop on Standards for the Interoperability of Simulators*. Orlando, FL: Institute for Simulation and Training.

Guckenberger, D., Uliano, K., & Lane, N. (1992). The application of above real-time training (ARTT) for simulators: Acquiring high performance skills. *Proceedings of the 14th Industry/Interservice Training Systems and Education Conference*. Orlando, FL: National Security Industrial Association.

Guckenberger, D., Uliano, K., Lane, N., & Stanney, K. (1993). The effects of above real-time training (ARTT) on three tasks in an F-16 part-task simulator. *Proceedings of the 15th Industry/Interservice Training Systems and Education Conference*. Orlando, FL: National Security Industrial Association.

Hoey, R. G. (1976). *Time compression as a means for improving the value of training simulators*. In P. Crane, D. Guckenberger, B. Schreiber, & R. Robbins (1997, August), *Above Real-Time Training (ARTT) Applied to Air Combat Skills: Final Report*, pp. 41–65 (Tech. Report AFRL-TR-1997-0104). Mesa, AZ: Air Force Research Laboratory.

James, W. (1890). *Principles of Psychology, Vol. 1*. New York: Holt.

Klein, G. A. (1989). Recognition primed decisions. In W. B. Rouse (Ed.), *Advances in man–machines systems research* (pp. 47–92). Greenwich, CT: JAI Press.

Kolf, J. (1973). Documentation of a simulator study of an altered time base. In P. Crane, D. Guckenberger, B. Schreiber, & R. Robbins (1997, August), *Above Real-Time Training (ARTT) Applied to Air Combat Skills: Final Report*, pp. 36–39 (Tech. Report AFRL-TR-1997-0104). Mesa, AZ: Air Force Research Laboratory.

Schneider, W. (1985). Training high performance skills: Fallacies and guidelines. *Human Factors, 27*, 285–300.

Schneider, W. (1989). Getting smarter quicker: Training more skills in less time. *Science and Public Policy Seminar*. Washington, DC: Federation of Behavioral, Psychological, and Cognitive Sciences.

Schneider, W., Vidulich, M., & Yeh, Y. (1982). Training spatial skills for air-traffic control. *Proceedings of the Human Factors Society 26th Annual Meeting*. Santa Monica, CA: Human Factors Society.

Vidulich, M., Yeh, Y., & Schneider, W. (1983). Time compressed components for air intercept control skills. *Proceedings of the Human Factors Society 27th Annual Meeting*. Santa Monica, CA: Human Factors Society.

Chapter 9

# How Can We Train Pilots to Make Better Decisions?

Gary Klein
*Klein Associates, Inc., Fairborn, OH*

The question guiding this chapter, how to train pilots to make better decisions, is simple and straightforward. It is linked to important issues of application because inadequate pilot decision making has been strongly linked to aviation accidents. It is also linked to important questions of basic research because the challenge of training decision-making skills is so difficult.

This chapter uses the Naturalistic Decision Making (NDM) framework (Flin, Salas, Strub, & Martin, 1997; Klein, Orasanu, Calderwood, & Zsambok, 1993) to explain how people function in a variety of challenging situations. It has been applied in many fields: firefighting, command and control, process control, medicine, and a number of others, including aviation (Flin, 1996; Klein, 1998). Although the NDM framework is primarily descriptive, it carries important implications for improving training. In order to explain the rationale for using the NDM framework, it may be useful to provide some background about the nature of NDM research.

According to Zsambok and Klein (1997), NDM studies the way people use their experience to make decisions in field settings. One key aspect of this definition is the emphasis on experience. The issue is how people develop and use experience, and the types of strategies that are adopted to take advantage of experience. Researchers have difficulty replicating those strategies when they perform controlled studies with novices who are only given a brief amount of practice. In natural settings, novices are rarely given responsibility for high stakes decisions. Therefore, we need to learn more about how experience is gathered and used. A second key aspect of the definition is the emphasis on field settings. A great deal of NDM research is conducted in the field in the form of observations and interviews, and even in retrospective accounts of previous incidents. Even when the research is conducted under more controlled conditions, it is usually in the context of an engaging simulation that carries functional fidelity to the task at hand, so that the results can generalize to the field settings.

NDM research also examines a range of factors that are not easily studied in context-restricted experiments. Orasanu and Connolly (1993) listed many of these, and Cannon-Bowers, Salas, and Pruitt (1996) updated that list, as presented in Table 9.1. The factors of interest are time pressure, shifting conditions, feedback loops to reflect the consequences of actions, meaningful consequences of actions, ill-structured tasks, multiple (and conflicting) goals, uncertainty, team interactions, organizational constraints, quantity of information, and experienced decision makers. These are the features of tasks, environments, and people that make decision making challenging. Most NDM studies incorporate several of these characteristics. One of the goals of NDM is to provide a better understanding of how these factors affect decision making.

Rasmussen (1997) offered a different perspective, one that is consistent with Table 9.1, but that approaches NDM from an historical perspective. He viewed NDM as part of an orderly progression found in other fields of inquiry. In several fields, the initial approach is based on a formal analysis of the process. In time, the limitations of this analysis become clear, and a second account develops that is a critique of the first. Eventually, researchers become less interested in the formal analysis and its limitations, and begin an exploration of the phenomenon itself. Rasmussen described this progression in the field of decision making: the initial research centered around formal analyses such as game theory and Bayesian statistics. Next, Tversky and Kahneman (1982) provided a critique of the normative approach by showing that people did not follow these normative principles, but instead relied on imperfect heuristics. The third phase, exemplified by NDM, is where researchers have little interest in techniques for defining optimal decision strategies. Instead, they explore how people adapt to the complexities of different environments that are marked by the features listed in Table 9.1.

It is important to note that the field of NDM investigates topics such as problem solving, planning, and situation awareness, as well as decision making. For most field settings, it is difficult to make fine distinctions between these processes. In contrast, many traditional decision researchers concentrated on the moment of choice, either between alternate courses of action or alternate hypotheses, and excluded consideration of problem solving and planning processes.

The term NDM was derived in 1989, at a conference of researchers from industry, government, and academia (Klein, Orasanu, Calderwood,

TABLE 9.1
Factors That Are Central to NDM Research

| | *Feature of* | | | |
|---|---|---|---|---|
| *Variable* | *Decision/ Task* | *Decision Maker* | *Environment* | *Comments* |
| Uncertain, dynamic task | • | | | Perhaps the single most important feature of interest |
| Multiple event feedback loops | • | | • | Related to the dynamic nature of many naturalistic decisions |
| Meaningful consequences | • | • | | Can be related to the decision maker's motivation and/or to the nature of the task |
| Ill-structured decision | • | | | Structured decisions are not typically complex or in need of attention using "naturalistic" methods |
| Multiple goals | • | | • | Multiple goals may or may not be shifting, competing, or unclear |
| Time constraints | • | | • | Time pressure affects the processes by which a decision is made and is a critical NDM factor |
| Decision complexity | • | | | Complex decisions are more interesting to investigate than simple ones in part because they pose greater challenges for aiding and training |
| Multiple players | • | | | Multioperator decisions are clearly of interest under NDM; however, complex individual decisions are important as well |
| Congruent organizational norms and goals | | • | • | Congruence between the decision maker and the organization can make a decision more difficult; situations in which goals are compatible are of interest as well |
| Quantity of information | • | | • | Knowledge-rich environments present added challenges to decision makers; information overload is an important naturalistic feature |
| Level of experience | | • | | While a major goal of NDM is to look at the impact of expertise on decision making, there is much to be gained by investigating processes employed across a wide range of expertise. |

*Note.* From Cannon-Bowers, Salas, & Pruitt (1996). Reprinted with permission from *Human Factors, Vol. 38, No. 2,* 1996, by the Human Factors and Ergonomics Society. 

& Zsambok, 1993). Researchers such as Hammond (1993), Rasmussen (1986), and Wohl (1981) had performed NDM research years earlier, and many of the investigators at the conference had already been doing NDM types of studies. That was why they were interested in identifying common aspects of their work. One of the discoveries at the 1989 conference was that they were relying on common methods and similar questions. The findings in one field appeared to be relevant to other fields, so there was a basis for learning from each other, and the applications, such as training and design, also appeared to generalize across domains.

One of the challenges facing NDM researchers is to demonstrate the "value added" of the approach. In order to have an impact, NDM studies must discover aspects of decision making that would otherwise be hidden, and NDM applications must be different from the recommendations made by practitioners working from other frameworks. This is one of the objectives of this chapter: to show why NDM makes a difference to the aviation community.

## HOW DO PILOTS MAKE DECISIONS?

Researchers believed they knew the answer to this question. However, the answer that was widely accepted turned out to be wrong, and the consequences for pilot training are important.

The original answer was that pilots made decisions by considering a range of options, evaluating these options on a set of evaluation categories, judging how each option scored on each evaluation category, perhaps assigning different weights to the categories to reflect their importance, totaling up the results, and finding the option with the highest score. This can be referred to as a Rational Choice strategy (Klein, 1998). It is widely taught in business schools and schools of engineering, and for many years it has been taught to pilots. One popular version is the DECIDE model (Benner, 1975). The six steps of the prescriptive model of decision making are to:

- Detect that an important change has occurred.
- Estimate how that change will impact the mission.
- Choose a safe outcome.
- Identify plausible courses of action.
- Do (execute) the action that was chosen.
- Evaluate the effect of the action.

The training was to consider all the relevant options, consider the relevant evidence, and find the best.

The Rational Choice method can be useful, and it has a number of strengths. It is a general method, so pilots can use it with different types of decisions. It is reliable—pilots can rely on getting the same answer if they run the method several times. It is comprehensive. The method requires pilots to consider a wide range of evidence and issues.

However, as researchers studied decision making in natural settings (Klein, 1998; Klein et al., 1993; Zsambok & Klein, 1997), we also learned that this Rational Choice method has some limitations. It takes a lot of time to perform the analysis, and in settings such as aviation, a pilot may not have that time. An aborted takeoff does not permit careful consideration of alternative courses of action. The Rational Choice method requires a great deal of data, and pilots often have to wrestle with uncertainty, such as weather conditions, the nature of a malfunction, the possibility of a go-around, and so forth. Another limitation is that a Rational Choice method is most helpful when the options are close together, rather than having one option clearly good and the others clearly bad. The closer together the options are, however, the less it matters to find the best. In fact, under time pressure, pilots often look for the first workable option and don't care about finding the absolute best. Klein (1989a) found that in a variety of natural settings, skilled decision makers rarely compared options. Even a watered-down version of the Rational Choice method may be used less than 5% to 10% of the time for critical decisions (Klein, 1998).

If pilots aren't comparing options, then how are they making decisions? We found that skilled decision makers can use their experience to generate an effective option as the first one they consider (Klein, Wolf, Militello, & Zsambok, 1995).

The Recognition-Primed Decision (RPD) model (Klein, Calderwood, & Clinton-Cirocco, 1986) attempts to explain how people can make decisions in field settings, under conditions that make it difficult to generate and compare multiple options. It must be emphasized that the RPD model is only one of many NDM models. Lipshitz (1993) reviewed nine different NDM models.

The basic assertions of the RPD model are that:

- People are able to use their experience to size up a situation as familiar.
- By recognizing a situation, people can generate a reasonable option as typical.
- Usually, this typical option is the first one they consider.
- If they are unsure of the situation, they may build stories to explain different events (although this may take a little time).
- If they are unsure of the option they generated, and have the time, they may build stories to evaluate how the option will be carried out in context.

The RPD model consists of three variations, shown in Fig. 9.1. In the simplest case, the decision maker judges the situation to be functionally identical to a prototype. A prototype is a typical, or representative, case; a robin and a swallow are prototype birds, whereas an ostrich is not. The match to a prototype allows a determination of the relevant cues, reasonable goals, expected developments, and a typical course of action. This is a form of rule-following behavior. However, we should not treat this as mechanistic. Skill is needed to judge whether the antecedent condition of the rule has been satisfied. Experience is needed to recognize a situation as typical, and to make an appropriate prototype match. All pilots know the rule not to fly into turbulent weather. It is a simple and important rule. However, merely telling a novice pilot the rule is no guarantee of safety. Skill is needed to determine what counts as turbulent weather. Many rules sound simple but turn out to be complex under field conditions.

Klein (1989b) reported a great deal of evidence from different domains (e.g., firefighters, tank platoon leaders, designers) that skilled decision makers rarely compare courses of action to each other. Variation 1 of the RPD model explains how that can be. If decision makers can use their experience to generate a reasonable course of action (i.e., one that is workable and appears to satisfy the required outcome) as the first one they consider, then there is little need to go through the exercise of generating a range of options. Under conditions such as time pressure, it can be critical to decide and act quickly.

There is some empirical evidence supporting this claim of the RPD model. Stokes, Kemper, and Kite (1997) conducted a study of pilot decision making and found that experienced pilots generally selected the first option they considered, even when they followed instructions and generated several options. Klein, Wolf, Militello, and Zsambok (1995) showed that for even moderately experienced chess players, the first option they considered was usually a good one. This explains why pilots don't need to use a Rational Choice method. Their experience base lets them identify a good option as the first one they consider, once they have sized up the situation.

Sometimes, pilots may be unsure about what is going on in a situation. They do not try to represent their uncertainty by estimating probabilities of different explanations. Rather, they try to build stories to explain the evidence, and they judge the plausibility of each story. Thus, when faced with a strange reading that one of the fuel tanks seems to be running lower than the others, a pilot might explain this away as a failure of the measuring instruments. An alternative explanation is that there might be a leak in that fuel tank, but this is more rare than a fault in a sensor. The pilot does not attempt to estimate the probability of each, but rather compares the plausibility of the two. It is easier to imagine a sensor going bad than a fuel tank suddenly developing a leak for no apparent reason. The

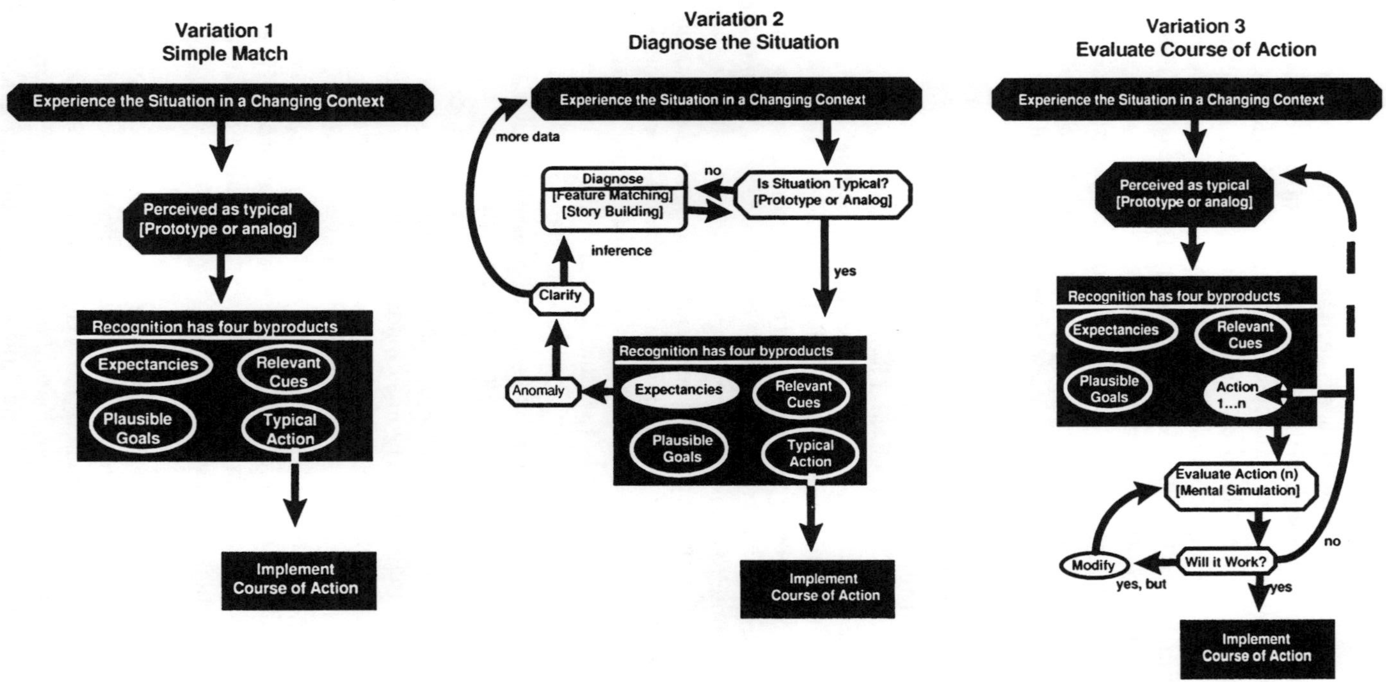


FIG. 9.1. Recognition-Primed Decision (RPD) Model

assessment relies on the heuristic of the availability of the two stories. This is described in Variation 2 of the RPD model. It takes experience to be able to build coherent stories (Cohen & Freeman, 1996; Klein & Crandall, 1995). Novices are unable to do this, particularly when the evidence is complex. This type of story building is often seen in diagnosis of equipment malfunctions (De Keyser & Woods, 1993). It is more difficult to build stories in time-pressured situations.

Pilots may need to evaluate a course of action once they recognize it. How can you evaluate an option without comparing it to others? We found that decision makers again build stories, to see how the option will be carried out. This is Variation 3 of the RPD model. We call this a process of mental simulation, thinking out how the action will proceed. If they do not find any difficulties, then they are fine. If they do spot a problem, they may find a way to improve the action plan. If they can't find a way to improve the action, then they may reject the option and look for a better one. For example, a captain preparing to land with engine problems might rehearse the sequence with the first officer. In so doing, they might identify points at which nonessential power requirements could be cut, to ensure sufficient power in case a go-around was necessary (a go-around is an aborted landing, followed by a route to attempt the landing a second time). If the uncertainty is too high (e.g., if the airport has been experiencing wind shear), then the pilots may reject the option of landing at that field and select another.

Clearly, there are times when pilots do need to compare several options rather than select the first that is satisfactory. For example, if a malfunction requires an airplane to divert (i.e., abandon the plan to land at a given airport in favor of an alternative), the pilot must consider which airports are in the vicinity and pick one of them. Here, we might see some sort of Rational Choice comparison. The RPD model was not developed to explain cases in which multiple options are compared. However, I believe it is more likely that the decision is made by mentally simulating each option, to see if there are any difficulties. In this way, some airports will be dropped from consideration. The remaining ones, all acceptable, may be compared at a global level, depending on which one provides the greatest comfort level. This explanation follows the work of de Groot (1978), who referred to the process of *progressive deepening* (essentially the same as mental simulation) whereby chess players imagine how a line of play might develop and determine a global reaction to the type of position they might expect, then progressively deepen another move, form an impression of it, and eventually select the move that offers the most positive reaction. In this way, different moves are compared. However, the comparison does not take place along a common set of evaluation dimensions (e.g., how much center control does each move provide, how strong is the king side attacking potential, and so forth). This type of strategy is not the same as noncompensatory strategies such as *elimination by aspects* (Tversky, 1972), in which all options are evaluated along the most important evaluation dimension, the inadequate options are dropped, then they are rated on the next most important dimension, until one option remains. De Groot's work showed that the chess players were imagining each move in the context of the game, not along common dimensions.

I suspect the same is true of pilots. It is possible that pilots who need to divert are using an elimination-by-aspects strategy (Tversky, 1972), comparing different airports on common dimensions: which airports are within 100 miles; then, if more than one remains, which airport has the longest runways; then which airport has the fewest terrain risks, etc. However, I hypothesize that the process is more along the lines of de Groot's concept of progressive deepening. A pilot might first identify the airports that are close enough to consider. Then, for each one, a progressive deepening would either cull it out (too much of a chance that they'll close that airport by the time I get there), or continue it (I can live with it, but I wish it had better connections for my passengers), or select it (I'm not crazy about flying in there because of the mountains, but the winds seem to be light enough so that I shouldn't have much trouble).

For the NDM accounts of aviation decision making, situation awareness is a key process: the way experienced pilots size up a situation tells them how they should respond. *Situation awareness* is the ability to identify the affordances in a task; a pilot with good situation awareness will be able to comprehend the various forces that are acting on the airplane and will be able to anticipate how these forces will shape the future course of the airplane. The simple variation of the RPD model shows a reaction entirely conditioned on the size-up of the situation. The variation about diagnosing the events is about how pilots would construct explanations and stories to form situation awareness and thereby know how to proceed. The variation about mentally simulating a course of action describes how pilots would project the current situation forward in time and imagine how a course of action would be carried out in the context of the current and projected situation. Choosing between options sometimes occurs, but it does not appear to require an analytical calculation of strengths and weaknesses. Pilots usually generate options one at a time until they find one that works. They are not attempting to find the absolute best option. When pilots do have to make choices between options, they do not seem to be balancing strengths and weaknesses, or taking stock of probabilities and utility values. Rather, they are running mental simulations of their mental models regarding the dynamics of the situation and the way alternative courses of action would be carried out. Their choices are based on their assessment of the situations that are likely to be produced. By inspecting the mental simulation for any given action, they can see if it leads to unintended consequences that are unacceptable.

## WHY DO PILOTS STRUGGLE WITH DECISIONS?

The most common reason for poor decisions is a lack of experience.

- It takes a high degree of experience to recognize situations as typical; to notice anomalies, it is first necessary to recognize typicality so that you can see when something unusual has happened.
- It takes a high degree of experience to build stories to diagnose problems and to mentally simulate a course of action.
- It takes a high degree of experience to prioritize cues, so workload won't get too high.
- It takes a high degree of experience to develop expectancies and to identify plausible goals in a situation.

In a review of several studies in different domains, Klein (1993) traced most poor decisions to a lack of experience in anticipating how a course of action could run into difficulties. The problem was not in the decision strategy, but in the lack of knowledge to appreciate the implications of the situation.

Acute stressors are another reason for poor decisions. Conditions such as time pressure, uncertainty, high workload, and increased risk of accidents all contribute to the stress level of pilots. However, Klein (1996) argued that these stressors do not themselves result in poor decisions. Their impact is more indirect. Time pressure reduces the opportunity to gather information and to build stories and mental simulations. Uncertainty reduces the quality of the data used to make the decision. Physiological changes (rapid heart rate and breathing) may be monitored and managed, adding a distracting task at a time when attention needs to be focused. So, for experienced decision makers, the decision strategy itself is not degraded by stress. The effect of stressors is on other factors that interfere with decision making.

Another reason for poor decisions is a tendency to explain away early signs of problems. In a variety of domains (e.g., Perrow, 1984), we find even skilled decision makers explaining away anomalies and cues that indicate difficulties. If a warning light comes on, it can be explained away as a malfunctioning indicator. Pilots may respond to it, but without giving it credibility. We observed this in a simulated environment, where aircrews responded to warnings of fuel loss in one of the tanks. This was an unlikely malfunction, and the initial response was to explain it away as a malfunctioning indicator rather than a problem with the fuel tank. More than 10 minutes later, supporting evidence was received, and the malfunction was treated more seriously.

Decisions can be difficult because of uncertainty. Uncertainty is a condition of doubt that threatens to block action (Lipshitz & Strauss, 1996). Pilots can be uncertain about the data, about inferences drawn from the data, and about how to use these inferences to anticipate future states. The uncertainty may be caused by missing information, by unreliable information, by ambiguous or conflicting information, and by complex information (Cohen & Freeman, 1996; McCloskey, 1996). Schmitt and Klein (1996) described the various aspects of uncertainty affecting United States Marine Corps commanders. Despite hopes that information technology will eventually dispel uncertainty, our conclusion was that uncertainty is inevitable in most operational environments. Technology may shift the nature of the uncertainty, rather than reducing it. For example, a commander who used to be frustrated by not knowing where all his troops are, is now frustrated by a blizzard of messages telling him more than he needs to know and making it harder to find the critical information. The fog of war is being replaced by a dust cloud of data. The hope in aviation is that skillful design of decision support systems and interfaces will make the pilot's job easier. Sarter and Woods (1991) were not optimistic. Their research into clumsy automation (i.e., the use of computers to perform a task in a way that interferes with skillful performance) showed that systems such as Flight Management Systems (FMS) are not improving the life of pilots. The FMS takes over the simple tasks in aviation (such as maintaining airspeed) and makes the pilot's job even more tedious, but during nonroutine events, the FMS creates its own confusions (such as the way the system "understands" the nonroutine events, and the reaction the system is "trying" to implement). To the uncertainty of events, the pilot now has to contend with uncertainty about the FMS and how it is going to respond.

Unsafe attitudes are another source of difficulty. Berlin et al. (1982) documented a set of common attitudes that can lead to accidents, and many of these can be traced to goal conflict (e.g., conflicts between safety and schedules). It is easy for airlines to issue guidelines for safety, but it is not as easy to transform these into consequences. The consequences for taking risks to meet a schedule are usually favorable. As long as no problems occur, pilots are rewarded for risk-taking behavior. When airlines begin to discipline pilots who successfully take risks, then we may see some change from the ideal of the pilot with the "right stuff" who can make it through regardless.

I should point out one factor that is not included on the list of reasons for pilots to struggle with decisions, and that is the influence of decision biases. Decision heuristics refer to strategies people use to obtain rough answers and solutions in cases where the algorithms are too complex to be computed. Sometimes, these heuristics generate misleading answers and may be seen as biasing the decision maker. Although the study of heuristics and biases has been active for more than 20 years, Christensen-Szalanki (1993), Cohen (1993), Fraser et al. (1989), Gigerenzer (1991), and many others have raised serious questions about whether these biases create difficulties outside of the laboratory setting, particularly when experienced personnel are involved.

## APPROACHES FOR IMPROVING THE DECISIONS OF PILOTS

This section explores different pathways for training pilots to improve their decision-making skills. Three general approaches are discussed: ways of asking better questions, applications of NDM models of decision making, and a specific decision skills training approach.

### Asking Better Questions

The first thing we can do to improve decision training is very simple: identify the critical decisions that need to be trained. In many flight manuals, much of the space is devoted to the task elements, such as releasing brakes, advancing power, and so forth, and the decisions and judgments are tucked away underneath, as adjuncts to unlikely sequences of events that should not happen if all the steps are carefully followed. Thus, one airline prepared a task analysis of aviation performance, listing the major tasks (e.g., 2.1: perform normal takeoff operations), then the second level tasks (2.1.2: perform takeoff roll procedures), the third level more specific tasks (2.1.2.1: position aircraft on runway centerline and stop . . . 2.1.2.5: set weather radar for takeoff . . . 2.1.2.13: monitor engine and flight instruments), and down to a very fine level of detail (2.1.2.13.1: if abnormality exists, captain must decide and initiate rejected takeoff). Needless to say, the emphasis in training was on the primary tasks, even though these were easier to teach and were fairly procedural. The decision-making tasks received little attention and virtually no training.

We need to ask better questions. We cannot help pilots make better decisions by teaching them some magical formula that will always apply. Instead, we need to find out which decisions are creating difficulties and make those the target of interest. Instead of asking "How can we teach pilots to make better decisions?", we should ask more specific questions, such as "How can we teach pilots to make more accurate assessments of the likely effects of weather conditions?" The first question seems hard, but it is just too vague. It gives us a forum to examine different ideas, but it does not allow a training specialist to determine what to do next. It is ill-defined; we do not know what counts as an effective answer. Also, it can lead to some misguided steps if a training specialist does try to give a simplistic answer (i.e., teaching a generic strategy such as Rational Choice, or teaching fundamentals of probability theory). There is no reason to believe that these steps will be useful. The second question is one that we can work on more productively. In pursuing it, we will undoubtedly learn that there are certain weather conditions that account for most of the problems, and we can focus the training even more clearly. We don't need to train for all weather conditions. If we can identify the most frequent tough conditions, we can start there. In this way, the question becomes more specific and more useful.

One way to proceed is to identify the decision requirements for pilots. Decision requirements are: the critical judgments and decisions, why they are difficult, what types of errors people make, and what types of strategies are used by more experienced personnel. Table 9.2 presents a decision requirements table from a study of en route air traffic controllers (Hutton, Olszewski, Thordsen, & Kaempf, 1997). It is not enough to assert that en route air traffic controllers need to be able to prioritize their scanning quickly. By using cognitive task analysis methods, Hutton et al. identified the reasons why the prioritization was difficult (e.g., having to anticipate the secondary effects), the potential errors (e.g., tunneling on one problem and missing the early signs of others), and the strategies that seem to help (e.g., provide hand-offs as early as possible). These types of descriptions offer a good basis for building a training program.

To summarize, if we want to improve the quality of decisions made by pilots, we should try to focus our efforts around specific types of decisions and around the factors that make these decisions problematic.

## Applications of NDM Models of Decision Making

The very fact that we would consider training decision making is a sign of progress. Kaempf and Klein (1994) reviewed the history of training decision skills for pilots. Twenty years ago, pilot training was about stick and rudder skills. Task analysis methods were used to identify behaviors, rather than to identify the judgments and decisions within a task. Then researchers found that decision errors, rather than procedural or manual errors, were implicated in a number of accidents. The reason for accidents changed from control errors to decision errors as aircraft design improved. In the early 1980s, aviation decision making was developed to address the problems of poor cockpit decisions. However, the training emphasized the Rational Choice method (e.g., the DECIDE model of aircrew decision making) and, as already explained, this does not match with the conditions faced by pilots. The time pressure, uncertainty, shifting conditions, and ill-defined goals all make it difficult to apply a Rational Choice strategy, and pilots therefore do not use such an analytical approach.

TABLE 9.2
Decision Requirements Table for Air Traffic Controllers

| *Cognitive Demand: Resolve Conflicts* | | | | |
|---|---|---|---|---|
| *Cues* | *Strategies* | *Why Difficult* | *Potential Errors* | *Tools/Aids* |
| • free altitudes<br>• free air space in sector for vectoring<br>• proximity to sector boundary<br>• aircraft type/capabilities (nav, climb/descent rates, turn rates)<br>• altitude<br>• aircraft type<br>• proximity of aircraft to be deconflicted<br>• proximity/heading of other aircraft<br>• distance from destination<br>• wind speed/direction | • depends on personal preference<br>• depends on what has worked well in the past (few "problems" are occurring for the first time; they are often repetitive problems with different nuances in terms of timing)<br>• look for a simple solution to reduce workload (vectoring requires more than one action, altitude change may only require on action)<br>• pilots may want to go higher for fuel conservation | • ripple effects may make a problem worse<br>• an involved solution may increase workload, leading to the controller "losing the bubble" or "going down the tubes"<br>• knowing when to give up on a course of action and use your "out"<br>• calculating speed, taking wind, altitude, and climb/descend into account<br>• accounting for aircraft capabilities (climb/turn rates)<br>• allowing for pilot tendencies to stay high and fast as long as possible<br>• allowing for nuances in maneuvering | • complex solution increases workload, where a simple solution would avert other problems<br>• may have to react to a failing course of action by asking for a rapid maneuver<br>• fixate on one course of action and make it work, even if it looks like it won't work<br>• complex solutions may require more effort in terms of communication (vectoring) and coordination (for point-outs or hand-offs)<br>• forgetting to complete a vector | • memory aids for vector resolution (J-ball, offset strips)<br>• leader lines to judge rate of closure and time available to effect a course of action |

For purposes of training, an instructional program may find it useful to have students compare options. Trainees can learn a great deal by seeing why one strategy works whereas another is inadequate. However, even if trainees are taught to contrast different options, I hypothesize that they will learn more by mentally simulating each option and gauging their comfort with the imagined outcome, than by trying to set up a standardized set of weighted evaluation criteria and ranking the options on the criteria.

Thus, if recognitional decisions depend on judging typicality, then a training requirement is to develop a sense of typicality for a variety of situations. This may include perceptual learning in order to make fine discriminations and to detect patterns. It could include building the sorts of mental models that will enable a decision maker to generate expectancies. It could include the development of a repertoire of typical reactions. Moreover, the mental models should be sufficiently rich to enable the decision maker to mentally simulate the consequences of different courses of action, as well as to construct stories that would explain observed events. Finally, training could help the decision maker learn to manage attentional resources in order to effectively seek new information at the appropriate times.

The Recognition-Primed Decision (RPD) model, along with other NDM models and complementary work in cognitive psychology, can be used to offer suggestions about where to focus the training so that pilots can learn to make better decisions. Figure 9.2 shows, on the left-hand side, a composite version of the RPD model, blending all three variations. The right-hand side translates the RPD model into suggestions for training.

The RPD begins with a recognition of typicality. Therefore, this is a training objective, to teach typicality, to teach common patterns and configurations and in so doing, to provide a basis for noticing deviations, for detecting anomalies. Problem detection is a critical skill for pilots. It requires a firm grasp of typicality. Novices may delay in judging an event as atypical, and in so doing, may lose valuable time for responding.

The recognition of typicality is an aspect of situation awareness, and so this becomes a cluster of skills to consider in training. Endsley (1995) described three levels of situation awareness; the awareness of data, the drawing of inferences based on those events, and the ability to project current states forward in time (e.g., mental simulation). All of these levels are potential targets for training interventions. In addition, I would add the ability to quickly determine shifts in a situation. De Groot (1978) noted additional skills here, such as the judgment of urgency, that is, when to prioritize so that a less threatening problem is put on the back burner and a subtle, but more threatening problem, is responded to quickly before it gets out of control. Waag and Bell (1997) also obtained evidence showing the linkage between performance and decision making, with situation awareness skills central to decision-making performance.

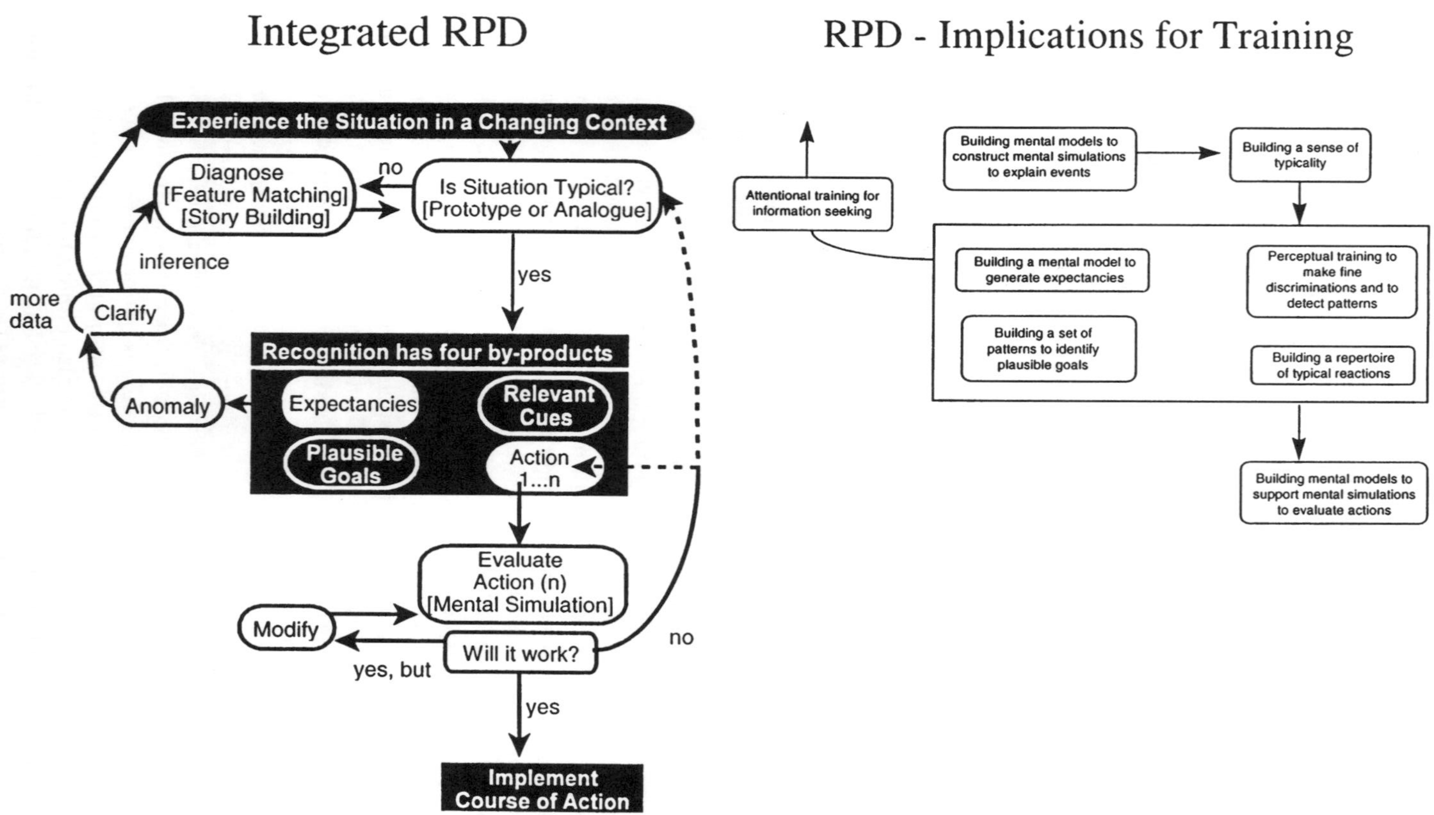


FIG. 9.2. Integrated RPD Model and Implications for Training

We can introduce the concept of mental models (e.g., Rouse & Morris, 1986) as a complementary concept linked to situation awareness and to the RPD model. With experience, a decision maker presumably develops a mental model of a task or of a piece of equipment. The mental model contributes to the judgments of typicality and anomaly. For example, Sarter and Woods (1991) studied the types of mental models that pilots have for the FMS. In the case of the FMS, the pilots' mental models were fairly impoverished; they were too often confused about the mode the system was in, and therefore were unable to project forward to anticipate what the system would do next. These are obvious targets for training (and also for improvements in design, particularly in the design of the human–computer interface). Smith, McCoy, and Layton (1993) showed the function of mental models and representations in the way pilots diagnose malfunctions. The schema used predicted the likelihood of success and the nature of the strategies employed.

For skills such as decision making, we might suggest that the training can be enhanced by providing trainees with information about the cognitive processes of highly proficient pilots engaged in the same maneuver. Charness (1989) showed that it was possible to represent the mental models of experts and novices faced with the same problem (this study used problems taken from the card game of bridge). This can be an aid to practice. Bloom and Broder (1950) showed that novices (college students having trouble with multiple choice tests) improved greatly when provided with think-aloud protocols of skilled test takers. In comparing these protocols to the ones they generated themselves, the novices were able to identify areas of weakness and to discover better strategies. Taking this further, we suggest that the value of case studies could be enhanced by selectively incorporating cognitive process representations of the skilled performers at key decision points. Moreover, if the trainee could generate some sort of protocol prior to inspecting the cognitive processes of the experts, then it might be possible for the novice to make a more meaningful examination of the differences.

Once a situation has been judged to be typical, a decision maker knows what the relevant cues are, so this is another suggestion for training: to differentiate relevant from irrelevant cues. This is particularly important with regard to eye-scanning patterns. Although it is tempting to teach systematic eye-scanning patterns to beginners, DeMaio, Parkinson, Leshowitz, Crosby, and Thorpe (1976) showed that instructor pilots do not follow the eye scan patterns they teach. During nonroutine events, they are faster and more accurate in sizing up the situation. (On the other hand, Hong & O'Neil [1992] noted that intermediate performers may not be able to benefit from the sophisticated mental models of experts.) Along with teaching relevant cues, we can consider teaching relevant goals and typical courses of action. These are commonly taught as standard operating procedures. However, for complex events, there may be conflicts between goals, and practice may be useful in resolving such conflicts.

Sometimes, expectancies are violated. The decision maker may determine that there is some sort of anomaly. One reaction is to seek more information, and another is to use the existing information to build stories to explain the events. Both of these can be the target of training. With regard to seeking more information, this is far from straightforward. Under time pressure and high workload, it may not be easy to judge which item of information is highly diagnostic and easily obtained. Skilled pilots appear able to recognize just what item of information to go after. Less-experienced pilots may waste time and effort gathering information with low value in the context of the situation, or may pass up opportunities because they do not appreciate the significance of a datum. They may misjudge the amount of effort needed, or the likelihood of succeeding in obtaining the data element. All of these difficulties need to be addressed through training.

Another aspect of building decision-making skills in pilots is to improve their ability to generate mental simulations for the purposes of evaluation, diagnosis, and planning. These mental simulations may be for evaluating a course of action to see how it will play out, or they may be for building stories to diagnose symptoms and explain puzzling events. Mental simulations can come into play for planning a flight. This is not simply a matter of going over the route to be flown, but an active scanning of conditions to anticipate where problems might arise and how those will be handled. This can be done prior to the flight, and also prior to difficult portions of a flight while en route. We know that the better crews do this type of mental simulation to prepare themselves for difficulties (Orasanu, 1993). Klein (1997) described a mental simulation technique in which a person reviews a plan by assuming that the plan has failed, and asks what were the most likely reasons for the failure. The exercise generates candidates' weaknesses in the plan, within the context of the mission. The exercise seeks to avoid defensiveness about the plan, because it challenges the person to use his or her expertise to find weaknesses, not to deny them. The mental simulation technique calls for counterfactual thinking (Anderson, & Sechler, 1986) in the form of mental simulations.

The RPD model did not address a range of cognitive processes, including memory, attention, metacognition, or the basis for pattern matching. It did not address the basis of mental simulation, although Klein and Crandall (1995) offered a more specific account of mental simulation. One reason that the RPD model avoided these processes is that the objective was to explain how skilled decision makers could make decisions without comparing options. Cognitive processes such as memory, attention, metacognition, and perception did not play a necessary role in this account. By expanding the RPD model to address these processes, the complexity would have detracted from achieving the objective. This does not imply that the cognitive processes I listed are unimportant, merely that they were not directly relevant to the purpose of presenting the RPD model. If the purpose is to generate concepts for training decision makers, then it becomes essential to go beyond the RPD model, to consider other models that do address a wider set of cognitive processes.

Means, Salas, Crandall, and Jacobs (1993) identified metacognition as a likely target for training. Means et al. reviewed the literature on training generic decision strategies, such as Rational Choice strategies, and found little evidence to support the effectiveness of such training. However, metacognitive skills did seem amenable to training. Cohen, Freeman, and Wolf (1996) discussed an approach to teaching metacognitive skills, particularly the skill of sizing up situations. Here, the metacognitive training is in learning to be sensitive to the possibility of mis-identifying a situation, and in learning strategies for generating stories to explain away inconvenient data, and stories that contrast with the original interpretation. The objective is to prevent fixation on the original interpretation, but not to leave decision makers in a permanently confused state, victims of every transient or accidental data point inconsistent with their views. The metarecognition model advanced by Cohen et al. also takes into account the variable of time pressure, and includes as part of the training strategy a concern for the time available and the feasibility of entering into deliberative activities. Cohen, Freeman, and Thompson (1998) developed this approach into a critical thinking training program and obtained empirical support for its effectiveness with naval officers performing simulated command and control tasks. Woods (1993) studied disturbance management (the attempt to correct for an unexpected deviation, often in parallel with an attempt to diagnose the cause for the deviation), and found that pilots need to simultaneously manage the aircraft and diagnose the cause(s) of the malfunction. Metacognitive skills are needed to coordinate different types of activities.

Martin-Emerson and Kramer (1995) and Gopher (1992) studied the way attentional control can be taught. This is another cognitive process that was not covered by the RPD model, and offers important possibilities for training, akin to metacognition. Kramer et al. (1994) studied the linkage between flight experience and instrument scan patterns. They used 17 students and 21 instructor pilots (12 with 200 to 1000 hours of flight time and nine with more than 1000 hours) flying segments in a flight simulator under time constraints. Performance measures did not do a good job of discriminating among the three groups of pilots (students, low-time instructors, high-time instructors). According to chance, discrimination levels between the three groups was 33%, and the performance measures for different flight segments averaged 45%. The eye movement data showed an average classification accuracy of 67%, and when both performance measures and eye scan measures were used in combination, the average discrimination accuracy increased to 92%. These data suggest that the more experienced pilots were able to fixate on the most relevant indicators, and that this ability to control attention was more marked than performance of controlling the aircraft.

Gopher (1992; Gopher, Weil, & Bareket, 1994) showed that it is possible to train attentional skills. Gopher et al. trained 33 cadet pilots in the Israeli Air Force using a complex computer game designed to improve the ability to cope with high processing and response demands and improve attention control. Ten hours of training resulted in higher levels of performance on subsequent test flights, compared to a group of 25 matched control subjects. The difference between cadets trained with the computer game and the controls was most pronounced for performance on the most complex flights. On the basis of these and related results, the computer game has been incorporated into the regular training program of the Israeli Air Force to develop and maintain attentional control. Gopher et al. estimated that the 10 hours of training increased the probability of graduation from the program by 30%. They speculated that "when subjects develop their experience by trial and error under high load, they tend to adopt suboptimal attention strategies. Moreover, these strategies tend not to change or be replaced by better ones with the process of experience . . . in contrast, by a systematic manipulation of emphasis on different task subelements, subjects . . . explore a wider range of attention strategies and improved their ability to cope with the high load of tasks" (p. 389).

Pilots need to be able to manage stress, and this is another area for developing training. Johnston and Cannon-Bowers (1996) described a stress resistance training program that appears to be effective and to generalize. This type of training can be valuable in aviation decision making. Schmitt and Klein (1996) examined one type of stressor, uncertainty, and found that it covers a variety of different configurations. People can be uncertain because of missing information, information perceived as unreliable, ambiguous and conflicting information, and complex information that cannot be integrated or comprehended. Moreover, the uncertainty can be about data elements, inferences drawn from the data, or ability to project the inferences into the future. For aviation, it may be useful to determine the dominant aspects and sources of uncertainty. Training can be introduced to help pilots prepare better for uncertainty, represent it more effectively, reduce it where appropriate, and manage it.

Thus far this subsection has been oriented toward the individual decision maker. We must also consider teams, particularly because cockpit crews function as teams, both within the cockpit and outside, in interactions with air traffic controllers and dispatchers. The field of aviation has already developed one effective strategy for training teams on a number of tasks: Crew Resource Management (CRM) (Helmreich & Foushee, 1993). One suggestion from the NDM perspective is to elaborate the CRM methods to emphasize shared situation awareness, so that all crew members have the same big picture where necessary, without providing excessive and distracting information. The work of Orasanu and Salas and their colleagues (Orasanu, 1993, 1994; Orasanu & Salas, 1993; Prince, Chidester, Bowers, & Cannon-Bowers, 1992; Swezey & Salas, 1992) demonstrated the importance of shared situation awareness. Cannon-Bowers, Tannenbaum, Salas, and Volpe (1995) defined team competencies in order to establish team training requirements. Zsambok (1993) described a model of team performance that centers around decision making. Moreover, Stout, Cannon-Bowers, and Salas (1996/1997) discussed approaches to develop and maintain shared situation awareness.

## Decision Skills Training

The previous section described a variety of methods and approaches that could be used to build a training program for pilots. These appear to be powerful methods for training specific aspects of decision making. They also require a fair level of front-end analysis to prepare the instructional programs.

We also developed a complementary approach to training, one that does not require so much preparation. We developed this in response to a request from the U.S. Marine Corps, who wanted to provide a program to improve decision making in squad leaders, but did not have sufficient lead time to perform the front-end analysis or to develop the instructional materials (Klein, McCloskey, Pliske, & Schmitt, 1997).

The Decision Skills Training program (Klein, 1997) that we developed was in keeping with the NDM assertion that effective decision making depends on experience, rather than on following procedures or normative strategies. Therefore, the approach we adopted was to help the squad leaders rapidly boost their experience level with regard to making difficult decisions. Instead of trying to teach them to make difficult discriminations, or to distinguish different types of situational dynamics, we instead attempted to provide them with the tools to boost their experience level. That included ways of preparing better for the training they received, ways of augmenting their training, and ways of reflecting on the lessons they learned during training.

To help them prepare better for training, we helped them identify the decision requirements of their missions: the difficult and critical judgments and decisions, along with the reasons for the difficulty, types of errors that are often made, and strategies that seem helpful. The squad leaders' missions were primarily reconnaissance and surveillance activities in tracking enemy forces. The decision requirements included being able to estimate the time it would take them to get into their positions once they were inserted by helicopter, the ability to identify good landing sites for helicopters sent to extract them, the ability to infer enemy force disposition from glimpses of a few vehicles, and so forth. This exercise helped them define the decision skills that they needed, and helped them determine how to get practice and feedback on these specific decision skills during their field training exercises. We also taught them to use a mental simulation strategy prior to beginning a field exercise in which they would imagine that their plans had fallen apart, and to try to discover what might have gone wrong. This helped them identify the weaknesses in their plans prior to execution. The key to successful practice includes the availability of feedback that is timely, accurate, and diagnostic. In addition, the experts in many fields engage in deliberate practice, such as setting goals for themselves in each session, rather than just putting in their time. Therefore, careful preparation prior to field exercises or training exercises can result in a greater level of learning during those exercises.

To help augment their training, we prepared tactical decision games (e.g., Schmitt, 1994) to provide low-level simulations. The tactical decision games offered high degrees of uncertainty in order to allow them to practice making decisions with key data elements missing or ambiguous. The games also were conducted under extreme time pressure, to prepare them to handle this type of stressor. The intent behind these tactical decision games was to provide them with a conditioning process in which they practiced making difficult decisions several times a week in preparation for their culminating field exercise. The program also taught methods for personnel to build their own tactical decision games in response to decision requirements that they needed to practice further.

To help them reflect on the training, we provided them with a decision checklist to help them debrief themselves after field exercises (and also after tactical decision games). The checklist asked them to identify areas of uncertainty in the exercise and how they handled the uncertainty; to identify difficult judgments and subtle cues and patterns, and how they discriminated these; and to reflect on the difficult judgments and decisions that they had just encountered.

The qualitative evaluations to the Decision Skills Training were very positive. Both attitude surveys and interviews showed enthusiasm for the training and its perceived benefit. As a result, the U.S. Marine Corps has requested that we further develop this program for use in different courses at the Marine Corps University. In one version, we were asked to include a component on coaching, to help the more experienced officers articulate what they are seeing in order to do a better job of providing on-the-job training to officers with less experience. Zsambok, Crandall, and Militello (1994) found in a variety of domains that on-the-job training accounts for the majority of skill development. However, most organizations do not have any program for performing on-the-job training, and most of the people who wind up providing on-the-job training have little idea of how to do it. They have a very small repertoire of training methods. Worse yet, they tend to think they are doing a decent job because they don't know any better and because their own supervisors are not able to give them feedback. Aviation offers many effective opportunities for using on-the-job training in training, and the advantage is that it is not expensive because it can be provided during flying operations. The training generalizes to the operational setting because it is presented in that setting. Currently, many trainers provide feedback about the actions taken, but not about the decision strategies used. We found that trainers can be shown how to pay attention to the processes of decision making, along with the actions. They can check what pilots were seeing, what cues and patterns they were noticing, which options they were considering, and so forth. Decision training can, and should, occur in the context of regular training opportunities. This helps to ensure that the learning will generalize, and it also keeps the costs down. By preparing instructors to provide feedback about processes, the regular training can become more valuable.

It appears that many aspects of this program could be useful for aviation training. In the past, instructors noted that pilots needed to do a good job of making decisions, and without differentiating types of decisions, were content to teach generic strategies. There is little evidence for the effectiveness of this approach. Instead, we may be able to do a much better job by:

- Helping pilots identify the specific judgments and decisions that they need to practice.
- Providing pilots with tools to use, and to build decision scenarios that allow practice in making decisions under uncertainty and time pressure.
- Designing debriefing checklists to enable pilots to identify the decision requirements they encounter during challenging situations, so that they can use these challenges to direct their own learning.
- Showing pilots that there are coaching methods for accelerating skill development.

In defining Naturalistic Decision Making as the study of how people use their experience to make decisions in field settings, the spotlight is placed on expertise and the way it develops. Instead of emphasizing generic decision strategies, NDM applications tend toward specifying the types of decisions that are of concern, and the training is designed to build up expertise for these types of decisions. This brings us to the linkage between NDM and the field of expertise. The NDM community needs to take advantage of research on expertise, such as the framework presented by Shanteau (1992) for different aspects of expertise. It appears that we can differentiate experts from novices in several different ways. These include the ability of the expert to adapt and improvise more freely than novices, to project ahead more skillfully than novices, as well as to focus attention more effectively than novices.

Ericsson and Charness (1994) took the strong position that high levels of expertise in many fields can be attained through practice. To become a chess grandmaster, Charness (1989) estimated that it takes approximately 32,000 hours, or 4 hours of practice a day, every day, for about 20 years. The Decision Skills Training program is an attempt to improve decision making through the accelerated development of expertise.

## CONCLUSIONS

In this chapter I used the NDM framework to generate a range of recommendations for training pilots to improve the way they make different types of decisions. One of my assumptions is that the NDM framework is a fruitful source of ideas and suggestions. Yet it is possible for a skeptic to challenge that assumption. A skeptic might claim that training consists of practice and feedback, and that the NDM framework has not changed that. Moreover, a skeptic might question whether any of the recommendations offered in this chapter are new or are based on the NDM framework. Surely it is possible that people working on the issue of pilot decision making might have generated these assumptions even if they had never heard about NDM.

In this section I show the value added by the NDM perspective, in order to counter such skepticism. The majority of suggestions are linked to the way the field of NDM views the issues. In fact, the recommendations generated by the NDM framework are a radical departure from previous approaches.

### The Questions Posed by NDM Are Different From the Questions Raised in the Past

As long as there was a belief in generic decision strategies, it made sense to ask how to teach pilots to make better decisions. Once that belief was challenged, the limitations of the question became clear. NDM is radical in rejecting the question about how to train pilots to make better decisions. It is radical in insisting that we substitute better questions, by specifying which decisions are the source of difficulty, why they are difficult, what types of errors are being made, and how proficient pilots avoid these errors. NDM posits the use of decision requirements in place of global and unfocussed questions. NDM claims that in many cases, the magic is in the question, not in the solution. Once the right question is posed, at the right level, the solution may appear mundane. Rather than being disappointed, we should acknowledge the value of a straightforward answer.

### The Solutions Posed by NDM Are Different From the Training Approaches Offered in the Past

One prominent decision researcher was once asked how he would train pilots to make better decisions and, without hesitation, he answered, "I would teach them probability theory." This is not an answer that would be given by NDM researchers. Generic strategies such as Rational Choice methods, Bayesian statistics, and debiasing procedures are all weak because generic methods must be weak and because they have a limited role for expertise. Even worse, as Erev, Bornstein, and Wallsten (1993), Wilson and Schooler (1991), and others have shown, when people apply analytical strategies, their performance may deteriorate, perhaps because pattern matching processes are interfered with. Sometimes, the generic strategies can be helpful, and NDM researchers are trying to learn the boundary conditions for the quantitative methods such as multiattribute utility analysis that decompose tasks, as well as the boundary conditions for holistic methods that depend on heuristics and expertise. Sometimes, NDM researchers do offer generic approaches, such as Cohen, Freeman, and Thompson's (1998) critical thinking program. But this is done with caution and with acknowledgment that the program is not always appropriate. Part of the program includes practice in determining when *not* to apply the critical thinking strategy. Yet another difference is in the way NDM researchers treat uncertainty, not as a judgment that informs the rating of a course of action, but as a dynamic feature of the environment and the decision maker that affects the strategies used. In short, NDM is radical in rejecting the multiattribute utility analysis, Rational Choice strategy, that has been the mainstay of training programs in aviation decision making and in many other applied domains.

### The Emphasis on Expertise Differentiates NDM From the Training Approaches in the Past

Research on heuristics (Tversky & Kahneman, 1974) has shown that decision makers relied on them even under conditions where they were misleading. These data were interpreted by many as providing evidence for biases, showing the fallibility of decision makers. However, Lopes (1982) argued that in the heuristics and biases research, the experimental conditions were arranged to make sure that the use of heuristics led to inferior performance. This was done as a way of demonstrating that people use heuristics even when they are not effective. However, a common perception is that heuristics generally result in biased performance and errors. Such a conclusion is unwarranted. In fact, for most settings, the heuristics provide useful shortcuts. The RPD model centers around the use of heuristics such as availability and representativeness for judging situations as typical and for noticing anomalies. The RPD model also depends on the simulation heuristic. Rather than seeing these heuristics as limiting factors, the RPD emphasizes them as a basis for proficiency. If we see the heuristics as helpful, and if we find that experienced subjects are able to make skillful use of the heuristics, then our job is not to discourage the use of heuristics, but to build the experience base for using heuristics more skillfully. We can determine the types of mental models and cognitive processes used by proficient pilots and present these for modeling by students.

### The Emphasis on Situation Awareness Contrasts With the Traditional Focus on Choosing Between Courses of Action

Neither Bayesian theory, with its emphasis on the fusion of different types of evidence, nor social judgment theory, with its emphasis on the Brunswik lens model, addresses the issue of extracting meaning from a situation, building stories to explain the evidence. Cohen (1993) argued that a Bayesian approach is fundamentally limited because it does not address the possibility of explaining away inconsistent evidence, which is sometimes a reasonable way to proceed and sometimes a dangerous way to proceed. Regardless of its merits, it stands as a common pattern, and a much richer one, than simply combining evidence streams. The treatment of situation awareness as sensemaking (Weick, 1995) is a radical departure from quantitative approaches to judgment, and the emphasis on situation awareness over choosing between courses of action is a radical departure from the fixation on the moment of choice (Berkeley & Humphreys, 1982). Moreover, at the team level, recent emphases on training shared situation awareness is a radical departure from the program of using Delphi methods or the nominal group technique to enable groups to gather impressions about the strengths and weaknesses of different courses of action.

### The Emphasis on Metacognition Includes a Concern for Managing Stress and for Controlling Attention

The approaches to training metacognitive skills as a way to improve decision making are a departure from conventional programs.

### Training Methods Are Being Introduced That Do Not Fit Into the "Practice and Feedback" Paradigm

Offering representations of experts' perceptions is not a simple matter of practice and feedback. In this sense, cognitive modeling is a radical departure from the "practice and feedback" cycle. The management of scenarios so that trainees can quickly gain a sense of typicality is a radical departure from the "practice and feedback" approach to training. Working with trainers and on-the-job training providers to help them identify decision strategies and processes that may be causing difficulties for trainees, and to help them articulate their own strategies, is not a part of the "practice and feedback" paradigm. It is unlikely that the NDM framework will ever produce prescriptive methods for ensuring effective decision making, for the reasons illustrated in Fig. 9.1. The NDM framework has abandoned normative theory, and with it, has abandoned the attempt to define generic strategies. In the years to come, we will need to see if NDM researchers arrive at a set of training methods that can substitute for normative, prescriptive methods.

## ACKNOWLEDGMENT

Robert Hutton provided very useful comments and suggestions on a draft of this manuscript.

## REFERENCES

Anderson, C. A., & Sechler, E. S. (1986). Effects of explanation and counter explanation on the development and use of social theories. *Journal of Personality and Social Psychology, 50*(1), 24–34.

Benner, L. (1975). D.E.C.I.D.E. in the hazardous materials emergencies. *Fire Journal, 69*(4), 13–18.

Berkeley, D., & Humphreys, P. (1982). Structuring decision problems and the 'bias heuristic.' *Acta Psychologica, 50*, 201–252.

Berlin, J. I., Gruber, E. V., Holmers, C. W., Jensen, P. K., Lau, J. R., Mills, J. W., & O'Kane, J. M. (1982). *Pilot judgment training and evaluation* (DOT/FAA/CT-82/56). Daytona Beach, FL: Embry-Riddle Aeronautical University (NTIS No. AD-A117 508).

Bloom, B. S., & Broder, L. J. (1950). *Problem solving processes of college students: An exploratory investigation*. Chicago: University of Chicago Press.

Cannon-Bowers, J., Tannenbaum, S., Salas, E., & Volpe, C. (1995). Defining team competencies and establishing team training requirements. In R. Guzzo & E. Salas (Eds.), *Team effectiveness and decision making in organizations*. San Francisco, CA: Jossey-Bass.

Cannon-Bowers, J. A., Salas, E., & Pruitt, J. S. (1996). Establishing the boundaries of a paradigm for decision-making research. *Human Factors, 38*(2), 193–205.

Charness, N. (1989). Expertise in chess and bridge. In D. Klahr & K. Kotovsky (Eds.), *Complex information processing: The impact of Herbert A. Simon* (pp. 183–208). Hillsdale, NJ: Lawrence Erlbaum Associates.

Christensen-Szalanski, J. J. J. (1993). A comment on applying experimental findings of cognitive biases to naturalistic environments. In G. A. Klein, J. Orasanu, R. Calderwood, & C. E. Zsambok (Eds.), *Decision making in action: Models and methods* (pp. 252–264). Norwood, NJ: Ablex.

Cohen, M. S. (1993). The naturalistic basis of decision biases. In G. A. Klein, J. Orasanu, R. Calderwood, & C. E. Zsambok (Eds.), *Decision making in action: Models and methods* (pp. 51–99). Norwood, NJ: Ablex.

Cohen, M. S., & Freeman, J. T. (1996). Thinking naturally about uncertainty. *Proceedings of the 40th Human Factors & Ergonomics Society*, 179–183. Santa Monica, CA: HFES.

Cohen, M. S., Freeman, J. T., & Thompson, B. (1998). Critical thinking skills in tactical decision making: A model and a training method. In J. Cannon-Bowers & E. Salas (Eds.), *Making decisions under stress: Implications for individual and team training* (pp. 155–189). Washington, DC: American Psychological Association.

Cohen, M. S., Freeman, J. T., & Wolf, S. (1996). Meta-recognition in time-stressed decision making: Recognizing, critiquing, and correcting. *Human Factors, 38*(2), 206–219. Santa Monica, CA: HFES.

de Groot, A. D. (1978). *Thought and choice in chess*. New York, NY: Mouton. (Original work published 1946)

De Keyser, V., & Woods, D. D. (1993). Fixation errors: Failures to revise situation assessment in dynamic and risky systems. In A. G. Colombo & A. Saiz de Bustamente (Eds.), *Advanced systems in reliability modeling*. Norwell, MA: Kluwer Academic.

DeMaio, J., Parkinson, S., Leshowitz, B., Crosby, J., & Thorpe, J. A. (1976). *Visual scanning: Comparisons between student and instructor pilots* (Tech. Report: AFHLR-TR-76-10). Air Force Human Resources Laboratory.

Endsley, M. R. (1995). Measurement of situation awareness in dynamic systems. *Human Factors, 37*(1), 65–84.

Erev, I., Bornstein, G., & Wallsten, T. S. (1993, June). The negative effect of probability assessments on decision quality. *Organizational Behavior and Human Decision Processes, 55*(1), 79–94.

Ericsson, K. A., & Charness, N. (1994). Expert performance: Its structure and acquisition. *American Psychologist, 49*(8), 725–747.

Flin, R. (1996). *Sitting in the hot seat: Leaders and teams for critical incident management*. Chichester, England: Wiley.

Flin, R., Salas, E., Strub, M., & Martin, L. (Eds.). (1997). *Decision making under stress: Emerging themes and applications*. Aldershot, England: Ashgate Publishing.

Fraser, J., Strohm, P., Smith, J., Galdes, D., Svirbely, J., Rudmann, S., Miller, T., Blazina, J., Kennedy, M., & Smith, P. J. (1989). Errors in abductive reasoning. *Proceedings of the 1989 IEEE Conference on Systems, Man and Cybernetics* (pp. 1136–1141). New York: IEEE.

Gigerenzer, G. (1991). How to make cognitive illusions disappear: Beyond "heuristics and biases." *European Review of Social Psychology, 2*, 83–115.

Gopher, D. (1992). The skill of attention control: Acquisition and execution of attention strategies. In D. E. Meyer & S. Kornblum (Eds.), *Attention and performance XIV: Synergies in experimental psychology, artificial intelligence and cognitive neuroscience*. Cambridge, MA: MIT Press.

Gopher, D., Weil, M., & Bareket, T. (1994). Transfer of skill from a computer game trainer to flight. *Human Factors, 36*(3), 387–405.

Hammond, K. R. (1993). Naturalistic decision making from a Brunswikian viewpoint: Its past, present, future. In G. A. Klein, J. Orasanu, R. Calderwood, & C. E. Zsambok (Eds.), *Decision making in action: Models and methods* (pp. 205–227). Norwood, NJ: Ablex.

Helmreich, R. L., & Foushee, H. C. (1993). Why crew resource management? Empirical and theoretical bases of human factors training in aviation. In E. L. Wiener, B. G. Kanki, & R. L. Helmreich (Eds.), *Cockpit resource management*. New York: Academic Press.

Hong, E., & O'Neil, H. F. Jr. (1992). Instructional strategies to help learners build relevant mental models in inferential statistics. *Journal of Educational Psychology, 84*(2), 150–159.

Hutton, R. J. B., Olszewski, R., Thordsen, M. L., & Kaempf, G. L. (1997). *En route air traffic controller decision making model and decision maker vulnerabilities* (Contract for NYMA Inc., Egg Harbor Township, NJ). Fairborn, OH: Klein Associates.

Johnston, J. H., & Cannon-Bowers, J. A. (1996). Training for stress exposure. In J. E. Driskell & E. Salas (Eds.), *Stress and human performance* (pp. 223–256). Mahwah, NJ: Lawrence Erlbaum Associates.

Kaempf, G. L., & Klein, G. (1994). Aeronautical decision making. In N. Johnston, N. McDonald, & R. Fuller (Eds.), *Aviation psychology in practice* (pp. 223–254). Aldershot, England: Avebury Press.

Klein, G. (1993). Sources of error in naturalistic decision making tasks. *Proceedings of the 37th Annual Meeting of the Human Factors and Ergonomics Society*, 368–371. Santa Monica, CA: HFES.

Klein, G. (1996). The effect of acute stressors on decision making. In J. E. Driskell & E. Salas (Eds.), *Stress and human performance* (pp. 49–88). Mahwah, NJ: Lawrence Erlbaum Associates.

Klein, G. (1997). Developing expertise in decision making. *Thinking and Reasoning, 3*(4), 337–352. East Sussex, UK: Psychology Press Ltd.

Klein, G. (1998). *Sources of power: How people make decisions*. Cambridge, MA: MIT Press.

Klein, G., McCloskey, M., Pliske, R., & Schmitt, J. (1997). Decision skills training. *Proceedings of the 41st Human Factors and Ergonomics Society*, 182–185. Santa Monica, CA: HFES.

Klein, G., Wolf, S., Militello, L., & Zsambok, C. (1995). Characteristics of skilled option generation in chess. *Organizational Behavior and Human Decision Processes, 61*(3). San Diego, CA: Academic Press.

Klein, G. A. (1989a). Strategies of decision making. *Military Review*, May, 56–64. Fort Leavenworth, KS: Department of the Army.

Klein, G. A. (1989b). Recognition-primed decisions. In W. B. Rouse (Ed.), *Advances in man–machine systems research, Vol. 5* (pp. 47–92). Greenwich, CT: JAI Press.

Klein, G. A., Calderwood, R., & Clinton-Cirocco, A. (1986). Rapid decision making on the fireground. *Proceedings of the 30th Annual Human Factors Society*, 576–580. Santa Monica, CA: HFES.

Klein, G. A., & Crandall, B. W. (1995). The role of mental simulation in naturalistic decision making. In P. Hancock, J. Flach, J. Caird, & K. Vicente (Eds.), *Local applications of the ecological approach to human–machine systems, Vol. 2* (pp. 324–358). Hillsdale, NJ: Lawrence Erlbaum Associates.

Klein, G. A., Orasanu, J., Calderwood, R., & Zsambok, C. E. (1993). *Decision making in action: Models and methods*. Norwood, NJ: Ablex.

Kramer, A., Tham, M., Konrad, C., Wickens, C., Lintern, G., Marsh, R., Fox, J., & Merwin, D. (1994). Instrument scan and pilot expertise. *Proceedings of the Human Factors and Ergonomics Society 38th Annual Meeting, 38(1)*, 36–40.

Lipshitz, R. (1993). Converging themes in the study of decision making in realistic settings. In G. A. Klein, J. Orasanu, R. Calderwood, & C. E. Zsambok (Eds.), *Decision making in action: Models and methods* (pp. 103–137). Norwood, NJ: Ablex.

Lipshitz, R., & Strauss, O. (1996). How decision-makers cope with uncertainty. *Proceedings of the Human Factors and Ergonomics Society 40th Annual Meeting*, 189–193. Santa Monica, CA: HFES.

Lopes, L. L. (1982). Doing the impossible: A note on induction and the experience of randomness. *Journal of Experimental Psychology: Learning, Memory, and Cognition, 8*(6), 626–636.

Martin-Emerson, R., & Kramer, A. F. (1995). Capture of attention by visual onsets. *Proceedings of the Human Factors and Ergonomics Society 39th Annual Meeting*, 1385–1389. Santa Monica, CA: HFES.

McCloskey, M. (1996). An analysis of uncertainty in the Marine Corps. *Proceedings of the Human Factors and Ergonomics Society 40th Annual Meeting*, 194–198. Santa Monica, CA: HFES.

Means, B., Salas, E., Crandall, B., & Jacobs, O. (1993). Training decision makers for the real world. In G. A. Klein, J. Orasanu, R. Calderwood, & C. E. Zsambok (Eds.), *Decision making in action: Models and methods* (pp. 306–326). Norwood, NJ: Ablex.

Orasanu, J. (1993). Decision-making in the cockpit. In E. Weiner, B. Kanki, & R. Helmreich (Eds.), *Crew resource management*. New York, NY: Academic Press.

Orasanu, J. (1994). Shared problem models and flight crew performance. In N. Johnston, N. McDonald, & R. Fuller (Eds.), *Aviation psychology in practice* (pp. 255–285). Aldershot, England: Avebury Press.

Orasanu, J., & Connolly, T. (1993). The reinvention of decision making. In. G. A. Klein, J. Orasanu, R. Calderwood, & C. E. Zsambok (Eds.), *Decision making in action: Models and methods* (pp. 3–20). Norwood, NJ: Ablex.

Orasanu, J., & Salas, E. (1993). Team decision making in complex environments. In G. A. Klein, J. Orasanu, R. Calderwood, & C. E. Zsambok (Eds.), *Decision making in action: Models and methods* (pp. 327–345). Norwood, NJ: Ablex.

Perrow, C. (1984). *Normal accidents: Living with high-risk technologies*. New York: Basic Books.

Prince, C., Chidester, T., Bowers, C., & Cannon-Bowers, J. (1992). Aircrew coordination—Achieving teamwork in the cockpit. In R. Swezey & E. Salas (Eds.), *Teams: Their training and performance* (pp. 329–353). Norwood, NJ: Ablex.

Rasmussen, J. (1986). *Information processing and human-machine interaction: An approach to cognitive engineering*. New York: North Holland.

Rasmussen, J. (1997). Merging paradigms: Decision making, management, and cognitive control. In R. Flin, E. Salas, M. Strub, & L. Martin (Eds.), *Decision making under stress: Emerging themes and applications* (pp. 67–81). Aldershot, England: Ashgate Publishing.

Rouse, W. B., & Morris, N. M. (1986). On looking into the black box: Prospects and limits on the search for mental models. *Psychological Bulletin, 100*(3), 349–363.

Sarter, N. B., & Woods, D. D. (1991). Situation awareness: A critical but ill-defined phenomenon. *The International Journal of Aviation Psychology, 1*(1), 45–57.

Schmitt, J. F. (1994). *Mastering tactics*. Quantico, VA: Marine Corps Association.

Schmitt, J. F., & Klein, G. (1996, August). Fighting in the fog: Dealing with battlefield uncertainty. *Marine Corps Gazette, 80*, 62–69.

Shanteau, J. (1992). Competence in experts: The role of task characteristics. *Organizational Behavior and Human Decision Processes, 53*, 252–266.

Smith, P. J., McCoy, E., & Layton, C. (1993). Design-induced error in flight planning. *Proceedings of the Human Factors and Ergonomics Society 37th Annual Meeting*, pp. 1091–1095.

Stokes, A. F., Kemper, K., & Kite, K. (1997). Aeronautical decision making, cue recognition, and expertise under time pressure. In C. Zsambok & G. Klein (Eds.), *Naturalistic decision making* (pp. 183–196). Mahwah, NJ: Lawrence Erlbaum Associates.

Stout, R. J., Cannon-Bowers, J. A., & Salas, E. (1996/1997). The role of shared mental models in developing team situational awareness: Implications for training. *Training Research Journal, 2*, 85–116.

Swezey, R. W., & Salas, E. (1992). *Teams: Their training and performance*. Norwood, NJ: Ablex.

Tversky, A. (1972). Elimination by aspects: A theory of choice. *Psychological Review, 79*(4), 281–299.

Tversky, A., & Kahneman, D. (1974). Judgment under uncertainty: Heuristics and biases. *Science, 185*, 1124–1131.

Tversky, A., & Kahneman, D. (1982). Causal schemas in judgments under uncertainty. In D. Kahneman, P. Slovic, & A. Tversky (Eds.), *Judgment under uncertainty: Heuristics and biases*. Cambridge, MA: Cambridge University Press.

Waag, W. L., & Bell, H. H. (1997). Situation assessment and decision making in skilled fighter pilots. In C. Zsambok & G. Klein (Eds.), *Naturalistic decision making* (pp. 247–254). Mahwah, NJ: Lawrence Erlbaum Associates.

Weick, K. (1995). *Sensemaking in organizations*. Thousand Oaks, CA: Sage.

Wilson, T. D., & Schooler, J. W. (1991). Thinking too much: Introspection can reduce the quality of preferences and decisions. *Journal of Personality and Social Psychology, 60*, 181–192.

Wohl, J. C. (1981). Force management decision requirements for Air Force tactical command and control. *IEEE Transactions on Systems, Man, and Cybernetics*, 618–639.

Woods, D. D. (1993). Process-tracing methods for the study of cognition outside of the experimental psychology laboratory. In G. A. Klein, J. Orasanu, R. Calderwood, & C. E. Zsambok (Eds.), *Decision making in action: Models and methods* (pp. 228–251). Norwood, NJ: Ablex.

Zsambok, C. E. (1993). Advanced team decision making in $C^2$ settings. *Proceedings of the 1993 Symposium on Command and Control Research* (pp. 45–52). McLean, VA: Information Systems Division, Science Applications International Corporation.

Zsambok, C. E., Crandall, B., & Militello, L. (1994). *OJT: Models, programs, and related issues* (Contract MDA903-93-C-0092 for the U.S. Army Research Institute for the Behavioral and Social Sciences, Alexandria, VA). Fairborn, OH: Klein Associates.

Zsambok, C. E., & Klein, G. (Eds.). (1997). *Naturalistic decision making*. Mahwah, NJ: Lawrence Erlbaum Associates.

Part II

# MODELS AND TOOLS FOR ASSESSMENT OF AIRCREW TRAINING

*Chapter* **10**

# Simulation Design for Training and Assessment

Stephen Alessi
*The University of Iowa*

This chapter has two goals. The first is to suggest what is needed in a theory of learning or assessment by simulation. The second is to frame a model for simulation design. Given the constraints of space, I discuss only the more important elements for both the theory and the model. Although neither the theory nor the model are comprehensive at this time, I believe the components discussed are among the most essential.

The theory section discusses the main variables affecting learner interaction with simulations. There are other simulation theories (de Jong & van Joolingen, 1998; Gibbons, Fairweather, Anderson, & Merrill, 1997; Reigeluth & Schwartz, 1989), but we have not yet devised one adequately encompassing the various types and uses of simulation. In the model section, I propose an approach that is based on the theory. The model does not include development, so I do not discuss evaluation, tools for development, or other production issues. My goal is to examine how a simulation theory should guide design. The audience for this chapter is people who design simulations for instruction or assessment. A primary consideration in both sections is simulation fidelity (or realism), a complex issue that affects all aspects of simulation design and that continues to need systematic research. In general, research on simulation has been insufficient to provide designers with the guidance they need. I conclude with suggestions for research to advance simulation theory and design.

## THEORY OF LEARNING WITH SIMULATIONS

Figure 10.1 illustrates the four critical components, and their relationship, that are necessary for a theory of learning with simulations: knowledge attributes, learner attributes, and simulation attributes, all of which affect the process of learners encoding, representing, and using knowledge.

### Knowledge Attributes

The first component is the attributes of the knowledge to be learned. The most important are the type of knowledge, its organization and complexity, and precision. For an alternative approach to classifying knowledge attributes, consider de Jong & Ferguson-Hessler (1996).

***Type of Knowledge.*** There are many knowledge taxonomies (e.g., de Jong & Ferguson-Hessler, 1996), but Anderson's (1980) broad division into declarative versus procedural knowledge is a useful one to begin with for simulation theory. The objectives of most simulations are either *how to do* something (procedural), to improve understanding *about* something (declarative), or a combination of both. The type of knowledge has some obvious and some less obvious implications for simulation theory. As an example of the latter, near transfer and learner actions are generally more important for procedural objectives, whereas far transfer, conceptual models, and motivation may be more critical for declarative objectives.

***Organization and Complexity of the Knowledge.*** Some domains are very hierarchical (mathematics), whereas others have many interconnections (history or economics). The procedures for an aircraft's preflight check may be fairly linear although cockpit crew management is not. Complexity is related to, but not equivalent to, organization. The domain may have a sim-

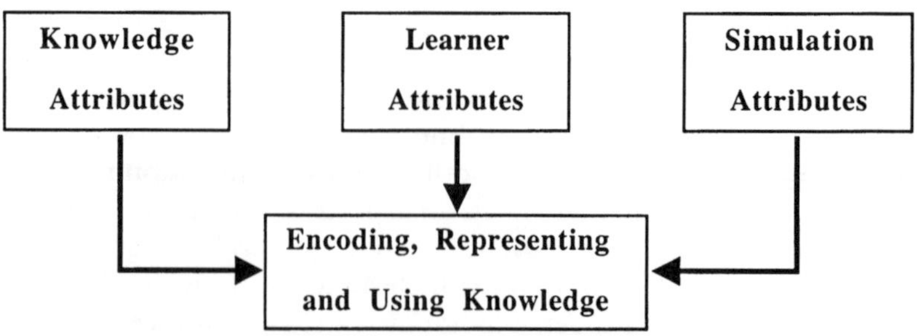


FIG. 10.1. Knowledge, learner, and simulation attributes determine learning and transfer.

ple hierarchy or a complex one. The skills of flying different aircraft vary in complexity, even if organization is similar. Organization and complexity impact instructional sequence, learning supports, and whether to reveal the simulation's underlying model, that is, whether to use a glass-box approach, in which the underlying model is visible, or a black-box approach, in which the underlying model is hidden (Wenger, 1987, p. 15).

***Precision of the Knowledge.*** Precision (Alessi & Trollip, 1991) means the accuracy and predictability of the actual phenomenon or objects to be simulated. Equipment constructed by humans has precise characteristics. In contrast, our understanding of human behavior is imprecise. Our understanding of some natural phenomena (mechanics and electrodynamics) are precise; our understanding of others (weather) are imprecise. Precision impacts transfer and fidelity. The unpredictable (or probabilistic) nature of imprecise domains makes design for transfer more difficult. Similarly, creating a realistic simulation of a phenomenon is more challenging when the domain is less precise. It can be aided through the use of randomness and probability distributions. For example, a simulation teaching radar operation (based on precise principles) can be very deterministic (the same inputs leading repeatedly to the same outputs), whereas a simulation to teach weather forecasting (based on less precise principles) will benefit from some random elements.

### Learner Attributes

The second essential component for our theory is learner attributes. Although research on individual differences suggests that cognitive ability has the greatest impact on learning, prior knowledge is more relevant to designers, followed by abilities, motivation, and learning style.

***Prior Knowledge.*** Prior knowledge, embodied largely in the novice–expert continuum, plays a central role in designing a simulation's features. Examples of considering prior knowledge include the use of metaphor, selection of delivery mode, and adjusting fidelity and learner control (Fitzgerald, Wilson, & Semrau, 1997; Park & Hannafin, 1993; Thurman, 1993).

***Cognitive Abilities.*** Cognitive ability is not a single entity. Some people learn math more easily than languages or art, whereas the reverse may be true for other people. Although we cannot easily alter learners' cognitive abilities, we can assess *which* cognitive abilities are critical to the domain, and given them, which methods of instruction or assessment to use—for example, the use of text versus voice as a function of learners' reading ability. For almost all simulations, the general cognitive abilities of memory, reading, and logical reasoning are important. Other abilities (such as listening, speaking, problem solving, planning, organizing, and scientific reasoning) are not always critical, but may be needed for simulations teaching particular domains.

***Metacognitive Abilities.*** Metacognition (Flavell, 1979; Gourgey, 1998; Mayer, 1998; Schraw, 1998; Sternberg, 1998; Veenman, Elshout, & Meijer, 1997; White & Frederiksen, 1998) refers to self-awareness about one's cognitive activities and ability to modify them consciously. This includes reflection, self-assessment, planning, and selecting fix-up strategies to correct problems. Strategies reflecting metacognition include taking notes, rereading text, and studying with partners (Derry, 1992; Novak, 1997; O'Neil & Abedi, 1996). Two opposing points may be made about metacognition and simulation. On one hand, because many simulations demand good metacognition (making decisions, planning, being reflective), they may be more appropriate for high than for low metacognitive learners. On the other hand, if success in simulations demands metacognition, then simulations represent a good environment for learning and exercising metacognitive skills (Veenman & Elshout, 1995; Veenman, Elshout, & Busato, 1994). Stressing the latter view, simulations may have a dual advantage for some learners, as they both teach the simulation's domain and provide the opportunity for practice of metacognitive skills, but only if there is support to teach those skills so they can be practiced.

***Motivation and Learning Style.*** Like cognitive abilities, the designer has little control over learners' initial motivation—the learner's prior interest in and desire to learn the content or skill that is taught by the simulation. Unlike cognitive abilities, designers can enhance motivation. Learning style (also difficult to change, though it may be accommodated) is used in reference to several things (Rayner & Riding, 1997). It can refer to preferred strategies. One may prefer note taking and use it whenever possible, even if it is not always the best strategy. Learning style is also used in reference to constructs about how different people learn. A frequently researched continuum of learning style (or cognitive style) is field-dependence versus field-independence (Tinajero & Páramo, 1997; Weller, Repman, Lan, & Rooze, 1995). A characteristic claimed of the former is the need for externally provided organization, whereas the latter is characterized by creating or finding organization. Simulations using a discovery approach may be better for the field-independent learners. Simulations with more learning support may be better for field-dependent (Veenman & Elshout, 1995; Weller, Repman, Lan, & Rooze, 1995). Learning style also refers to conscious preferences for learning activities (Dunn, Dunn, & Price,

1984), such as a preference for learning from books versus computers. An implication of learning style research is that different learners' approaches are better accommodated if we provide a variety of learning resources and give learners choices. Motivation and learning style probably have less importance for assessment simulations than for instructional simulations, although prior knowledge and ability are clearly important for both.

## Simulation Attributes

The third component of the theory is simulation attributes. Once again, I only discuss a few of the many attributes and do so within eight categories: the underlying model, the scenario, presentation modes, delivery mode, methodology, instructional support, motivators, and fidelity.

***Underlying Model.*** Attributes of a model affect individual learners in different ways. Abstract models, for example, are more difficult for those with little prior knowledge. Model attributes are also affected by domain attributes. More precise domains permit greater resolution and precision in a model. Whether the model is more mathematical versus more logical or whether it is statistical versus algebraic impacts students differently depending on their cognitive abilities. Making the model more visible to the learner may be based on domain complexity and whether the goal is a discovery-oriented simulation or an expository or practice-oriented simulation. Designers of discovery-oriented simulations frequently use a black-box approach in which the underlying model is hidden and must be determined by the learner. In contrast, it has been suggested that for learning complex models, a glass-box approach is more successful.

***Scenario.*** The scenario is the situation or storyline through which learners perceive and interact with the underlying model. Some simulations have a natural and common scenario, such as a passenger airline pilot flying from Boston to Los Angeles. Others have fantasy or less common scenarios, such as battle simulations for fighter pilots. The scenario is critical because a good scenario enhances motivation (Choi, 1997; Keegan, 1995) and lays the basis for the simulation methodology and interactions.

***Presentation Modes.*** Presentation modes include visual versus auditory information, static versus dynamic presentations, and single versus multiple representations of the phenomenon. Presentation modes are affected by the domain (is it verbal, mathematical, or more visual), by the learners (can they read, do they have prior experience, will they have misconceptions), and, in interaction with those attributes, will impact learning, transfer, and test performance (Kozma & Russell, 1997; Moore, Burton, & Myers, 1996).

***Delivery Mode.*** Delivery modes include independent microcomputer-based simulations, distributed simulations with networked users, virtual reality rooms with headsets, and large-scale physical simulators such as are used for aviation training. Delivery mode affects fidelity, learner acceptance, learning activities, encoding, transfer, and test validity in assessment simulations (e.g., O'Neil, Chung, & Brown, 1997; O'Neil, Mayer, Herl, & Niemi, 1998).

***Methodology.*** There are many approaches to simulation and several classification schemes. Different methods and distinctions include microworlds, virtual reality, discovery versus expository simulation, laboratory (popular in the sciences), role-playing (popular in law), and operator-in-the-loop simulation (typical of aviation training). For several classification schemes see Gibbons, Fairweather, Anderson, and Merrill (1997), Towne (1995), Alessi and Trollip (1991), van Joolingen and de Jong (1991), or Reigeluth and Schwartz (1989). The choice of such methods may be a function of instructional philosophy, subject matter, and other prior considerations, but to the extent a designer has choice of methodology, it has impact on learning, motivation, and transfer.

***Instructional Support.*** Instructional support (termed the *instructional overlay* by Reigeluth & Schwartz, 1989) includes coaching, collaborative learning, making models visible (the glass-box approach), providing artificial feedback, giving learners intermediate goals and assignments, and metacognitive support. Instructional support is often avoided in simulations because it tends to be negatively related to fidelity. Increasing instructional support generally lowers the sense of realism in a simulation. But instructional support facilitates learning processes (Boyle, Stevens-Wood, Feng, & Tikka, 1996; de Jong et al., 1999; de Jong & Njoo, 1992; de Jong & van Joolingen, 1998; Goodyear, Njoo, Hijne, & van Berkum, 1991; Kang, 1996; Veenman & Elshout, 1995), so designers must strike a balance between the two.

***Motivators.*** A common assumption is that simulation is inherently more motivating than other instructional methodologies such as reading a book or listening to a lecture (Duchastel, 1994; Gibbons & Fairweather, 1998; Keegan, 1995). This is a misleading assumption. Learners do not compare simulations to books and lectures but to "doing the real thing." Airline pilots compare flying a simulator to flying an aircraft, and unless design takes motivation into consideration, the simulation may be judged as unacceptable by them. Most critical to motivation are the scenario, delivery mode, and fidelity (Choi, 1997; de Jong & van Joolingen, 1996; Keegan, 1995). Scenarios may include game or fantasy elements to increase motivation, but for most adults, realism is the main factor affecting their acceptance of a simulation as worthwhile (de Jong & van Joolingen, 1996; Thurman, 1993). This is true for both learners and their instructors.

***Fidelity.*** Although I discuss it last, fidelity is one of the most important simulation attributes. Fidelity is the degree to which the simulation replicates reality. I avoid the alternative word *realism* because to many people it evokes only the *perceptual* aspects of fidelity—how much the simulation looks, feels, and sounds like the real device or phenomenon. Fidelity goes beyond perceptual fidelity to include *functional* fidelity (which includes how we control or operate the simulation as well as its behaviors, both natural and in response to our actions), and *model* fidelity (the extent to which the underlying math or logical model replicates the intricacies of the real device or phenomenon). Fidelity impacts learning, test validity, transfer, acceptance, anxiety, motivation and the cost of simulation development and delivery (de Jong & van Joolingen, 1996; Detterman & Sternberg, 1993; Franken & O'Neil, 1994; Gagné, 1954; O'Neil, Allred, & Dennis, 1997). The concept of fidelity seems simple, but it is deceptive in that regard. Following are four of the more complex issues related to simulation fidelity. First is the assumption, held for years, that higher fidelity is better, even though studies have proven that notion as simplistic. (Andrews, Carroll, & Bell, 1995; Detterman & Sternberg, 1993; Duchastel, 1994; Hays & Singer, 1989). This assumption is still exhibited today in claims that virtual reality will be the *best* training method or that "the real world is the best teacher." Second is the belief that fidelity is a general or overall simulation quality. Rather, fidelity varies for different parts of a simulation. A good simulation may require high fidelity for some parts and low fidelity for others, termed *selective fidelity* (Andrews, Carroll, & Bell, 1995). We must separately consider fidelity of time, the underlying model, presentations, learner actions, and system feedback. Each of these has subcomponents—for example, the fidelity of pictures, sounds, and of animations when considering presentations. The third issue is that higher fidelity facilitates some valuable outcomes, whereas lower fidelity facilitates others, thereby requiring compromise. High fidelity increases acceptance among learners familiar with the domain (e.g. experienced pilots), has been proposed as relatively more important for assessment simulations (Gagné, 1954), and as *sometimes* more important for transfer (Alessi, 1988; Guckenberger & Guckenberger, 1996). Lower fidelity, in contrast, may be more acceptable to novices (Watson, 1995), relatively better for instruction than assessment (Gagné, 1954), better for initial learning than transfer (Alessi, 1988), and is less expensive (Baum et al., 1982). The fourth issue is that learners' *perception* of fidelity is more critical than actual fidelity. Novices sometimes perceive medium fidelity simulations as more realistic than either low or high fidelity simulations (Dittrich, 1977; Thurman, 1993). Providing learners control of fidelity may increase the perception that it is high because they can adjust it to what they consider realistic. Perception of fidelity is also relative to the complexity of the phenomenon, to a learner's prior simulation experience, and to what is possible given the delivery mode. For example, the same visuals may seem very simple and low in fidelity on a large-scale flight simulator (i.e., a physical simulator of a military or commercial aircraft with a full cockpit and motion base) but sophisticated and high in fidelity on a microcomputer.

Elsewhere I proposed a theoretical framework for fidelity analysis (Alessi, 1988, 1995; Alessi & Trollip, 1991). Some of the more salient aspects of that framework are: Higher fidelity is more important for advanced learners, transfer, and assessment, whereas lower fidelity is more important for novices, initial learning, and performance improvement. For declarative knowledge, fidelity of the underlying model, time, and presentations are more relevant. For procedural knowledge, fidelity of learner activities and system feedback are more relevant. Lastly, the complexity of fidelity and the compromises it demands may be reconciled by emphasizing *perceived* fidelity, by using *dynamic* fidelity, and by giving learners *control* of fidelity.

My colleagues and I engaged in several studies of simulation fidelity and its impact on initial learning, transfer of learning, motivation, and simulator sickness. Procedural learning and transfer was assessed in Alessi (1995). Learners in this study used one of three simulations of a volt–ohm meter: one was of fixed high fidelity, one of fixed low fidelity, and one had dynamic fidelity. Fidelity was varied for the visuals and for the multimeter's functions, but the underlying model and the feedback were always high in realism. After studying the simulation, learners were given a transfer test in which they operated a real volt–ohm meter to measure the values of 20 batteries and resistors. The hypotheses were that low and dynamic fidelity would yield best learning *during* the simulation, whereas dynamic and high fidelity would yield best *transfer*. The second hypothesis was confirmed but not the first. During the simulation, all students learned equally. Many students remarked that they expected the simulation to help them more when they made errors. That prompted a second investigation in which feedback fidelity was also high (realistic reactions from the volt–ohm meter), low (text explanations and suggestions), or dynamic (starting with text explanations but fading them over time). This did affect initial learning, and learners in the low fidelity simulation made significantly fewer errors than in the high fidelity simulation. Learners receiving dynamic fidelity performed in between the other groups, but more similar to those with low fidelity. However, the same pattern was found for the transfer test. Improved learning during the simulation may have facilitated transfer more than the higher fidelity.

Declarative learning was investigated in Quinn and Alessi (1994). The simulation allowed the student to manipulate and study four variables affecting the spread of influenza in a community: time to illness, duration of the illness, number of contacts a person has with other people per week, and length of the immune period after illness. Of interest were students' hypothesis generation and testing strategies and success at minimizing the impact of the flu epidemic. One group of students started with only two variables to investigate and progressed to three and later four variables when they felt they were ready. A second group of students started with three variables, progressing to four when they felt ready. A third group started with all four variables to investigate. Additionally, half of each group was encouraged to generate multiple hypotheses before experimenting and half were not. The results were mixed. The benefit from dynamically increasing complexity only occurred for students generating multiple hypotheses when going from low complexity (two variables to manipulate) to medium complexity (three variable) tasks. The time-to-illness variable was counterintuitive in nature, and students had more success determining its effect when starting with a simpler task and generating multiple hypotheses about the relationship. In a follow-up study, Quinn and Klein (1996) found that presenting the counterintuitive variable first and on its own improved learning efficiency (the number of simulation runs required). In another follow-up study, Quinn, Peña, and McCune (1996) found that dynamic complexity interacted with cooperative learning (students working alone or in cooperative groups). Individual learners benefited from dynamic (increasing) complexity; group learners benefited from starting the simulation at its full level of complexity. Although these studies demonstrated small benefits from dynamic fidelity, they mostly raised questions. Is *learner choice* a good basis for changing fidelity level? Will dynamic fidelity only be a benefit given certain types or amounts of complexity? How much does it depend on other aspects of learner activity and context, such as group or individual learning?

Research on the Iowa Driving Simulator (IDS) project is particularly meaningful to other large-scale physical simulation such as for aircrew training. The IDS (Kuhl, Evans, Papelis, Romano, & Watson, 1995) is a ground vehicle simulator that allows various car, truck, and military vehicle cabs to be mounted on a motion base surrounded by projected visuals and sound. Visuals may be either nonreactive (roadways, buildings, trees) or reactive (other vehicles, people, animals). Different scenarios may be run, such as rural roadways, interstate highways, or city streets. Development is also underway at the University of Iowa on the National Advanced Driving Simulator (NADS), which will include higher fidelity motion and visuals (Haug, 1990).

A major concern with vehicle simulators such as the IDS and NADS is simulator sickness and its effect on operator performance and validity. Watson (1995) investigated simulator sickness as a function of varying levels of visual fidelity, kinesthetic fidelity, and automobile driving experience. Although significant effects were not found for visual or kinesthetic fidelity, novice drivers (high school students who had not yet learned to drive) had significantly less simulator sickness than did experienced drivers, and they perceived the simulator as more realistic than did experienced drivers. This supports the value of designing fidelity based on learner experience and of emphasizing perceived fidelity more than actual fidelity. It also reminds us that experienced learners may demand a high level of realism to consider a simulator acceptable. In subsequent research, Watson (1998) provided evidence that although high visual and kinesthetic fidelity decreased simulator sickness, repeated use of the simulator was the most important variable for reducing simulator sickness, and simulator sickness did not significantly affect driving performance. Thus, although fidelity may not affect simulator sickness appreciably, simulator sickness is significantly less for users with either little experience on the real device (in this case driving automobiles) or with repeated experience in the simulator. The implication of this is that simulators may be best suited for either novice learners or for assessment because users being given an assessment are likely to have had prior repeated experience during instruction on the simulator.

An issue we are now beginning to investigate is the effect of fidelity on *implicit* learning, sometimes called intuitive learning (Jarz, Kainz, & Walpoth, 1997; Moore & Greeno, 1991; Schacter, Cooper, & Delaney, 1990; Schacter, Cooper, Delaney, Peterson, & Tharan, 1991; Swaak, 1998; Swaak, van Joolingen, & de Jong, 1998). Implicit learning is subconscious or semiconscious learning that we cannot put into words, but that affects our behavior in observable ways. When learning to operate a device (such as an automobile or an aircraft), there appears to be implicit learning of psychomotor skills that we cannot easily explain to other people but that clearly contributes to proficient operation. One familiar example (for drivers of manual transmission vehicles) is the simultaneous and smooth coordination of the accelerator and the clutch to start moving forward. Another example is the ability to coordinate a vehicle's motion with other vehicles seen in the side or rear-view mirrors, even though their directions (right and left) are reversed. Research on object recognition (Schacter, Cooper, & Delaney, 1990; Schacter, Cooper, Delaney, Peterson, & Tharan, 1991) and three-dimensional graphics (Hamel & Ryan-Jones, 1997) suggests that implicit learning of device operation may depend on a visual memory system that only functions if the images are sufficiently realistic. High perceptual or functional fidelity may be essential (at a neural level) for implicit learning of complex psychomotor skills such as piloting an aircraft, especially if those skills require a high degree of automaticity.

## Encoding, Representing, and Using Knowledge

The fourth component in our theory includes the processes of learning and transfer. This component can be summarized as follows. As a function of knowledge, learner, and simulation attributes, the learner first develops component or partial procedural knowledge (motor or psychomotor) and declarative knowledge (concepts, rules, principles). By means of functions such as practice, compilation (Neves & Anderson, 1981), and procedure modification heuristics (Bibby & Payne, 1996), learners develop automated skills and mental models. Through processes such as assimilation and accommodation, learners develop schemata and complex skills (Thurman, 1993). What is learned must then be transferred for application in other situations. This summary is what the last 30 or more years of cognitive psychology has been all about (Winn & Snyder, 1996). I discuss only mental models and transfer of learning, both of which are closely linked to the concept of simulation fidelity and require continued examination in that regard.

***Mental Models.*** Different meanings for the term *mental model* are found in work by Mayer (1997), Greeno (1989), White and Frederiksen (1998), and others (several of whom are discussed in Park & Gittelman, 1995). Common to each is the notion of a mental model as an entity in working memory that represents a process, phenomenon, or procedure. The learner can *run* a mental model to make predictions or solve problems. One might have a mental model of how electricity flows in a wire, how a computer's central processor executes commands in a loop, or how an aircraft flies in a holding pattern. A mental model may have visual, mathematical, or verbal components. A learner's mental model may be correct, partially correct, or completely incorrect. An advantage claimed for simulations is that they serve as interactive *conceptual* models (conceptual models being defined as instructional models external to the learner, such as textbook diagrams or computer animations) that help students to create or modify their own internal mental models. Gibbons (1998), for example, contends that most learning is model building and that instruction should be designed to foster that. Kieras & Bovair (1984) illustrated that a correct mental model can significantly enhance procedural learning. Simulation models that can be seen, manipulated, and run, are claimed to be more concrete for learners and thus more likely to facilitate good mental model formation.

It is insufficient to say that simulations may help develop mental models. What is critical in learning from simulations is the development of correct and useful mental models. For example, Larkin (1983) suggested that the models of learners new to a domain are *naive representations*, whereas experts' models are more useful and correct *physical representations*. Many researchers (e.g., White & Frederiksen, 1998) investigated mental model formation and provided evidence that erroneous model formation is a frequent impediment to learning. Simulation design should emphasize techniques, such as animation and experimentation, to help learners create good internal models. The choice of animations, metaphors, and interactions can dramatically affect the internal models that a learner develops. Gentner and Gentner (1983) for example, reported that giving learners different analogies led to different learning, some correct and some incorrect. Rieber (1998) contended that the choice of visual representation is at the heart of which mental model will develop. Hagmann, Mayer, and Nenninger (1998) suggested that novices require more direct instruction about a conceptual model than do experts. From this evidence, we may conclude that a theory of learning from simulation must better explain how correct versus incorrect mental models form as a function of a simulation's visual representations, interactions, and other attributes.

***Transfer of Learning.*** Transfer of learning is a central concern in the design and use of simulation (Alessi, 1988; Allen, Hays, & Buffardi, 1986; Detterman & Sternberg, 1993; Guckenberger & Guckenberger, 1996; Lintern, Sheppard, Parker, Yates, & Nolan, 1989; Pappo, 1998; Thomson, 1989). A primary justification for simulation as a training technology, despite its cost compared to other instructional methods and media, is the goal of deriving transfer benefits compared to those other methods and media. For many years, the goal of transfer was pursued via an emphasis on high fidelity and the theory of identical elements. Today's simulation designers realize that although fidelity is one factor affecting transfer, there are others. For example, in many applications where we say transfer is desired, generalization is the issue. Generalization is enhanced by increasing *variability* of presentations and practice activities rather than by trying to create a specific learning situation identical to the application environment. Understanding of transfer must consider near transfer, for which similarity is more relevant, and far transfer, for which variability is more relevant (Greeno, Smith, & Moore, 1993). It must consider transfer between procedural and declarative knowledge as well as within each (Brooks & Dansereau, 1987). It must consider transfer efficiency as well as transfer effectiveness (Rose, Evans, & Wheaton, 1987). Although there has been increased interest and research on transfer in recent years (Cormier & Hagman, 1987; de Croock, van Merriënboer, & Paas, 1998; Detterman & Sternberg, 1993; Dienes, Altmann, Gao, & Goode, 1995; Gentner, Rattermann, & Forbus, 1993; Harvey & Anderson, 1996; Pennington, Nicolich, & Rahm, 1995; Taylor et al., 1997; van Merriënboer, de Croock, & Jelsma, 1997), we still lack a complete picture of how to optimize transfer from learning in simulations.

### Summary

This section proposed a framework for a theory of learning from simulations. Only a few of the many relevant variables were discussed. These variables are divided into four categories: knowledge attributes, learner attributes, simulation attributes, and the encoding, representation, and use of knowledge by the learner. Clearly there are many interrelationships and dependencies between these categories and their components. A complete theory must describe those interrelationships and dependencies in more detail, a task for much future research. One reason for outlining the theory is to shape a research agenda. The other, and the object of the next section of this chapter, is to guide a model of simulation design.

## MODEL FOR SIMULATION DESIGN

Based on the preceding theory, I now suggest a framework for a model of simulation design. Its purpose is to encourage careful consideration of all critical design issues and decisions in an appropriate sequence. Most existing design models are specific to particular types of simulation, for example, simulations for procedural learning (Pappo, 1998) or conceptual discovery learning (de Jong & van Joolingen, 1998; van Joolingen & de Jong, 1997). As illustrated in Fig. 10.2, informed by the theory, the designer should begin with consideration of the four primary dimensions (knowledge, the learner, purpose, importance) that drive most subsequent decisions. Those decisions begin with fairly general and important ones that I simply call *first level decisions*, and progress to successively more detailed and dependent decisions.

### The Knowledge Dimension

Following from the theoretical discussion, the knowledge dimension is whether instructional objectives are to learn *about* something or to learn *how to do* something, corresponding to the declarative versus procedural distinction. Consideration of this dimension includes analyzing the domain and the instructional objectives to identify which are declarative and which are procedural. In part this should be an analysis of those objectives to be accomplished within the simulation (initial learning), but

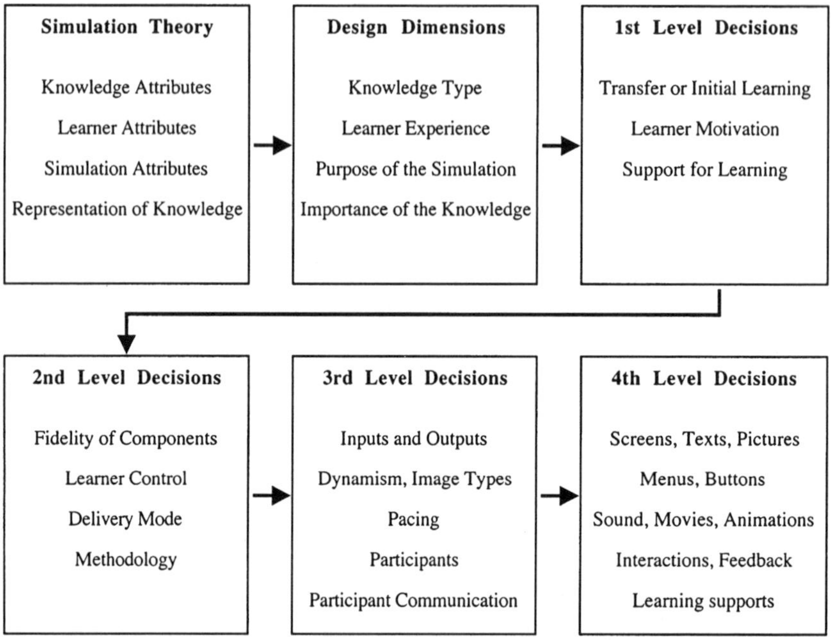


FIG. 10.2. The primary considerations of design that follow from simulation theory.

equally and often more important is analyzing the transfer environment to see what procedural and declarative knowledge one needs when working in that environment.

## The Learner Dimension

Learners have many characteristics (as discussed in the section on learner attributes) with logistical implications for simulation design. The main characteristic I wish to emphasize is current expertise (prior knowledge) in the domain, which may be subdivided into expertise for particular objectives, some procedural and some declarative. The learner dimension may be considered a continuum from novice to expert. This dimension is an interesting one because instructional simulations need to accomplish two things. First, they must accommodate *different* learners who are at different points (novice to expert) in this dimension. Second, they must accommodate *individual* learners as they progress from novice to becoming an expert. Expertise in a large domain, such as aviation, probably changes slowly over a relatively long period of time. For that purpose, it may be sufficient for accommodation to take place across larger instructional events (lessons) rather than within them. For example, successive events may become progressively more complex, sophisticated, detailed, and realistic, as suggested by Reigeluth's (1992) Elaboration Theory of Instruction, just as aviation students start with ground school (lectures, books, and videos), before progressing to simulators and then real aircraft. In addition to considering overall expertise in the larger domain, we must consider increasing expertise for component skills or knowledge, such as the objectives of a particular lesson. Those skills will develop more quickly, even within the duration of a lesson activity. Therein lies a compelling rationale for dynamically changing a simulation's characteristics as learners progress through it.

### The Purpose Dimension

Purpose is whether a simulation is intended for instruction or assessment, and instruction may be subdivided into initial learning or practice. Although a particular simulation may need to satisfy different points on the knowledge or learner dimensions, it can usually be designed for a single purpose. We can design different simulations for different purposes, even if content is the same in each. Many of the issues I discussed in this chapter are relevant primarily to instructional simulations. Designers must recognize that when a simulation is to be used for assessment, the factors of importance will be very different. In assessment the main general concerns are test reliability and validity. To enhance reliability, we must seek to reduce simulator sickness, assess knowledge and performance with multiple exposures (the equivalent of increasing test length in traditional tests), and standardize administration procedures. We should view simulations as a type of adaptive exam and use statistical models appropriate to them. Test validity is essentially the issue of transfer and is usually sought via high fidelity. But fidelity only increases *content* or *face* validity. Ultimately we are more interested in *predictive* validity, the simulation's accuracy at predicting performance in the transfer environment. This requires validation studies in which simulation performance is correlated with transfer performance and the simulation revised to improve that correlation. Some interesting questions remain to be addressed concerning assessment simulations. Should "rare events," which are presented in instructional simulations more frequently than in reality (such as aircraft engine failure), be similarly frequent in assessment, or will that decrease validity? How can the equivalent of "increasing the number of items in a test" be accomplished in simulation without decreasing realism? Should random or probabilistic events, useful for teaching imprecise domains, be used during assessments or will they decrease the standardization of administration procedures? Is test anxiety a factor in assessment simulations? Finally, should the many techniques useful and common in instructional simulations (instructional support, user control, glass-box models, motivators, multiple representation modes, or learning style accommodation) be eliminated partially or completely during assessment simulations?

### The Importance Dimension

By *importance* I mean whether the consequences of error are severe (a patient dies or an aircraft crashes) or if the consequences of good performance are very positive (a company earns a large profit or an enemy aircraft is eliminated). The more important the content, the more critical are decisions about learning support, transfer, motivation, and delivery mode. Most simulation design models, and for that matter most instructional design models, do not address importance. Although it is probably in the back of a designer's mind, especially when the content *is* very important, it should be more explicitly considered during design. Guckenberger and Guckenberger (1996) suggested that the stress of real environments with important consequences improves performance, and that simulations lacking real consequences will exhibit lowered transfer unless other design techniques (they suggested requiring performance at slightly faster than real-time) are employed.

### First Level Decisions

Being aware of the four dimensions (knowledge, learner, purpose, and importance) helps the designer make *deliberate* decisions about simulation characteristics. Because there are many, it is useful to sequence those decisions based on how they influence one another. Those I call *first level decisions* are most directly influenced by the four dimensions and in turn influence most other decisions. The first level decisions are transfer versus initial learning, motivation, and support.

Although design always considers both learning and transfer, one or the other may need to be emphasized due to the dimensions of the simulation, and which one is emphasized will impact later decisions. For example, transfer will tend to be more critical given procedural knowledge, more advanced learners, more important knowledge, and for assessment (where transfer relates to test validity). When making *second* level decisions, a transfer emphasis may require greater fidelity, less (or different) learner control, and particular delivery modes such as a physical simulator.

Motivation has multiple components, some a function of the simulation dimensions and some not. More important skills may increase intrinsic motivation and thus require *less* design to enhance it. More experienced learners have higher demands for relevance during instruction, so design to enhance that aspect of motivation may be more critical. A primary reason for high fidelity motion in aircraft simulators has been student pilots' (and instructors') insistence that simulators feel like real aircraft. Novice learners will probably demand realism less, as in Watson (1995), where non-drivers rated the Iowa Driving Simulator as more realistic than did experienced drivers.

The need for learning support increases for novices and as importance increases, although it decreases for assessment simulations. Support for learning includes metacognitive support. For example, an instructional simulation for novice students might pause periodically and encourage note-taking or reflecting on simulation behavior. An assessment simulation might suggest that students double-check their answers when they indicate they are done and if there is time to do so.

### Second Level Decisions

The *second level decisions* (those that should be based on the simulation design dimensions and the first level decisions) include fidelity of the various components, learner control, delivery mode, and methodology. Fidelity is considered a second level decision, despite its great importance to simulation theory and design, because it depends on and follows from the dimensions and first level decisions. Following are examples of fidelity guidelines based on the theory and assuming particular dimensions for a simulation. For assessments, fidelity should be high (Alessi & Trollip, 1991; Gagné, 1954; Gerson, 1997), especially when the information or skills are important. For procedural tests it is important that examinee actions and system reactions be very realistic and there be little or no unnatural feedback. In contrast, for instructional simulations, fidelity and complexity should be lower for novice learners and when initial learning is the primary concern, but higher for more advanced learners and where practice and transfer are of greater concern (Gerson, 1997; Watson, 1995). A way to accommodate the needs of both novice and experienced learners, or the individual learner's progress from novice to expert, is to use dynamic variation of fidelity and complexity, with changes being a function primarily of performance but in part based on learner or instructor choice (Alessi, 1995). During initial learning, procedural objectives may best be facilitated by providing learning supports such as corrective feedback or decreasing the number of actions learners must contend with, then removing the supports and increasing the number of actions as the learners gain experience and the need for transfer becomes more critical (Alessi, 1995). For declarative learning (understanding concepts, principles, or relationships), initial learning may be optimized by starting with simple models and presentations, then adding details to the model and variables the learner must attend to as experience increases (Quinn & Klein, 1996). To maintain motivation and learner attitude towards the use of simulation, the perception of realism is more important than actual realism (Dittrich, 1977; Thurman, 1993). Learner acceptance may be facilitated by providing high fidelity for the more salient features of a simulation (motion, visuals, actions frequently taken) and lower fidelity for less salient features such as the underlying model or uncommon actions (Andrews, Carroll, & Bell, 1995). An interesting example of fidelity design occurs when parts of the domain lack precision. In that case, perception of fidelity may be better if we use random numbers or probability distributions to trigger events, rather than strict algebraic formulas. For example, in a flight simulator, the controls and behaviors of the aircraft (which have high precision) may be programmed algebraically, and system reactions will follow predictably from learner actions and other inputs. In contrast, weather and its prediction (a less precise domain) is based on a combination of algebraic formulas, probability distributions, and random numbers. Weather in the simulation should not be too predictable, just as it is not in reality. Lastly, simulated cockpit crew behavior (a very imprecise domain) are based on a combination of logical rules, probability distributions, and perhaps random numbers. A simulated copilot may react calmly and rationally in one emergency but irrationally in another.

Many designers today take an idealistic approach to learner control—it is "learner oriented" and therefore good. Others take a more empirical approach, citing many research studies showing that too much learner control hinders achievement (Singhanayok & Hooper, 1998; Steinberg, 1989; Williams, 1996). Neither may be optimal. In simulation design, decisions about learner control must consider the learner and purpose dimensions, motivation, support, and fidelity. More control is needed for experienced learners. Fewer and different controls should be provided in assessment simulations. Although some types of control may hinder learning, they may also improve motivation, and the two must be carefully balanced. Greater fidelity, which might be necessitated by the need for transfer, suggests providing more natural types of controls for the learner.

Delivery mode (a large-scale physical simulator, a network of microcomputers for distributed simulation, a virtual reality room or headset) should be a function of importance, fidelity, motivation, and the need for transfer. Physical simulators are often required when extreme high fidelity is necessary for assessment or for teaching very important skills. In the future, if dynamic fidelity culminating in very high fidelity is desired, virtual reality devices might be better than physical simulators. However, motivation and learner acceptance may still demand the latter.

The simulation methodology (e.g., a diagnostic simulation, an interpersonal role-playing simulation, or a war-game simulation) is mostly a function of the domain. When teaching aircraft mechanics to fix engines, a diagnostic simulation makes more sense than a war-game simulation.

But for teaching military personnel to operate a jet fighter's weapons, a war-game methodology is probably more effective. Although the domain will largely suggest an appropriate methodology, the methodology may need to be modified or replaced if the simulation's purpose is assessment (for example, a game methodology may not be considered appropriate because it decreases a test's construct validity), if the consequences of error are severe (for example, a nuclear reactor simulator demands such high transfer of learning that extremely realistic methodologies may be required), or if the domain's complexity is so great that extensive learning support is required (such as part-task trainers that break the complete activities down into manageable components).

### Third and Fourth Level Decisions

Space does not allow an extended discussion of third and fourth level decisions. The third level requires detailed decisions about simulation inputs and outputs, whether presentations are dynamic or static, the types of images required (schematic, photographic, video), pacing, the number of participants, and the types of communication between them. Examples might be the following: For assessment, the examinee may need to be the only participant to ensure test reliability. If high fidelity is deemed necessary, an instructional simulation may use dynamic visual images, fast pace, and realistic modes of input and output. The many fourth level decisions include the design of each individual screen, text, picture, menu, button, sound, movie, animation, interaction, feedback, and learning supports such as hints, cues, or coaching (Acovetti & Gamble, 1997; Rieber, 1996). At this point, the impact of prior decisions (fidelity, purpose, importance, learner experience) become fairly obvious. For example, high fidelity interpersonal interactions may require using voice or video. Novice learners require extensive learning support features during instruction but may need fewer or no support features during assessment.

## RESEARCH FOR THE FUTURE

The simulation theory I outlined suggests a framework for a simulation research agenda as depicted in Table 10.1. The main effects and interactions of the independent variables in the left-hand column should be assessed for their effects on intermediate variables in the middle column and the outcome variables listed in the right-hand column. Independent variables (whether natural or determined by the designer) are those not constrained by other variables. Intermediate variables are constrained or determined by independent variables. Outcome variables are those we are

TABLE 10.1
Framework for a Research Agenda on Learning and Assessment With Simulation

| *Independent Variables* | *Intermediate Variables* | *Outcome Variables* |
|---|---|---|
| Knowledge Attributes | mental models | transfer |
| procedural versus declarative | other mental representations | near transfer |
| organization and complexity | implicit learning | far transfer |
| Learner Attributes | motivation and learner attitude | transfer effectiveness |
| novice versus expert | acceptance by learners | transfer efficiency |
| initial motivation | acceptance by instructors | retention |
| ability, metacognition, style | initial learning and error rates | automaticity |
| Simulation Attributes | learning efficiency | metacognition |
| fidelity | navigation and control success | test predictive validity |
| glass versus black-box approach | test reliability | acceptance by learners |
| delivery modes and methods | test content validity | acceptance by instructors |
| instructional supports | test construct validity | cost |
| motivators | | |
| user controls | | |

ultimately most interested in because they reflect cost and performance quality and are in turn constrained or determined by intermediate variables and independent variables.

Initial research should not seek to uncover optimal combinations because the combinations are far too numerous for such research to be viable. Optimization should be the goal of design and accomplished through prototyping, evaluation, and revision. Research should assess more general questions such as the following: Do novice learners benefit from more natural simulations or those with greater (and less natural) learning supports? Will mental models of complex systems be facilitated more by glass-box or black-box approaches? Will test construct validity (facilitated by higher fidelity) lead to greater test predictive validity? The important area of research on fidelity has had weak and inconsistent results. Possible reasons for that, which must be corrected, include the following: Different fidelity levels considered high and low by researchers are often, when compared to reality, all very low. A wider range of fidelity variation is needed with the high end closer to reality. Future research must consider fidelity *relative* to reality, learners' *perceived* fidelity rather than actual fidelity, and the fidelity of specific simulation components.

## CONCLUSION

Simulations, especially large-scale physical simulators for devices like aircraft, are often designed in an atheoretical context with the emphasis on logistic factors such as cost, safety, and user acceptance. In the case of aircraft simulators, Federal Aviation Administration regulations (U.S. Department of Transportation, Federal Aviation Administration, 1991) determine in large part the design and construction of those simulators. Although logistic and regulatory considerations are important, advancement of simulation-based learning and assessment depends on a richer underlying theory, supported by research, to guide initial design, evaluation, and refinement. This is especially important because on the horizon are the new virtual reality approaches. Past experience and the predictions of this chapter's theory suggest both pitfalls and benefits in virtual reality approaches. On one hand (potential pitfalls), virtual reality proponents make the same erroneous claim that has been repeated for years, that reality is the best teacher and that the higher the realism the better the outcomes. That belief must be tempered (or supported) with empirical evidence. On the other hand, a potential advantage that some forms of virtual reality may provide is the ability to dynamically vary fidelity and complexity. Physical simulators such as most current aircraft simulators include part of the actual aircraft (the cockpit) and although fidelity of some components (system feedback, views outside windows) may be varied, fidelity of other components (cockpit instruments and controls) are essentially fixed. Headset and glove-based virtual reality may provide the mechanism for varying all the necessary visual, functional, and kinesthetic variables. It is essential that a clearer research agenda be developed regarding such issues (for example, how fidelity should be varied for particular learners, knowledge, and purposes) and that agenda should be used to guide research and design for virtual reality and other new simulation approaches.

## REFERENCES

Acovetti, M., & Gamble, M. (1997). A coaching agent for learners using multimedia simulations. *Educational Technology, 37*(2), 44–48.

Alessi, S. M. (1988). Fidelity in the design of instructional simulations. *Journal of Computer-Based Instruction, 15*(2), 40–47.

Alessi, S. M. (1995). *Dynamic versus static fidelity in a procedural simulation*. Paper presented at the Annual Meeting of the American Educational Research Association, San Francisco, CA.

Alessi, S. M., & Trollip, S. R. (1991). *Computer-based instruction: Methods and development* (2nd ed). Englewood Cliffs, NJ: Prentice-Hall.

Allen, J. A., Hays, R. T., & Buffardi, L. C. (1986). Maintenance training simulator fidelity and individual differences in transfer of training. *Human Factors, 28*(5), 497–509.

Anderson, J. R. (1980). *Cognitive psychology and its implications*. San Francisco: W. H. Freeman.

Andrews, D. H., Carroll, L. A., & Bell, H. H. (1995). The future of selective fidelity in training devices. *Educational Technology, 35*(6), 32–36.

Baum, D. R., Smith, D. A., Hirshfeld, S. F., Klein, G. A., Swezey, R. W., & Hays, R. T. (1982). *Specification of training simulator fidelity: A research plan*. (ERIC Document Reproduction Service No. ED 241 793).

Bibby, P. A., & Payne, S. J. (1996). Instruction and practice in learning to use a device. *Cognitive Science, 20*(4), 539–578.

Boyle, T., Stevens-Wood, B., Feng, Z., & Tikka, A. (1996). Structured learning in a virtual environment. *Computers and Education: An International Journal, 26*(1–3), 41–49.

Brooks, L. W., & Dansereau, D. F. (1987). Transfer of information: An instructional perspective. In S. M. Cormier & J. D. Hagman (Eds.), *Transfer of learning: Contemporary research and applications* (pp. 121–150). San Diego: Academic Press.

Choi, W. (1997). Designing effective scenarios for computer-based instructional simulation: Classification of essential features. *Educational Technology, 37*(5), 13–21.

Cormier, S. M., & Hagman, J. D. (1987). *Transfer of learning: Contemporary research and applications*. San Diego: Academic Press.

de Croock, M. B. M., van Merriënboer, J. J. G., & Paas, F. G. W. C. (1998). High versus low contextual interference in simulation-based training of troubleshooting skills: Effects on transfer performance and invested mental effort. *Computers in Human Behavior, 14*(2), 249–267.

de Jong, T., & Ferguson-Hessler, M. G. M. (1996). Types and qualities of knowledge. *Educational Psychologist, 31*(2), 105–113.

de Jong, T., Martin, E., Zamarro, J.-M., Esquembre, F., Swaak, J., & van Joolingen, W. R. (1999). The integration of computer simulation and learning support: An example from the physics domain of collisions. *Journal of Research in Science Teaching, 36*(5), 597–615.

de Jong, T., & Njoo, M. (1992). Learning and instruction with computer simulations: Learning processes involved. In E. De Corte, M. C. Linn, H. Mandl, & L. Verschaffel (Eds.), *Computer-based learning environments and problem solving* (pp. 411–427). Berlin: Springer-Verlag.

de Jong, T., & van Joolingen, W. R. (1996). *Discovery learning with computer simulations of conceptual domains* (technical report IST-MEMO-96-02). Enschede, The Netherlands: University of Twente.

de Jong, T., & van Joolingen, W. R. (1998). Scientific discovery learning with computer simulations of conceptual domains. *Review of Educational Research, 68*(2), 179–201.

Derry, S. J. (1992). Metacognitive models of learning and instructional systems design. In M. Jones & P. H. Winne (Eds.), *Adaptive learning environments: Foundations and frontiers* (pp. 257–286). Berlin: Springer-Verlag.

Detterman, D. K., & Sternberg, R. J. (Eds.). (1993). *Transfer on trial: Intelligence, cognition, and instruction*. Norwood, NJ: Ablex.

Dienes, Z., Altmann, G., Gao, S.-J., & Goode, A. (1995). The transfer of implicit knowledge across domains. *Language and Cognitive Processes, 10*(3/4), 363–367.

Dittrich, J. E. (1977). Realism in business games: A three-game comparison. *Simulation & Games, 8*(2), 201–210.

Duchastel, P. (1994). Learning environment design. *Journal of Educational Technology Systems, 22*(3), 225–233.

Dunn, R., Dunn, K., & Price, G. E. (1984). *Learning Style Inventory*. Lawrence, KS: Price Systems.

Fitzgerald, G. E., Wilson, B., & Semrau, L. P. (1997). An interactive multimedia program to enhance teacher problem-solving skills based on cognitive flexibility theory: Design and outcomes. *Journal of Educational Multimedia and Hypermedia, 6*(1), 47–76.

Flavell, J. H. (1979). Metacognition and cognitive monitoring. *American Psychologist, 34*, 906–911.

Franken, J., & O'Neil, H. F. Jr. (1994). Stress induced anxiety of individuals and teams in a simulator environment. In H. F. O'Neil Jr. & M. Drillings (Eds.), *Motivation: Theory and research* (pp. 201–218). Hillsdale, NJ: Lawrence Erlbaum Associates.

Gagné, R. M. (1954). Training devices and simulators: Some research issues. *The American Psychologist, 9*, 95–107.

Gentner, D., & Gentner, D. (1983). Flowing waters or teeming crowds: Mental models of electricity. In D. Gentner & A. Stevens (Eds.), *Mental models* (pp. 99–127). Hillsdale, NJ: Lawrence Erlbaum Associates.

Gentner, D., Rattermann, M. J., & Forbus, K. D. (1993). The roles of similarity in transfer: Separating retrievability from inferential soundness. *Cognitive Psychology, 25*(4), 524–575.

Gerson, C. W. (1997). Situation awareness and dynamic performance training systems: Some reflections on the literature. *Journal of Educational Technology Systems, 25*(4), 373–407.

Gibbons, A. S. (1998, April). *Model-centered instruction.* Paper presented at the annual meeting of the American Educational Research Association, San Diego, CA.

Gibbons, A. S., & Fairweather, P. G. (1998). *Computer-based instruction: Design and development.* Englewood Cliffs, NJ: Educational Technology Publications.

Gibbons, A. S., Fairweather, P. G., Anderson, T. A., & Merrill, M. D. (1997). Simulation and computer-based instruction: A future view. In C. R. Dills & A. J. Romiszowski (Eds.), *Instructional development paradigms* (pp. 769–804). Englewood Cliffs, NJ: Educational Technology Publications.

Goodyear, P., Njoo, M., Hijne, H., & van Berkum, J. J. A. (1991). Learning processes, learner attributes and simulations. *Education & Computing, 6*, 263–304.

Gourgey, F. (1998). Metacognition in basic skills instruction. *Instructional Science, 26*(1–2), 81–96.

Greeno, J. G. (1989). Situation models, mental models, and generative knowledge. In D. Klahr & K. Kotovsky (Eds.), *Complex information processing: The impact of Herbert A. Simon* (pp. 285–318). Hillsdale, NJ: Lawrence Erlbaum Associates.

Greeno, J. G., Smith, D. R., & Moore, J. L. (1993). Transfer of situated learning. In D. K. Detterman & R. J. Sternberg (Eds.), *Transfer on trial: Intelligence, cognition, and instruction* (pp. 99–167). Norwood, NJ: Ablex.

Guckenberger, D., & Guckenberger, E. (1996). *Above real-time training theory: Implications of neuroscience, cognitive psychology, & learning theory in simulator training.* Orlando, FL: Institute for Simulation and Training.

Hagmann, S., Mayer, R. E., & Nenninger, P. (1998). Using structural theory to make a word-processing manual more understandable. *Learning and Instruction, 8*(1), 19–35.

Hamel, C. J., & Ryan-Jones, D. L. (1997). Using three-dimensional interactive graphics to teach equipment procedures. *Educational Technology Research and Development, 45*(4), 77–87.

Harvey, L., & Anderson, J. (1996). Transfer of declarative knowledge in complex information-processing domains. *Human–Computer Interaction, 11*(1), 69–96.

Haug, E. J. (1990, March). *Feasibility study and conceptual design of a National Advanced Driving Simulator.* Final Report DOT HS 807 596. Washington DC: U.S. Department of Transportation.

Hays, R. T., & Singer, M. J. (1989). *Simulation fidelity in training system design: Bridging the gap between reality and training.* New York: Springer.

Jarz, E. M., Kainz, G. A., & Walpoth, G. (1997). Multimedia-based case studies in education: Design, development, and evaluation of multimedia-based case studies. *Journal of Educational Multimedia and Hypermedia, 6*(1), 23–46.

Kang, S.-H. (1996). The effects of using an advance organizer on students' learning in a computer simulation environment. *Journal of Educational Technology Systems, 25*(1), 57–65.

Keegan, M. (1995). *Scenario educational software: Design and development of discovery learning.* Englewood Cliffs, NJ: Educational Technology Publications.

Kieras, D. E., & Bovair, S. (1984). The role of a mental model in learning to operate a device. *Cognitive Science, 8*, 255–273.

Kozma, R. B., & Russell, J. (1997). Multimedia and understanding: Expert and novice responses to different representations of chemical phenomena. *Journal of Research in Science Teaching, 34*(9), 949–968.

Kuhl, J. G., Evans, D., Papelis, Y. E., Romano, R. A., & Watson, G. S. (1995). The Iowa Driving Simulator: An immersive environment for driving-related research and development. *IEEE Computer, 28*(7), 35–41.

Larkin, J. H. (1983). The role of problem representation in physics. In D. Gentner & A. Stevens (Eds.), *Mental models* (pp. 75–98). Hillsdale, NJ: Lawrence Erlbaum Associates.

Lintern, G., Sheppard, D. J., Parker, D. L., Yates, K. E., & Nolan, M. D. (1989). Simulator design and instructional features for air-to-ground attack: A transfer study. *Human Factors, 31*(1), 87–99.

Mayer, R. E. (1997). Multimedia learning: Are we asking the right questions? *Educational Psychologist, 32*(1), 1–19.

Mayer, R. E. (1998). Cognitive, metacognitive, and motivational aspects of problem solving. *Instructional Science, 26*(1–2), 49–63.

Moore, D. M., Burton, J. K., & Myers, R. J. (1996). Multiple-channel communication: The theoretical and research foundations of multimedia. In D. H. Jonassen (Ed.), *Handbook of Research for Educational Communications and Technology* (pp. 851–875). New York: Macmillan.

Moore, J. L., & Greeno, J. G. (1991). Implicit understanding of functions in quantitative reasoning. *Proceedings of the 13th Annual Conference of the Cognitive Science Society* (pp. 221–226). Hillsdale, NJ: Lawrence Erlbaum Associates.

Neves, D. M., & Anderson, J. R. (1981). Knowledge compilation: Mechanisms for the automatization of cognitive skills. In J. R. Anderson (Ed.), *Cognitive skills and their acquisition* (pp. 57–84). Hillsdale, NJ: Lawrence Erlbaum Associates.

Novak, J. D. (1997). Ausubel's assimilation theory and metacognitive tools as a foundation for instructional design. In C. R. Dills & A. J. Romiszowski (Eds.), *Instructional development paradigms* (pp. 401–415). Englewood Cliffs, NJ: Educational Technology Publications.

O'Neil, H. F. Jr., & Abedi, J. (1996). Reliability and validity of a state metacognitive inventory: Potential for alternative assessment. *Journal of Educational Research, 89*(4), 234–245.

O'Neil, H. F. Jr., Allred, K., & Dennis, R. A. (1997). Validation of a computer simulation for assessment of interpersonal skills. In H. F. O'Neil Jr., *Workforce readiness: Competencies and assessment* (pp. 229–254). Mahwah, NJ: Lawrence Erlbaum Associates.

O'Neil, H. F. Jr., Chung, G. K. W. K., & Brown, R. S. (1997). Use of networked simulations as a context to measure team competencies. In H. F. O'Neil Jr., *Workforce readiness: Competencies and assessment* (pp. 411–452). Mahwah, NJ: Lawrence Erlbaum Associates.

O'Neil, H. F. Jr., Mayer, R., Herl, H., & Niemi, C. (1998). *Report of VR experimental research study*. (Technical Deliverable 3). Sherman Oaks, CA: Advance Design Information.

Pappo, H. A. (1998). *Simulations for skills training*. Englewood Cliffs, NJ: Educational Technology Publications.

Park, I., & Hannafin, M. J. (1993). Empirically-based guidelines for the design of interactive multimedia. *Educational Technology Research and Development, 41*(3), 63–85.

Park, O.-C., & Gittelman, S. S. (1995). Dynamic characteristics of mental models and dynamic visual displays. *Instructional Science, 23*(5–6), 303–320.

Pennington, N., Nicolich, R., & Rahm, J. (1995). Transfer of training between cognitive subskills: Is knowledge use specific. *Cognitive Psychology, 28*(2), 175–224.

Quinn, J., & Alessi, S. (1994). The effects of simulation complexity and hypothesis-generation strategy on learning. *Journal of Research on Computing in Education, 27*(1), 75–91.

Quinn, J., & Klein, D. (1996). *The affects of complementary dual goals in an instructional simulation*. Paper presented at the 1996 National Convention of the Association for Educational Communications and Technology, Indianapolis, IN.

Quinn, J., Peña, C., & McCune, L. (1996). The effects of group and task structure in an instructional simulation. *18th Annual Proceedings of selected research and development presentations at the 1996 National Convention of the Association for Educational Communications and Technology* (pp. 571–581). Ames, IA: Iowa State University.

Rayner, S., & Riding, R. (1997). Towards a categorisation of cognitive styles and learning styles. *Educational Psychology, 17*(1,2), 5–27.

Reigeluth, C. M. (1992). Elaborating the elaboration theory. *Educational Technology Research and Development, 40*(3), 80–86.

Reigeluth, C. M., & Schwartz, E. (1989). An instructional theory for the design of computer-based simulations. *Journal of Computer-Based Instruction, 16*(1), 1–10.

Rieber, L. P. (1996). Animation as feedback in a computer-based simulation: Representation matters. *Educational Technology Research and Development, 44*(1), 5–22.

Rieber, L. P. (1998). *Metaphors as graphical representations within open-ended computer-based simulations*. Paper presented at the Annual Meeting of the American Educational Research Association, San Diego, CA.

Rose, A., Evans, R., & Wheaton, G. (1987). Methodological approaches for simulator evaluations. In S. M. Cormier & J. D. Hagman (Eds.), *Transfer of learning: Contemporary research and applications* (pp. 261–278). San Diego: Academic Press.

Schacter, D. L., Cooper, L. A., & Delaney, S. M. (1990). Implicit memory for unfamiliar objects depends on access to structural descriptions. *Journal of Experimental Psychology: General, 119*(1), 5–24.

Schacter, D. L., Cooper, L. A., Delaney, S. M., Peterson, M. A., & Tharan, M. (1991). Implicit memory for possible and impossible objects: Constraints on the construction of structural descriptions. *Journal of Experimental Psychology: Learning, Memory, and Cognition, 17*(1), 3–19.

Schraw, G. (1998). Promoting general metacognitive awareness. *Instructional Science, 26*(1–2), 113–125.

Singhanayok, C., & Hooper, S. (1998). The effects of cooperative learning and learner control on students' achievement, option selections, and attitudes. *Educational Technology Research & Development, 46*(2), 17–33.

Steinberg, E. R. (1989). Cognition and learner control: A literature review, 1977–1988. *Journal of Computer-Based Instruction, 16*(4), 117–121.

Sternberg, R. J. (1998). Metacognition, abilities, and developing expertise: What makes an expert student? *Instructional Science, 26*(1–2), 127–140.

Swaak, J. (1998). *What-if: Discovery simulations and assessment of intuitive knowledge*. Doctoral dissertation. Enschede, The Netherlands: Twente University.

Swaak, J., van Joolingen, W. R., & de Jong, T. (1998). Supporting simulation-based learning: The effects of model progression and assignments on definitional and intuitive knowledge. *Learning and Instruction, 8*(3), 235–252.

Taylor, H. L., Lintern, G., Hulin, C. L., Talleur, D., Emanuel, T., & Phillips, S. (1997). *Transfer of training effectiveness of personal computer-based aviation training devices. Report No. DOT/FAA/AM-97/11*. Washington, DC: U.S. Department of Transportation, Federal Aviation Administration.

Thomson, D. R. (1989). Transfer of training from simulators to operational equipment—Are simulators effective? *Journal of Educational Technology Systems, 17*(3), 213–218.

Thurman, R. A. (1993). Instructional simulation from a cognitive psychology viewpoint. *Educational Technology Research and Development, 41*(4), 75–89.

Tinajero, C., & Páramo, M. F. (1997). Field dependence–independence and academic achievement: A re-examination of their relationship. *British Journal of Educational Psychology, 67*(2), 199–212.

Towne, D. M. (1995). *Learning and instruction in simulation environments*. Englewood Cliffs, NJ: Educational Technology Publications.

U.S. Department of Transportation, Federal Aviation Administration. (1991). *Airplane simulator qualification. Advisor circular number AC 120-40B*. Washington, DC: Author.

van Joolingen, W. R., & de Jong, T. (1991). Characteristics of simulations for instructional settings. *Education & Computing, 6*, 241–262.

van Joolingen, W. R., & de Jong, T. (1997). An extended dual search space model of scientific discovery learning. *Instructional Science, 25*(5), 307–346.

van Merriënboer, J. J. G., de Croock, M. B. M., & Jelsma, O. (1997). The transfer paradox: Effects of contextual interference on retention and transfer performance of a complex cognitive skill. *Perceptual and Motor Skills, 84*, 784–786.

Veenman, M. V. J., & Elshout, J. J. (1995). Differential effects of instructional support on learning in simulation environments. *Instructional Science, 22*(5), 363–383.

Veenman, M. V. J., Elshout, J. J., & Busato, V. V. (1994). Metacognitive mediation in learning with computer-based simulations. *Computers in Human Behavior, 10*, 93–106.

Veenman, M. V. J., Elshout, J. J., & Meijer, J. (1997). The generality vs domain-specificity of metacognitive skills in novice learning across domains. *Learning and Instruction*, 7(2), 187–209.

Watson, G. (1995). *Simulator effects in a high fidelity driving simulator as a function of visuals and motion*. Orlando, FL: Institute for Simulation and Training.

Watson, G. (1998). *Simulator sickness adaptation in a high-fidelity driving simulator as a function of scenario intensity and motion cueing*. Doctoral dissertation, The University of Iowa, Iowa City, Iowa.

Weller, H. G., Repman, J., Lan, W., & Rooze, G. (1995). Improving the effectiveness of learning through hypermedia-based instruction: The importance of learner characteristics. *Computers in Human Behavior, 11*(3–4), 451–465.

Wenger, E. (1987). *Artificial intelligence and tutoring systems: Computational and cognitive approaches to the communication of knowledge*. Los Altos, CA: Morgan Kaufmann Publishers.

White, B. Y., & Frederiksen, J. R. (1998). Inquiry, modeling, and metacognition: Making science accessible to all students. *Cognition and Instruction, 16*(1), 3–118.

Williams, M. D. (1996). Learner-control and instructional technologies. In D. H. Jonassen (Ed.), *Handbook of research for educational communications and technology* (pp. 957–983). New York: Macmillan.

Winn, W., & Snyder, D. (1996). Cognitive perspectives in psychology. In D. H. Jonassen (Ed.), *Handbook of research for educational communications and technology* (pp. 112–142). New York: Macmillan.

Chapter **11**

# Principles of Performance Measurement for Ensuring Aircrew Training Effectiveness

Daniel J. Dwyer
*Naval Air Warfare Center Training Systems Division, Orlando, FL*

Eduardo Salas
*University of Central Florida, Orlando, FL*

Performance measurement is paramount to training. Without it, administrators do not know if training is having the intended effect, researchers cannot be certain which instructional strategies are most effective, instructors cannot accurately identify deficiencies in need of remediation, and trainees cannot answer the question "How are we doing?" (Hallman & Campbell, 1997). Performance measures allow training researchers and practitioners to gauge progress, to properly focus follow-on training, to evaluate different training techniques, and to assess how much "bang" they are getting for their training "buck." In short, performance measurement allows us to make informed decisions concerning the training function (Brannick & Prince, 1997). Despite the critical role that performance measurement should play in training, the development and implementation of valid, reliable, and meaningful performance measures is surprisingly absent (Komaki, 1997). Why?

Some researchers attributed the absence of performance measures to the difficulty, challenge, and complexity surrounding the issue (Cannon-Bowers & Salas, 1997), suggesting that it may be easier to simply ignore measurement. Others argued that there has been a scarcity of research devoted to developing measures (Dickinson & McIntyre, 1997), thereby making it difficult for practitioners to easily develop performance measurement systems. Still others addressed the time and expense associated with performance measurement, suggesting that these types of resources might also be to blame (Brannick & Prince, 1997). In the education sector, similar problems and rationale exist for performance measurement (Baker, Abedi, Linn, & Niemi, 1996). Whatever the reason, the efficient use of valuable training resources is often hindered by the lack of performance measurement systems within training environments. If performance is not properly measured, it cannot be effectively managed. Given the tremendous resources devoted to training, it makes economic sense to systematically assess its effectiveness. Performance measurement allows this assessment.

In recognition of the importance of performance measurement for training, researchers began focusing attention in this area. Theories of team performance are beginning to form a foundation on which performance measurement can be based (Baker & Salas, 1997). Based on these theoretical developments, as well as on a body of empirical research, Cannon-Bowers and Salas (1997) identified four basic tenets that they believed should guide the selection of performance measures and the development of performance measurement systems for use in training environments. We now briefly describe these tenets.

First, Cannon-Bowers and Salas (1997) argued that the development of performance measures for training should consider multiple levels of measurement. For example, an instructor may want a training session to focus on teaching either specific individual skills or competencies, generic team competencies that are applicable to any team setting, specific team competencies that are tied to a specific team, or some combination (Cannon-Bowers & Salas, 1997). Once the focus is specified and the training regimen is created, measurement tools should be developed to tap into the focus area(s), whether it is one area or a combination.

The second tenet suggests that any performance measurement system for training should address both outcome and process measures. *Outcome measures* (e.g., did the missile hit the target) tend to answer the question "What happened?", yet this type of measure is not very diagnostic (Dwyer, Oser, Salas, & Fowlkes, 1999). For example, a positive outcome (e.g., bombs delivered on target at the specified time-on-target) may have been achieved due to the actions of a team member who salvaged the operation in the last minute (e.g., correction of a latitude–longitude location error that could have otherwise led to fratricide). In this case, simply looking at the positive outcome masks flaws in the underlying processes—flaws that created a situation in which someone (luckily) detected the urgency and had to intervene very late in the mission. The failure to intervene could have had dire consequences. *Process measures*, on the other hand, tend to answer the question "Why did this happen?" (Johnston, Smith-Jentsch, & Cannon-Bowers, 1997) and can provide a much more information-rich description of performance. Process measures provide a means for capturing the moment-to-moment behaviors that are exhibited during a training exercise. As such, they can be very diagnostic and serve as a powerful source of information for feedback. However, one of the weaknesses of process measures is that they do not provide an assessment of overall performance—an assessment that most trainees and instructors would like to know. Process measures, when used in conjunction with outcome measures, can paint a very thorough picture of training performance.

The third tenet states that performance measures for training must be able to describe, evaluate, and diagnose performance. The ability of a performance measurement system to accurately diagnose performance and to pinpoint specific shortfalls is perhaps its most critical function. Without proper diagnosis, problem areas may very well propagate themselves, producing deficiencies elsewhere. This diagnostic capacity is a critical driver for delivering feedback and identifying follow-on training requirements.

Cannon-Bowers and Salas' (1997) final tenet argued that performance measures for training must provide a foundation for remediation. The goal of any training system is to systematically improve performance. The purpose of follow-on training is to advance a training audience to a higher plateau; however, if the participants are not ready for that progression, remedial training is typically warranted. Performance measures must provide information on which to make such decisions and to help determine where remediation should be focused.

Cannon-Bowers and Salas (1997) summarized their concepts in a framework for measuring training performance (Fig. 11.1). As can be seen in the figure, the different cells of the matrix represent different focus areas. Thus, if the training intent is to focus on competencies associated with team processes, measures that assess these competencies should be developed. On the other hand, if the focus of a training exercise is to enhance the skills associated with individual processes, the competencies associated with individual processes should be targeted within the performance measurement system. Finally, if the training interest is focused on multiple cells within the matrix, measurement tools should be developed for each so that performance can be systematically traced to the areas that contribute to performance shortfalls.

The purpose of this chapter is to build on these tenets raised by Cannon-Bowers and Salas (1997) and to lay the groundwork for developing comprehensive performance measurement systems in aircrew training environments. We address recent developments in performance measurement in several domains that may be leveraged and applied to aircrew training. Then, we discuss what we believe are critical principles (and guidelines) that must be taken into account in order to properly develop performance measures in a manner that ensures aircrew training effectiveness. Finally, we close with a discussion of what we believe the future holds in the area of performance measurement for aviation training.

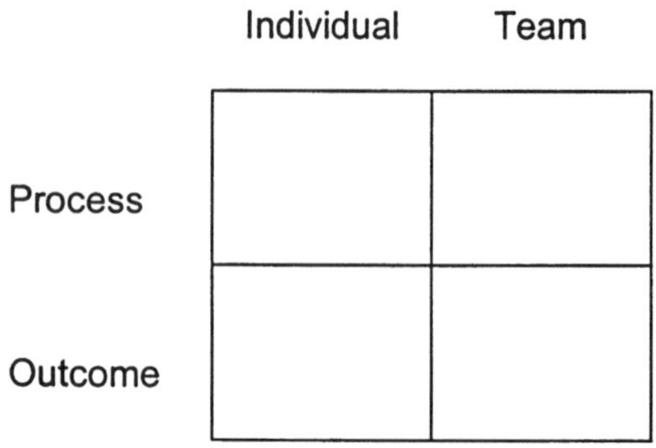


FIG. 11.1. Conceptual framework for measuring training performance.

## DEVELOPMENTS IN PERFORMANCE MEASUREMENT

Several performance measurement tools have emerged in recent years, mostly in the research and development community. An examination of these tools suggests that they address different pieces of the performance measurement "puzzle." Some for example, focus on team outcomes (Meliza & Tan, 1996), whereas others address team processes (Smith-Jentsch, Johnston, & Payne, 1998). Others use paper-and-pencil formats (Muñiz, Stout, Bowers, & Salas, 1998), whereas some take advantage of computer-based media (Pruitt, Burns, Wetteland, & Dumestre, 1997). Some use rating scales (Silverman, Spiker, Tourville, & Nullmeyer, 1997); their counterparts opt for behaviorally oriented checklists (Dwyer et al., 1999). Despite the differences, all of these approaches have something to offer. We believe that performance measurement in the aircrew training environment can benefit from these efforts by leveraging from these various approaches. Next, we briefly review some of the more current efforts involved in the development of performance measures—efforts that can be leveraged and used in measuring aircrew performance.

### The Targeted Acceptable Responses to Generated Events or Tasks (TARGETs) Measurement Methodology

The TARGETs measurement tool is a taskwork-oriented measure that has been used in a number of domains (Dwyer, Fowlkes, Oser, Salas, & Lane, 1997; Dwyer et al., 1999; Fowlkes, Lane, Salas, Franz, & Oser, 1994; Pierce, Hallion, Fowlkes, McCluskey, & Dwyer, 1998). A sample section from a TARGETs checklist is provided in Fig. 11.2. The TARGETs performance measurement approach uses a behaviorally oriented checklist to

| Expected Actions | Score |
|---|---|
| Fighter authenticates | 1 |
| Fighter provides brief to Forward Air Controller:<br>- Mission Number | 1 |
| - Number/type of aircraft | 0 |
| - Ordnance | X |
| - Time on Station | 1 |
| - Abort Code | 1 |
| 9-Line briefing provided to fighters:<br>- Initial Point (IP) | 0 |
| - Heading (IP to target) | X |
| - Distance (IP to target) | 1 |
| - Target description | 1 |
| - Friendly locations | 1 |

| Expected Actions | Score |
|---|---|
| Fighters informed of situation | 0 |
| Fighters told to orbit at safe distance | X<br>1 |
| Fighter plan time considered | 1 |
| Informal Airspace Coordination Area (ACA) established/re-established | 1 |
| Alternate ACA considered | 1 |
| ACA:<br>- Protects fighters | 1 |
| - Allows fighter maneuverability | 0 |
| - Fire Support Officer ensures indirect fires conform to ACA | X |
| - Clearly defined (e.g., suitable landmarks identified) | 1 |
| Time on Target revised | 1 |
| Availability of Suppression of Enemy Air Defense established | 0 |

Comments:______________________________

**1** = Observed/Performed Satisfactorily; **0** = Omitted/Failed to Perform Satisfactorily;
**X** = No Opportunity to Perform/Not Required

FIG. 11.2. Sample section from a TARGETs checklist for trigger event "heavy enemy artillery."

record the occurrence and nonoccurrence of expected actions performed in response to *trigger events* that occur during an exercise. Trigger events are cues that are embedded or injected into a scenario in order to prompt the training participant(s) to action—action related to the training objectives of interest. The trigger events provide specific and controlled opportunities for training participants to demonstrate their ability to perform a task or set of tasks. One of the defining characteristics of the TARGETs measurement tool is that the expected actions for each trigger event are identified a priori through task analyses and subject matter expert consultation. Each of the items within the checklist is scored as either present, absent, or not observed. Depending on how the data are aggregated, a TARGETs checklist can provide both process measures and outcome measures. To examine outcome, the proportion of checklist items performed correctly by the team can be computed (e.g., 72 behaviors performed correctly out of a total possible of 100 would be an outcome score of 72%). This type of analysis provides an overall indication of how teams performed in general terms but is not diagnostic. As a process measure, the checklist items can be grouped into meaningful clusters of behaviors (e.g., those items involving communications about enemy movements). These clusters can then be analyzed separately to provide an indication of how well teams performed in that area of interest. Various derivatives of the TARGETs format (e.g., the Tactically Relevant Assessment of Combat Events—TRACE—measurement tool) have recently emerged (McCluskey, Fowlkes, Pierce, & Dwyer, 1998; Pierce et al., 1998).

### The Situational Awareness Linked Instances Adapted to Novel Tasks (SALIANT) Measurement Approach

A recent adaptation of the TARGETs methodology is a measurement tool known as SALIANT (Muñiz et al., 1998). A sample section from a SALIANT measurement tool is provided in Fig. 11.3. SALIANT uses an event-based approach and focuses on behavioral indicators of team situational awareness. It is used to examine behavioral patterns exhibited during training, and uses those patterns as an indication of the level of team situational awareness present during the exercise. As is seen in Fig. 11.3, the SALIANT methodology requires a tight coupling between the measurement component and the scenario design process. The methodology requires that cues (i.e., events) be built into training exercises and that these events be designed such that they are noticeable by training participants. For each event, acceptable responses—that should be performed in response to that event—are identified. The extent to which the acceptable responses are carried out is taken as an indication of team situational awareness. The occurrence of an acceptable response is scored as a "hit" by marking an X in the

| Scenario Segment | Event | Acceptable Response | Code | Hit |
|---|---|---|---|---|
| From Point Golf To India | Team lost radio contact with Air Mission Commander | Team member discussed possibility of radio contact loss | 1.1 | X |
| | | Team member suggested to climb to higher altitude to re-establish contact | 1.2 | X |
| | | Team member mentioned that he will be climbing to a higher altitude | 1.3 | |

FIG. 11.3. Sample section from a SALIANT measurement form.

appropriate box. Preliminary data collected in commercial aviation demonstrated a high level of interrater agreement (Cohen's kappa ranged from .70 to .96) and demonstrated some success in predicting performance in team settings, in some cases accounting for 73% of the variance (Bowers, Barnett, Weaver, & Stout, 1998).

**The Teamwork Observation Measure (TOM)**

Another measurement instrument that has met with success in team training environments is a tool known generically as the TOM (Dwyer et al., 1999). TOM is a team process measure that taps into key dimensions of teamwork such as communication, coordination, and backup behavior. The approach involves the identification of team competencies through detailed task analyses—competencies that are believed to be critical to successful performance in a team setting. Smith-Jentsch et al. (1998) for example, identified communication, information exchange, supporting behavior, and team leadership/initiative as the key team competencies within air defense warfare (ADW) teams. Each competency was then decomposed into its component factors; for example, communications was divided into improper phraseology, inaudible communications, excess chatter, and incomplete reports. Within the team dimensional training (TDT) strategy, performance on each of these component factors is systematically assessed during the execution of a scenario (Smith-Jentsch et al., 1998). The measurement tool, which was a derivative of TOM, was called ATOM (Antiair warfare TOM). A small sample section from a much more comprehensive ATOM measurement tool is provided in Fig. 11.4. As can be seen in this figure, the frequency of occurrence for each of the four components of communication is selected by evaluators based on their observations. Specific examples of behavioral strengths and weaknesses are recorded before assessing the frequency of occurrence and for later use to support feedback. In its most recent configuration, the measurement tool is housed within a portable, hand-held computer (Pruitt et al., 1997). The data recording includes instructor observations and notes, plan view displays, and voice communications—made in real-time—from the actual exercise (training) environment. These data are saved in a repository to support subsequent feedback following the exercise.

**The Team-Mission Observation Tool (T-MOT)**

The Air Force was also involved in the development of measurement tools aimed at assessing individual performance and team coordination (Silverman et al., 1997; Tourville, Spiker, Silverman, & Nullmeyer, 1996). One tool

**Communication Behavior Ratings**

| | Frequency of Occurrence | | |
|---|---|---|---|
| Improper Phraseology | 0 | 1-3 | >3 |
| Inaudible Communication | 0 | 1-3 | >3 |
| Excess Chatter | 0 | 1-3 | >3 |
| Incomplete Reports | 0 | 1-3 | >3 |

FIG. 11.4. Sample section from Antiair Teamwork Observation Measure (ATOM).

(the T-MOT) was developed for use in tactical team resource management training (Silverman et al., 1997). This approach uses a Likert scale format along with subject matter expert (SME) observations to assess the extent to which coordination processes are demonstrated in five areas: situation awareness; function allocation of workload; tactics employment; time management; and command, control, and communications. This measurement approach yields both qualitative (SME annotations) and quantitative (rating) data for both individual and team behaviors. Figure 11.5 shows a sample section that is used for assessing situation awareness (SA). Similar forms are used for assessing the remaining coordination areas, each of which contains multiple items similar to that in Fig. 11.5. Figure 11.6 provides the T-MOT assessment matrix used for providing overall ratings of the five coordination processes for each aircrew position. Raters used a 5-point scale to make their assessments, and used a "by exception" rule whereby ratings where provided only in cases where performance was viewed as either better or worse

**1.0** **(SA) The Aircrew Commander (AC) will (typically) pay strict attention to overall mission preparation details (e.g., knowing everybody's job, and checking other aircrew members' work efforts).**

**a** **Did the AC pay attention to detail?............................................................ YES / NO**
**(Explain)** ____________________

**b** **Did the AC promote an environment for "open" team communications? ........................................................................ YES / NO**
**(Explain)** ____________________

**c** **Did non-operational factors (such as social Interaction) interfere with any Crewmember's time to perform necessary tasks? ........................ YES / NO**
**(Explain)** ____________________

FIG. 11.5. Sample section from Team-Mission Observation Tool (T-MOT) for qualitative assessment of situation awareness (SA).

| MISSION PREPARATION | AC | CP | LN | RN | FE | CSO | CREW |
|---|---|---|---|---|---|---|---|
| 1. Situation Awareness (SA) | | | | | | | |
| 2. Function Allocation (FA) | | | | | | | |
| 3. Tactics Employment (TE) | | | | | | | |
| 4. Time Management (TM) | | | | | | | |
| 5. Command, Control & Communication (C3) | | | | | | | |

NOTE. AC = aircrew commander; CP = co-pilot; LN = left navigator; RN = right navigator; FE = flight engineer; CSO = communications system officer; CREW = entire flight crew.

FIG. 11.6. Team-Mission Observation Tool (T-MOT) assessment matrix for rating coordination by aircrew position.

than "average." Initial tryouts were encouraging with the authors reporting statistically significant correlations (Pearson product moment) as high as .86 between overall process ratings and overall mission performance.

### The Team-Mission Performance Tool (T-MPT)

Another measurement tool recently developed by the Air Force is the T-MPT—an assessment tool for evaluating overall crew mission performance (Silverman et al., 1997). T-MPT, which can be categorized as an outcome measure, uses multiple behaviorally anchored rating scales to rate the quality of mission products developed at both the individual and team levels. The ratings are supplemented with observer notations in critical areas such as the number of briefings each crew conducts, briefing content, and so on. Moreover, the T-MPT is partly completed on-line (as the exercise is being run), but follow-up completion is also done post hoc with help from instructor operator station pages, flight plans, charts, etc. This multifaceted approach is intended to paint a thorough picture of what happened during mission execution.

### The Unit Performance Assessment System (UPAS)

Finally, the Army has developed a measurement tool known as UPAS (Meliza & Tan, 1996). UPAS—a computer-based system—provides outcome measures such as the number of bombs released, the number of enemy targets hit, percentage of bombs impacting targets, and battle damage assessment (Orlansky, Taylor, Levine, & Honig, 1997). The results of those data can also be displayed in the form of plan view displays, battle flow, timelines, tables, and graphs for use during after-action reviews (Meliza, Bessemer, & Hiller, 1994). The ability of this system to monitor a training exercise and to capture exercise data in networked training environments is an innovative way for automatically collecting critical performance data very efficiently. UPAS-based data have been useful in identifying potential problem areas in need of more in-depth coverage during after-action reviews (Moses, 1996).

Each of these different measurement approaches has a different focus area, due in part to different areas of research interest. Next, we attempt to distill from these (and other) measurement approaches (Smith-Jentsch et al., 1998) what we believe are the critical principles that must be taken into account when developing comprehensive performance measurement systems.

## PRINCIPLES OF PERFORMANCE MEASUREMENT FOR ENSURING AIRCREW TRAINING EFFECTIVENESS

Based on our review of the theoretical and empirical literature in the areas of team training and performance, we derive what we believe to be critical considerations that must be taken into account if we are to ensure training effectiveness through proper performance measurement. These considerations, presented as principles, are discussed next.

### Principle 1: A Theoretical Foundation Should Be the Basis of a Performance Measurement System for Aviation Training (Baker & Salas, 1992; 1997)

Borrowing from Lewin (1951) and Marrow (1969), Baker and Salas (1992) argue that, in order to understand teamwork, "There is nothing more practical than a good theory," thus our first principle.

From the early work of Hall and Rizzo (1975) to the present work by Cannon-Bowers and her colleagues (Cannon-Bowers, Salas, Tannenbaum, & Mathieu, 1995), theory has played a central role in team training research and development. Morgan, Glickman, Woodard, Blaiwes, and Salas (1986), for example, identified a "common set" of seven teamwork dimensions that they believed were characteristic of successful teams. These dimensions were team spirit, coordination, cooperation, communication, giving suggestions, adaptability, and accepting suggestions. Subsequent work refined these dimensions, adapted them, and applied them in additional areas. For example, Smith-Jentsch et al. (1998) identified four teamwork dimensions of Navy combat information center teams based on an analysis of the competencies necessary to perform successfully in an ADW environment. These dimensions included communication, information exchange, supporting behavior, and team leadership/initiative. In their Team Evolution and Maturation (TEAM) Model, Morgan, Salas, and Glickman (1994) identified taskwork tracks (e.g., interactions with equipment, technical components of the job) and teamwork tracks (e.g., interactions, interdependencies, cooperation among team members)—tracks that develop in parallel with one another, eventually merging into a single track as a team matures. Other models, such as the Teamwork Effectiveness Model (Salas, Dickinson, Converse, & Tannenbaum, 1992), expanded on earlier theories by describing where various training interventions might best impact training performance.

The point is that the theoretical foundation on which training is to be based should also drive the types of performance measures that are to be developed. For example, the "two track" theory of Morgan et al. (1994)—if this is one's theoretical foundation and focus of the training intervention—would require the development of two performance measurement protocols, one for the assessment of taskwork skills and one for the assessment of teamwork skills. Measurement tools such as TARGETs (Dwyer et al., 1997) may very well support the assessment of taskwork skills, whereas a measurement tool such as ATOM (Smith-Jentsch et al., 1998) could support teamwork assessment. Once the theoretical drivers are specified, the training approach and performance measurement system should flow from that theoretical basis.

### Principle 2: The Purpose of Human Performance Measurement Should Be Defined (Baker, O'Neil, & Linn, 1993; Cannon-Bowers et al., 1989)

Depending on the purpose of measurement, the assessment protocols may differ. Baker et al. (1993) provided an extensive list of such purposes. However, for training, the primary purposes (from Baker et al., 1993) center on diagnosis and remediation, formative evaluations of instructional improvement, and grading. Measures developed for grading performance will be different from measures developed for delivering feedback. Measures developed for feedback will differ from measures used for generating a normative database or conducting a training effectiveness evaluation. For example, if the intent is to diagnose performance deficiencies in order to identify follow-on training needs, measures of performance (MOPs)—measures that assess processes—are needed (Johnston et al., 1997). They must be conducive to generating information for feedback and directly tied to the critical tasks to be trained. Thus, they must be specific and detailed (Dwyer et al., 1999). On the other hand, if the purpose of measurement is to assess performance trends over a period of time, to determine overall performance, or to grade performance, a different format may be warranted (Oser, Cannon-Bowers, Dwyer, & Salas, 1997). In this case, measures of effectiveness (MOEs)—or outcome measures—may be derived. Here, global indicators of overall performance may meet the intended need. In short, it is important to determine how your performance data are to be used and to then develop your tools in a manner that allows you to collect the types of data needed.

### Principle 3: Team and Cognitive Task Analyses Are Necessary Tools for Determining Aviation Training Requirements and Performance Measurement Criteria

Another critical consideration for ensuring effective performance measurement focuses on the conduct of front-end analyses that define what is to be trained and how it is to be measured. Team and cognitive task analyses define the critical knowledge, skills, and abilities needed for successful performance, the critical tasks to be trained, and the competencies to be measured. The information provided by these analyses form the foundation on which the measurement approach is based. Without this foundation, we cannot be sure that we are measuring "the right stuff." Although it is often tempting to forego such analyses (due to budget cuts, time constraints, etc.), team and cognitive task analyses are absolutely critical and the payoffs will far exceed the cost.

### Principle 4: An Event-based Strategy for Training and Performance Measurement Is Required in Complex Aviation Environments (Dwyer et al., 1999; Fowlkes, Dwyer, Oser, & Salas, 1998; Johnston et al., 1997)

Arguments are frequently heard concerning the desirability of "free play" versus "scripted" scenarios. Military personnel often demand free play training environments to create the "fog of war" and to replicate real world operations in which the future is unknown. On the other hand, researchers argue for the ability to maintain control over what trainees experience in the training environment rather than "leaving training opportunities to chance." We maintain that this is a false dichotomy and that these are independent functions—both sides can simultaneously have their desires met. A free play scenario (in the eyes of the training audience) can still be a scripted scenario. In these types of scenarios, the training audience is free to perform in any manner desired in order to successfully complete the mission in the exercise. Aircraft can fly north, turn right, release ordnance, or fire missiles—anything the crew deems necessary to best meet its mission. Within that "free play" environment, however, trigger events could and should be embedded or injected, events that provide the training audience with "scripted" (i.e., preplanned) opportunities to perform critical tasks that have been targeted by the training. If properly done, these scripted events are transparent to the training audience and they appear as real-world occurrences that they may likely encounter in operational settings. Thus, regardless of whether the aircrew turns right or left, the instructor (or exercise developer) should ensure that the flight crew encounters key events that prompt them to action. For example, if one of the critical tasks of interest concerns the use of backup communications, an event that introduces a lost communications link could be incorporated in the scenario. When that trigger event occurs, the crew can perform their backup procedures for lost communications, rather than leaving that performance to chance. Equally as important, the event-based approach allows observers (assessors) to focus attention on the upcoming event, to systematically observe what happens, and to record how well the training participants performed during the event. Thus, an event-based approach allows instructors to provide specific and controlled training opportunities to the training audience and allows assessors to measure that performance, while the training audience perceives the scenario to be "free play" in nature. Event-basing a scenario, if done properly, ensures proper task coverage and exposure, and replicates real-world encounters.

### Principle 5: Measurement Systems in Aviation Must Use a Multilevel Focus (Baker & Salas, 1992)

A multilevel approach allows instructors to examine performance beyond the obvious outcome. Although team performance is obviously important, it is clear that that performance is determined, at least in part, by the competencies of the individual members (Salas, Bowers, & Cannon-Bowers, 1995; Stout, Salas, & Carson, 1994). For example, a measure such as 6 out of 10 bombs within a circular area of 50 meters (a 60% hit rate) provides an overall team outcome measure. However, such a measure is not diagnostic nor does it explain why four bombs missed the target. A supplemental process measure may help identify a late exchange of information between the pilot and the bombardier–navigator, a team process problem. In turn, individual process measures may indicate that the forward air controller was overtasked with multiple targets and was slow to convert a latitude–longitude location to a grid coordinate. In this example, an individual error may have been propagated, ultimately manifesting itself as a poor team outcome. Thus in training exercises, it may be important to assess performance on several levels to diagnose the true and complete nature of the deficiency. By using a multilevel focus and systematically working from the top down, instructors can develop a more detailed and diagnostic assessment of what really happened.

### Principle 6: Performance Measurement Protocols Need to Support the Diagnosis of Learning (Cannon-Bowers & Salas, 1997)

Our next principle for success is to design measurement protocols that describe, evaluate, and diagnose performance (Cannon-Bowers & Salas, 1997). Obviously, performance measures must accurately describe the behaviors that occurred during training. This is critical not only from a user acceptance viewpoint, but also from a psychometric perspective. Furthermore, in applications where expert teams are participating in training, performance changes may be very small. In these cases, measurement tools must be sensitive enough to accurately and reliably detect small—but important—performance changes. Careful consideration of the psychometric properties of the measurement tools facilitates this.

Related to the accurate description of performance is the ability of measurement tools to evaluate performance. This implies meeting some standard of acceptable performance (e.g., perform task X within 3 minutes of event A). In dynamic environments, however, fixed standards that are acceptable under all conditions are extremely difficult to specify (e.g., 3 minutes to perform task X in one situation may suffice, but in other situations 45 seconds may be critical for survival). In these settings, standards may need to be set with boundary conditions because the standard of acceptable performance may actually "slide" as a function of the constantly changing environment. Here, subject matter expertise becomes critical in determining where within the boundary limits performance should have occurred.

In addition, as we mentioned previously, measures must be able to accurately diagnose strengths and weaknesses in performance. It is absolutely essential that measurement tools are able to accurately uncover the specific causes of performance breakdown if follow-on training is to be conducted efficiently and effectively. Performance measures must identify performance shortfalls with sufficient accuracy so that deficient areas can be targeted and remediated.

### Principle 7: The Measurement Tools Must Be Practical to Use

To make the evaluator's job of data collection simple, logical, and straightforward, the data collection tools must be easy to use. Behavioral checklists that flow with a scenario as it unfolds have been accepted by subject matter experts involved in performance measurement; however, rating scales sometimes meet a more limited acceptance (Mirabella, Sticha, & Morrison, 1997). Measurement tools that use the evaluators' language and jargon should enhance acceptance (Mirabella et al., 1997), whereas logical, easy-to-use scoring, data entry, and data extraction procedures should facilitate usability. Moreover, measurement tools that can quickly and easily convert training performance data to proper formats that support feedback should be endorsed by users (Bell et al., 1997).

### Principle 8: Involve the User (Of Course!)

Developing reliable, valid, and useful performance measures demands the creation of a partnership between researchers and subject matter experts. Both groups bring critical pieces to the table. Researchers typically do not possess the subject matter expertise of military domains—consequently, that expertise is a necessary commodity. Conversely, military experts are often not well versed in the methodologies necessary for developing psychometrically sound measurement protocols. Forming a partnership between these two types of experts significantly increases the chance of developing measurement tools that are not only reliable and valid, but also meaningful and accepted by military personnel.

### Principle 9: Provide the Tools and Training for the Instructor

The final component that is necessary to ensure aircrew training effectiveness via performance measurement involves instructor training. We believe that performance measures—to be truly diagnostic—will always require some amount of observation by humans. Performance measures (particularly process measures) will not likely be fully automated in the near future. Consequently, training instructors to properly use performance measurement tools will remain a critical component. Typically, military personnel are not trained as instructional technologists. Despite their interest and motivation, well-intentioned personnel may not be familiar with the subtle nuances and techniques involved in developing performance measurement tools and in collecting reliable and valid measures of performance. As a result, those personnel who fill the role of "instructor" or "assessor" must be prepared to properly use the techniques and methodologies for data collection to support performance measurement. Carrying this concept a step further, the TDT program has focused on training instructors to be "facilitators" who guide (rather than direct) the feedback process through guided self-correction (Smith-Jentsch et al., 1998). The requirements for data collection in this successful work have a strong instructor-training component. Instructors trained in this methodology report high levels of satisfaction and a large-scale incorporation of the TDT principles in existing training programs (Smith-Jentsch et al., 1998).

We summarize the principles in Table 11.1. In addition, we offer some brief guidelines for how to implement the principles, and discuss the implications for performance measurement and training.

## THE FUTURE

Great strides have been taken in the area of performance measurement in the past several years, especially in aviation environments. Theoretical foundations have been proposed, empirical examinations have been conducted, and performance measurement systems have been tested and used in field applications. Paper-and-pencil formats have begun migrating toward computer-based media to streamline portions of the performance measurement data collection function. Performance measurement in military training environments is becoming commonplace. We have truly moved in a positive direction over the last decade. So, what does the future hold?

One of the most interesting and most recent movements in the area of performance measurement has been in the area of automated performance measurement. Vreuls and Obermayer (1985) addressed the challenges of developing automated performance measures many years ago. Today, many of those challenges remain; however, several substantial inroads have also been made. Although we do not believe that complete automation of the performance measurement function is achievable or desirable (assessments from subject matter experts should never be completely relinquished to a computer), some components of the performance measurement task can and have been automated.

For example, automated measurement in the form of pattern analysis of operator keystroke data is currently being developed in Navy settings in an attempt to enhance the ease of data collection (Zachary, Bilazarian, Burns, & Cannon-Bowers, 1997). The intent is to develop software algorithms that look for patterns in the keystroke data (from an operator's interactions with a console) to determine whether a particular task (e.g., a target has been labeled) has been performed. Similarly, the Army's UPAS monitors training data within a simulation and automatically captures information on key accomplishments (e.g., number of target kills). Although these techniques will significantly ease the burden of collecting performance measurement data, they must still be supplemented with human observations.

Other promising work is focused on developing hand-held, portable, computer-based measurement tools that require human input based on observations made during training. Evaluators can roam throughout a training environment in order to be "in the middle of the action" and provide a better assessment of performance. New systems with radio frequency (RF) links allow multiple evaluators to communicate digitally with each other and with exercise controllers, and allow them to transmit perform-

TABLE 11.1
Summary of Principles for Aviation Performance Measurement With Implications and Guidelines

| *Principles* | *Implication* | *Guidelines* | *Sources* |
|---|---|---|---|
| 1. A theoretical foundation should be the basis of a performance measurement system for aviation training. | • Theory of human behavior drives measurement content | • Adopt a theory of human performance that relates to aviation environments | Salas & Cannon-Bowers (in press) |
| 2. The purpose of human performance measurement should be defined. | • The purpose of measurement determines the kind of data to be collected and what instruments should be developed | • Establish the purpose of measurement early and clearly<br>• Ensure measurement tools support intended purpose | Brannick, Salas, & Prince (1997)<br>Vreuls & Obermayer (1985)<br>Baker et al. (1993) |
| 3. Team and cognitive task analyses are necessary tools for determining aviation training requirements and performance measurement criteria. | • These tools provide the content for assessment and training by defining required aviation individual and team competencies | • Conduct team task analyses<br>• When appropriate, conduct cognitive-oriented task analyses<br>• Specify the required competencies (e.g., KSAs) | Klein (1995)<br>Cannon-Bowers et al. (1995)<br>Gordon & Gill (1997) |
| 4. An event-based strategy for training and performance measurement is required in complex aviation environments. | • This approach allows for specific and controlled learning opportunities and a manageable context to measure complex aviation performance | • Adopt an event-based approach to training<br>• Implement key events (linked to competencies) in scenario to create opportunities to measure aviation performance | Fowlkes et al. (1998)<br>Dwyer et al. (1997) |
| 5. Measurement systems in aviation must use a multilevel focus. | • Focus is needed to determine not only "what happened" but "why it happened." | • Develop measures that tap both individual and team performance as well as process and outcomes to thoroughly assess performance | Baker & Salas (1992)<br>Prince, Brannick, Prince, & Salas (1997)<br>Brannick, Prince, Prince, & Salas (1995) |

TABLE 11.1
(Continued)

| *Principles* | *Implication* | *Guidelines* | *Sources* |
|---|---|---|---|
| 6. Performance measurement protocols need to support diagnosis of learning. | • Remediation of performance needs diagnostic analysis of behavior | • Create protocols to capture moment-to-moment aviation and behavior<br>• Capture outcome performance<br>• Links moment-to-moment action and behavior to outcome performance<br>• Establish causes of performance failure or successes | Fowlkes et al. (1994)<br>Brannick et al. (1997) |
| 7. The measurement tools must be practical to use. | • Ease of use facilitates the collection, evaluation, and synthesis of performance information for debriefs | • Create tools that are simple, straightforward, and task relevant<br>• Use automated data recording technology when feasible<br>• Prepare instruction for the assessors/instructors | Pruitt et al. (1997)<br>Johnston et al. (1997) |
| 8. Involve the user (Of course). | • Key to use, acceptance, and effectiveness of measurement system | • Create partnership between SMEs and human performance experts | Smith-Jentsch et al. (1998) |
| 9. Provide tools and training for the instructor. | • Needed for consistency of use and utility of performance measurement system | • Create training package for instructors<br>• Provide all required support (e.g., guide sheets, definitions, scenario events) to initiate and to facilitate data collection | Smith-Jentsch et al. (1998)<br>Prince & Salas (1993) |

ance data back to a central repository while they remain in the training environment. These "semi-automated" measurement tools can speed up data collection, facilitate aggregation and deaggregation of the performance data, and, by tapping into the scenario script, prompt the evaluator when and where to focus attention in order to measure performance on critical events. One such device, known as the Shipboard Mobile Aid for Training and Evaluation (ShipMATE), has been successfully demonstrated in Navy settings (Pruitt et al., 1997). Similar applications in the Army have also been conducted (McCluskey et al., 1998; Pierce et al., 1998).

Although automated and semi-automated performance measurement techniques will relieve some of the data collection burden, they are not panaceas. We believe that performance measurement will continue to require human observation. The goal is to strike a proper balance between what automation offers and what humans are best suited to do. Regardless of the data collection format, the considerations we addressed will help pave the way for performance measurement development in aircrew training environments.

## SUMMARY

We proposed several principles that researchers and practitioners should take into account in order to ensure effective training through the development and implementation of performance measurement systems in any domain. Our recommendations are based on the theoretical as well as empirical literature, and from both research and practitioner applications. Applying these principles (and guidelines) in a comprehensive performance measurement system, rather than implementing them in a piecemeal manner, will go a long way toward enhancing aircrew training performance. Aircrew training programs should embrace these principles to ensure training effectiveness.

## ACKNOWLEDGMENT

The views expressed here are those of the authors and do not reflect the official position of the organizations with which they are affiliated.

## REFERENCES

Baker, D. P., & Salas, E. (1992). Principles for measuring teamwork skills. *Human Factors, 34*, 469–475.

Baker, D. P., & Salas, E. (1997). Principles for measuring teamwork: A summary and look toward the future. In M. T. Brannick, E. Salas, & C. Prince (Eds.), *Team performance assess-*

*ment and measurement: Theories, methods, and applications* (pp. 331–355). Mahwah, NJ: Lawrence Erlbaum Associates.

Baker, E. L., Abedi, J., Linn, R. L., & Niemi, D. (1996). Dimensionality and generalizability of domain-independent performance assessments. *Journal of Educational Research, 89*(4), 197–205.

Baker, E. L., O'Neil, H. F., & Linn, R. L. (1993). Policy and validity prospects for performance-based assessment. *American Psychologist, 48*(12), 1210–1218.

Bell, H. H., Dwyer, D. J., Love, J. F., Meliza, L. L., Mirabella, A., & Moses, F. L. (1997). *Recommendations for planning and conducting multi-service training with distributed interactive simulation technology* (Research Product). Alexandria, VA: US Army Research Institute for the Behavioral and Social Sciences.

Bowers, C. A., Barnett, J., Weaver, J., & Stout, R. J. (1998). Empirical validation of the SALIANT methodology. *Proceedings of the First Human Factors & Medicine Panel Symposium on Collaborative Crew Performance in Complex Operational Systems*. Neuilly-sur-Seine, France: Research and Technology Agency.

Brannick, M. T., & Prince, C. (1997). An overview of team performance measurement. In M. T. Brannick, E. Salas, & C. Prince (Eds.), *Team performance assessment and measurement: Theories, methods, and applications* (pp. 3–16). Mahwah, NJ: Lawrence Erlbaum Associates.

Brannick, M. T., Prince, A., Prince, C., & Salas, E. (1995). The measurement of team process. *Human Factors, 37*, 641–651.

Brannick, M. T., Salas, E., & Prince, C. (1997). *Team performance assessment and measurement: Theory, methods & applications*. Mahwah, NJ: Lawrence Erlbaum Associates.

Cannon-Bowers, J. A., Prince, C., Salas, E., Owens, J. M., Morgan, B. B. Jr., & Gonos, G. H. (1989). Determining aircrew coordination training effectiveness. *Proceedings of the 11th Annual Meeting of the Interservice/Industry Training Systems Conference* (pp. 128–136). Washington, DC: National Security Industrial Association.

Cannon-Bowers, J. A., & Salas, E. (1997). A framework for developing team performance measures in training. In M. T. Brannick, E. Salas, & C. Prince (Eds.), *Team performance assessment and measurement: Theories, methods, and applications* (pp. 45–62). Mahwah, NJ: Lawrence Erlbaum Associates.

Cannon-Bowers, J. A., Salas, E., Tannenbaum, S. I., & Mathieu, J. E. (1995). Toward theoretically-based principles of training effectiveness: A model and initial empirical investigation. *Military Psychology, 7*, 141–164.

Dickinson, T. L., & McIntyre, R. M. (1997). A conceptual framework for teamwork measurement. In M. T. Brannick, E. Salas, & C. Prince (Eds.), *Team performance assessment and measurement: Theories, methods, and applications* (pp. 19–43). Mahwah, NJ: Lawrence Erlbaum Associates.

Dwyer, D. J., Fowlkes, J. E., Oser, R. L., Salas, E., & Lane, N. E. (1997). Team performance measurement in distributed environments: The TARGETs methodology. In M. T. Brannick, E. Salas, & C. Prince (Eds.), *Team performance assessment and measurement: Theory, methods, and applications* (pp. 137–153). Hillsdale, NJ: Lawrence Erlbaum Associates.

Dwyer, D. J., Oser, R. L., Salas, E., & Fowlkes, J. E. (1999). Performance measurement in distributed environments: Initial results and implications for training. *Military Psychology, 11*, 189–215.

Fowlkes, J. E., Dwyer, D. J., Oser, R. L., & Salas, E. (1998). Event-based approach to training (EBAT). *International Journal of Aviation Psychology, 8*(3), 209–221.

Fowlkes, J. E., Lane, N. E., Salas, E., Franz, T., & Oser, R. (1994). Improving the measurement of team performance: The TARGETs methodology. *Military Psychology, 6*, 47–61.

Gordon, S. E., & Gill, R. T. (1997). Cognitive task analysis. In C. E. Zsambok & G. Klein (Eds.), *Naturalistic decision making* (pp. 131–140). Hillsdale, NJ: Lawrence Erlbaum Associates.

Hall, E. R., & Rizzo, W. A. (1975). *An assessment of U.S. navy tactical training* (TAEG Rep. No. 18). Orlando, FL: Training Analysis and Evaluation Group.

Hallman, G., & Campbell, D. (1997). The measurement of team performance with a standardized survey. In M. T. Brannick, E. Salas, & C. Prince (Eds.), *Team performance assessment and measurement: Theories, methods, and applications* (pp. 155–171). Mahwah, NJ: Lawrence Erlbaum Associates.

Johnston, J. H., Smith-Jentsch, K. A., & Cannon-Bowers, J. A. (1997). Performance measurement tools for enhancing team decision-making training. In M. T. Brannick, E. Salas, & C. Prince (Eds.), *Team performance assessment and measurement: Theories, methods, and applications* (pp. 311–327). Mahwah, NJ: Lawrence Erlbaum Associates.

Klein, G. (1995). The value added by cognitive task analysis. *Proceedings of the Human Factors and Ergonomics Society 39th Annual Meeting* (pp. 530–533). Santa Monica, CA: The Human Factors and Ergonomic Society.

Komaki, J. L. (1997). Behind the scenes: Field-testing a measure of effectiveness for theater teams. In M. T. Brannick, E. Salas, & C. Prince (Eds.), *Team performance assessment and measurement: Theories, methods, and applications* (pp. 227–251). Mahwah, NJ: Lawrence Erlbaum Associates.

Lewin, K. (1951). *Field theory in social science*. New York: Harper & Row.

Marrow, A. J. (1969). *The practical theorist: The life and work of Kurt Lewin*. New York: Basic Books.

McCluskey, M. R., Fowlkes, J. E., Pierce, L. G., & Dwyer, D. J. (1998). Measurement of command/control staff performance in tactical training environments. *Proceedings of the 20th Interservice/Industry Training, Simulation and Education Conference* (pp. 826–834). Washington DC: National Security Industrial Association.

Meliza, L. L., & Tan, S. C. (1996). *SIMNET unit performance assessment system (UPAS) version 2.5 user's guide* (ARI Research Product 96-05). Alexandria, VA: U.S. Army Research Institute for the Behavioral and Social Sciences.

Meliza, L. L., Bessemer, D. W., & Hiller, J. H. (1994). Providing unit training feedback in the distributed interactive simulation environment. In R. F. Holtz, J. H. Hiller, & H. H. McFann (Eds.), *Determinants of effective unit performance* (pp. 257–280). Alexandria, VA: US Army Research Institute for the Behavioral and Social Sciences.

Mirabella, A., Sticha, P., & Morrison, J. (1997). *Assessment of user reactions to the multi-service distributed training testbed (MDT2) system* (ARI Tech. Rep. No. 1061). Alexandria, VA: U.S. Army Research Institute for the Behavioral and Social Sciences.

Morgan, B. B. Jr., Glickman, A. S., Woodard, E. A., Blaiwes, A. S., & Salas, E. (1986). *Measurement of team behaviors in a navy environment* (NTSC Tech. Rep. TR-86-014). Orlando, FL: Naval Training Systems Center.

Morgan, B. B. Jr., Salas, E., & Glickman, A. S. (1994). An analysis of team evolution and maturation. *Journal of General Psychology, 120*(3), 277–291.

Moses, F. M. (1996). The challenge of distributed training. *Proceedings of the 7th International Training Equipment Conference* (pp. 358–364). Arlington, VA: ITEC Ltd.

Muñiz, E. J., Stout, R. J., Bowers, C. A., & Salas, E. (1998). A methodology for measuring team situational awareness: Situational awareness linked indicators adapted to novel tasks (SALIANT). *Proceedings of the First Human Factors & Medicine Panel Symposium on Collaborative Crew Performance in Complex Operational Systems*. Neuilly-sur-Seine, France: Research and Technology Agency.

Orlansky, J., Taylor, H. L., Levine, D. B., & Honig, J. G. (1997). *The cost and effectiveness of the multi-service distributed training testbed (MDT2) for training close air support* (IDA Paper P-3284). Alexandria, VA: Institute for Defense Analyses.

Oser, R. L., Cannon-Bowers, J. A., Dwyer, D. J., & Salas, E. (1997). Establishing a learning environment for JSIMS: Challenges and considerations [CD-ROM]. *Proceedings of the 19th*

*Interservice/Industry Training, Simulation and Education Conference* (pp. 144–153). Washington, DC: National Security Industrial Association.

Pierce, L. G., Hallion, K. J., Fowlkes, J. E., McCluskey, M. R., & Dwyer, D. J. (1998). Defining requirements for cost efficient collective training for dispersed staffs. *Proceedings of the Command & Control Research & Technology Symposium*, Monterey, CA, 334–346.

Prince, A., Brannick, M. T., Prince, C., & Salas, E. (1997). The measurement of team process behaviors in the cockpit: Lessons learned. In M. T. Brannick, E. Salas, & C. Prince (Eds.), *Team performance assessment and measurement: Theory, methods and applications* (pp. 289–310). Mahwah, NJ: Lawrence Erlbaum Associates.

Prince, C., & Salas, E. (1993). Training and research for teamwork in the military aircrew. In E. L. Weiner, B. G. Kanki, & R. L. Helmreich (Eds.), *Cockpit resource management* (pp. 337–366). Orlando, FL: Academic Press.

Pruitt, J. S., Burns, J. J., Wetteland, C. R., & Dumestre, T. L. (1997). SHIPMATE: Shipboard mobile aid for training and evaluation. *Proceedings of the Human Factors and Ergonomics Society 41st Annual Meeting* (pp. 1113–1117). Santa Monica, CA: The Human Factors and Ergonomics Society.

Salas, E., Bowers, C. A., & Cannon-Bowers, J. A. (1995). Military team research: 10 years of progress. *Military Psychology, 7*, 55–75.

Salas, E., & Cannon-Bowers, J. A. (in press). The anatomy of team training. To appear in L. Tobias & D. Fletcher (Eds.), *Training and retraining: A handbook for business, military, industry, and government*. New York: Macmillan.

Salas, E., Dickinson, T. L., Converse, S. A., & Tannenbaum, S. I. (1992). Toward an understanding of team performance and training. In R. W. Swezey & E. Salas (Eds.), *Teams: Their training and performance* (pp. 3–29). Norwood, NJ: Ablex.

Silverman, D. R., Spiker, V. A., Tourville, S. J., & & Nullmeyer, R. T. (1997). Team coordination and performance during combat mission training [CD-ROM]. *Proceedings of the 19th Interservice/Industry Training, Simulation and Education Conference* (pp. 154–163). Washington, DC: National Security Industrial Association.

Smith-Jentsch, K. A., Johnston, J. H., & Payne, S. C. (1998). Measuring team-related expertise in complex environments. In J. A. Cannon-Bowers & E. Salas (Eds.), *Making decisions under stress: Implications for individual and team training* (pp. 61–87). Washington, DC: American Psychological Association.

Stout, R. J., Salas, E., & Carson, R. (1994). Individual task proficiency and team process behavior: What is important for team functioning? *Military Psychology, 6*, 177–192.

Tourville, S. J., Spiker, V. A., Silverman, D. R., & Nullmeyer, R. T. (1996). An assessment methodology for team coordination in combat mission training [CD-ROM]. *Proceedings of the 18th Interservice/Industry Training, Simulation and Education Conference* (pp. 79–89). Washington, DC: National Security Industrial Association.

Vreuls, D., & Obermayer, R. W. (1985). Human-system performance measurement in training simulators. *Human Factors, 27*(3), 241–250.

Zachary, W., Bilazarian, P., Burns, J., & Cannon-Bowers, J. A. (1997). Advanced embedded training concepts for shipboard systems [CD-ROM]. *Proceedings of the 19th Interservice/Industry Training, Simulation and Education Conference* (pp. 670–679). Washington, DC: National Security Industrial Association.

Chapter 12

# Assessment of Teamwork Skills Using Computer-Based Teamwork Simulations

Harold F. O'Neil, Jr.
*University of Southern California*

Shuling Wang
*Hua-Fan University*

Gregory Chung
Howard E. Herl
*University of California, Los Angeles*

## ASSESSMENT OF TEAMWORK SKILLS USING COMPUTER-BASED TEAM SIMULATIONS

Most of the U.S. Department of Defense activities (e.g. tactical missions, maintenance functions) are completed by teams of persons. Moreover, the ultimate bottom line, combat effectiveness, is accomplished by effective teams from multiple services and countries. Although the measurement of these team outcomes is problematic, there has been progress in using simulation environments (Fletcher, in-press; Dwyer & Salas, chapter 11, this volume). Less is known about how to measure team processes that lead to effective team outcomes. Traditional research on the measurement of team processes uses trained observers who rate such behavior in live or virtual environments (see Nullmeyer & Spiker, chapter 7, this volume).

Such approaches are people-intensive and time consuming and result in delayed reporting. What would be desirable is to measure team processes in real time and in a context that facilitates such research. The purpose of this chapter is to suggest an approach to measure team processes that would be useful in computer-based environments. The tasks that we used have ecological validity in the civilian sector (e.g., a union–manager negotiation simulation). Moreover, the distinctions suggested by Salas and his colleagues (Morgan, Salas, & Glickman, 1993) between team task skills and teamwork skills suggest some public transfer of our findings to the defense section. We discuss this theme later in the chapter.

Teamwork is a common, desired characteristic of many work environments, particularly high-performance, high-paying ones. Research on workforce skills in the civilian sector indicates that teamwork skills are judged to be essential (see O'Neil, 1997). These skills are important, the studies emphasize, because as responsibility is shifted further down the management hierarchy to groups of workers, the average worker needs to communicate and cooperate with other members of the organization to an increasing degree. Thus, the measurement of teamwork for selection, credentialing, promotion, and training outcomes is important.

In the present studies, we examined teamwork processes and teamwork skills using two different computer-based group tasks. The tasks were based on earlier studies and involved (a) a group of students negotiating with a simulated management team and (b) a group of students collaborating to jointly construct a knowledge map. The teamwork processes that participants used while they engaged in the tasks were measured using the communication among team members. The current work, together with our past efforts (e.g., Chung, O'Neil, & Herl, in press; O'Neil, Chung, & Brown, 1997; O'Neil, Wang, Chung, & Herl, 1998), suggests a promising approach toward the development of a real-time measurement system for teamwork skills.

Our expectation was that these two tasks differed along several important dimensions, and thus we were interested in how the task demands influenced team processes. A second purpose for these studies was to examine the stability of the measures across different populations. In our prior work (see Chung et al., in press; O'Neil, Chung, & Brown, 1997), we used participants from middle and high schools. In the current studies, we sampled the adult population.

## Teamwork Processes

We drew on the work of Morgan et al. (1993), Salas, Dickinson, Converse, and Tannenbaum (1992), Burke, Volpe, Cannon-Bowers, and Salas (1993), and others (O'Neil, Baker, & Kazlauskas, 1992; Webb, 1993; Webb & Palincsar, 1996) to aid our development of teamwork process measures. Morgan et al. provide insight into the nature of teams. In their model of team development, they postulate two tracks of team processes, a taskwork track and a teamwork track. The taskwork track accounts for specific activities unique to the task. Taskwork team skills influence how well a team performs on a particular task (e.g., how well a team of students is able to jointly construct a knowledge map). Taskwork skills are domain-dependent, task-related activities. The teamwork track and teamwork skills influence how effective an individual member will be as part of a team and are domain-independent. Teamwork skills encompass skills such as adaptability, coordination, cooperation, and communication. Effective teams develop competence along both tracks. Members of effective teams possess the basic skills required for the task and know how to coordinate their activities, communicate with each other, and respond effectively to changing conditions.

Six teamwork processes were measured: (a) adaptability—recognizing problems and responding appropriately, (b) communication—the exchange of clear and accurate information, (c) coordination—organizing team activities to complete a task on time, (d) decision making—using available information to make decisions, (e) interpersonal skills—interacting cooperatively with other team members, and (f) leadership—providing structure and direction for the team. Our prior research showed statistically significant positive relationships between decision making and team performance on a negotiation task (final contract offer and negotiation style), and a significant negative correlation between interpersonal processes and whether an agreement was reached (O'Neil, Chung, & Brown, 1997). A detailed discussion of these teamwork processes can be found in O'Neil, Chung, and Brown (1997) for the negotiation task and Chung et al. (in press) for the knowledge mapping task.

To measure teamwork processes, we provided participants with a set of pre-defined messages that they used to communicate with each other. Each message belonged to one of six teamwork processes (Chung et al., in press; O'Neil, Chung, & Brown, 1997). We assumed that the use of a message reflected the use of the associated team process; thus, our technique for measuring team processes was based on message usage.

### Overview of Studies

***Group Tasks.*** In the two studies reported here, we examined teamwork processes and performance of groups working on two different tasks. These studies were replications of earlier work, with slight modifications to the message set and a different population. In the current work, we collected data from a large number of groups and sampled an adult population, which provided a wider range of individual differences in terms of age, education, and work experience. Our earlier studies focused on students in middle and high schools. Note that the data for the current studies were collected originally to cross-validate a paper-and-pencil teamwork questionnaire (O'Neil et al., 1998). However, the data and analyses reported here are limited to team processes and outcomes, which is a continuation of our line of research on the use of computer-based networked group tasks.

In Study 1, we used a modified version of the simulation software developed for O'Neil, Chung, and Brown (1997). The only differences were that the number of messages available to participants was reduced from 51 to 31 based on a reliability analysis (O'Neil, Brown, & Chung, 1996), and the user interface was modified to reflect this change. Briefly, the task was a union–management negotiation. Participants were informed that they were part of a union team that was responsible for negotiating a contract with the management team. The management team was actually simulated by the computer and programmed to reflect priorities that were opposite to the union's team. (Participants were unaware of the simulation during the session and were debriefed at its conclusion.) Participants communicated with each other by sending predefined messages. Only one group member (the leader) could actually send an offer to the simulated management team. The remaining two members were instructed to advise the leader on what to offer. Leadership was randomly assigned and did not rotate.

In Study 2, participants worked in groups to jointly construct a knowledge map on environmental science. The terms and links of the knowledge map were provided to the group. As with the negotiation task, participants communicated by sending a predefined, but different, set of messages. In addition, leadership rotated among the group members. For the knowledge mapping task, leadership meant control of the knowledge map (i.e., the capability to add, delete, or modify the map). See Chung et al. (in press) for a more detailed description of the task.

***Similarities and Differences Between Group Tasks in Study 1 and Study 2.*** Because this study reports on the team processes in two different team tasks, the following discussion briefly outlines similarities and differences between the tasks that were expected to impact the team processes or team outcomes. Table 12.1 summarizes the differences.

Four characteristics were common to both tasks. First, both tasks used weak team tasks; that is, the task conceivably could be carried out by one member. However, the task instructions emphasized working together, and the user interface of the computer systems displayed all messages that all team members sent. A second characteristic of both tasks is that the participants were required to use a set of predefined messages for communication. This communication mode is the only feasible way to measure teamwork processes in real-time. We acknowledge the limitations of this method, but available and affordable technology (i.e., technology accessible to educational or workplace settings) does not allow for the accurate recognition of verbal input (text input by typing or speaking). Third, both tasks were open-ended; there was no single correct answer. The strategies a team adopted to complete the task were determined solely by the team. Finally

TABLE 12.1
Summary of Similarities and Differences Between the Union–Management Negotiation Task and the Collaborative Knowledge Mapping task

| | *Union–Management Negotiation Simulation* | *Collaborative Knowledge Mapping* |
|---|---|---|
| Theoretical framework for the development of the teamwork process categories and messages | Same | Same |
| Strength of team task—can task be completed only with participation of all members? | Weak | Weak |
| Content knowledge required to perform well on outcome measure | Low | High |
| User-interface attentional demands | Moderate. Participants had only to look at messages on screen and messages on handout | High. Users had to attend to messages on screen, handout, plus knowledge map configuration |
| Feedback on task performance | Some, from the counteroffer from simulated management team. | None |
| Task time | Limited to 8 rounds of negotiation. No time limit on the length of each round | Fixed to 36 minutes |

and most importantly, the messages available to participants in both tasks were based on the same theoretical framework. A common set of teamwork processes (i.e., adaptability, coordination, decision making, interpersonal skills, leadership, and communication) was used to derive the messages participants used to communicate with each other.

There were major differences between the two tasks. First, the amount of time available to complete the tasks differed. For the negotiation task, team members had eight rounds to finish, and the duration of each round was based on when the union team sent an offer instead of a fixed time limit. Thus, the time to complete the negotiation task typically ranged between 20 and 40 minutes. For the knowledge mapping task, there was a fixed time limit for the entire task (36 minutes) and for the amount of time any one person had control of the map (6 minutes). Another difference was in the availability of feedback. In the negotiation task, feedback was in the form of the response from the simulated management team. Because the simulated management team responded systematically to participants' offers, participants had available feedback about how well their negotiating strategy was working (or not working). The knowledge map, in contrast, offered no explicit feedback. There was no real-time scoring of the map to indicate to participants how well they were doing on the map. A third difference was in the amount of content knowledge required to complete the tasks. The knowledge requirement for the negotiation task was presumably sparse compared to that of the knowledge-mapping task. The crucial information participants needed to know was their negotiating priorities, and this information was provided to them. Participants then had to induce the management team's priorities from the management team. In contrast, the knowledge-mapping task was highly knowledge driven—participants had to know the nature of the relationship between concepts in order to perform well on the task. A fourth difference was that the leadership role rotated in the knowledge-mapping task, but not in the union–management task.

Both studies were done in the context of validation of a teamwork questionnaire. These results are reported elsewhere (O'Neil et al., 1998). We organized two sets of research activities into two virtual studies. The data for both studies were collected concurrently, and some of the participants in Study 1 also participated in Study 2. The task of Study 1 was a union–management negotiation scenario, whereas the task of Study 2 was a computer-based knowledge map in environmental science.

## STUDY 1: NETWORKED NEGOTIATION TASKS

Our previous work using our computerized negotiation task suggested that it was feasible, reliable, and valid in measuring individual negotiation skills (O'Neil, Allred, & Dennis, 1997a, 1997b). Our findings were relatively consistent across different negotiation contexts and samples. Further, we have some information on the feasibility of a network-based team version of this negotiation task.

Our computer-based negotiation task was a union–management negotiation of a contract. The participants played a union team that negotiates a contract with a management team, which was simulated by the computer. To facilitate team communication, we provided a set of predefined messages that map onto our taxonomy of teamwork processes (i.e., adaptability, coordination, decision making, interpersonal skills, and leadership). These predefined messages allowed participants to communicate with their team members. During the negotiation, based on our prior research, we expected the teamwork process of decision making to facilitate positive negotiation outcomes. The messages and their associated categories are described in greater detail in the discussion of methodology.

In Study 1, we focused on assessing team processes that were important during a negotiation task and how these team processes affected team performance.

## Networked Negotiation System

***User Interface.*** The user interface is shown in Fig. 12.1. The display is divided into three major sections. The upper left section consists of numbered message buttons. Each button corresponds to a particular message on the participants' message handout. Participants can click on a button and the message corresponding to that button will be sent to the other team members' computers. The lower left section of the screen displays the messages sent by the team members. The messages are listed in the order that they are sent. The right-hand section displays the issue chart, current union and management offer, and buttons to send and accept offers. Only the participant who is the leader can send or accept offers. The team's current offer is selected by clicking on one of the issue values. The selection appears on all team members' computers. For certain messages, user input is required. These messages are operated by using dialog boxes, but most messages require simple point-and-clicking.

***Message Types.*** In Study 1, there was a set of messages for each of five teamwork categories to measure team process. A total of 31 messages were used for the networked negotiation team process. These messages were

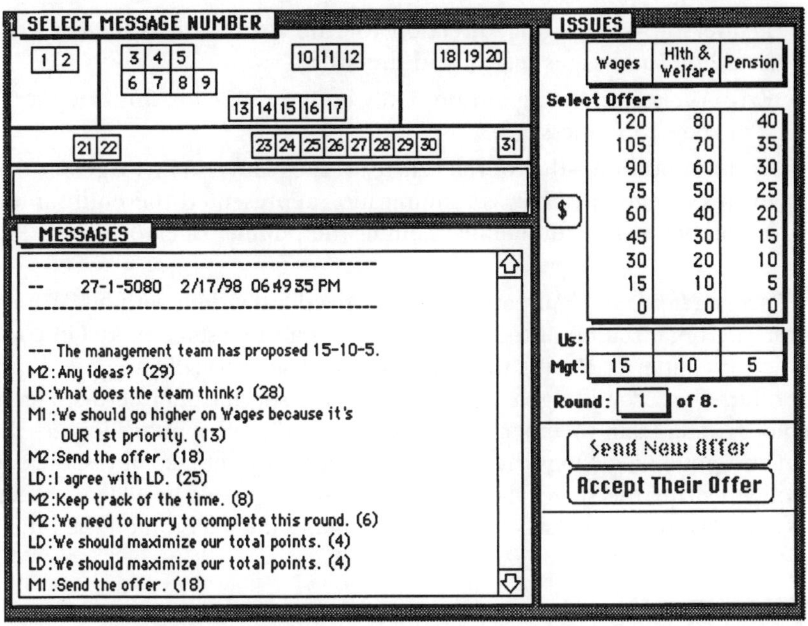


FIG. 12.1. User interface for the team leader in the networked negotiation task. Advisory team members' screens do not have the buttons for sending or accepting the offers.

selected from 51 messages with reasonable reliability from a previous study (O'Neil et al., 1996). An error message was provided to signal that a previously sent message was mistakenly sent by the participant. This message was excluded from the message categories. Communication skills were not measured by a specific message set; instead, because all of the messages were selected and accurately sent via the network, we considered all of the team process messages to be examples of communication. The message types are described below:

- Adaptability—recognizing problems and responding appropriately. This category consisted of 6 messages (e.g., Should we accept their offer?).
- Coordination—organizing team activities to complete a task on time. This category consisted of 6 messages (e.g., Keep track of the time).
- Decision making—using available information to make decisions. This category consisted of 6 messages (e.g., We should go higher on wages because it's our 1st priority).
- Interpersonal skills—interacting cooperatively with other team members. This category consisted of 5 messages (e.g., We have to work together on this).
- Leadership—providing direction for the team. This category consisted of 7 messages (e.g., Send the offer).
- Error—click on wrong button. Only one message for this category (I sent the wrong message).
- Communication—the overall activity of the team. No message was tied to this category; rather, communication represented the entire interaction between team members minus the number of error messages.

***Message Handout.*** All participants received a handout with all 31 messages. The organization of the handout follows a three-stage model of communication patterns. During the first stage, the leader usually suggests a specific offer. The second stage is a discussion of the specific offer between team members. The team members also discuss how the offer might impact the team and potential actions the management team might take in response to the offer. During this stage, alternative offers are suggested by all team members. The last stage consists of the leader making the decision on the offer.

### Measures

***Team Outcome Measures.*** The negotiation team outcome measures were based on the following negotiation measures: (a) final counteroffer, (b) whether or not an agreement was reached, (c) if an agreement was reached, the type of agreement (i.e., integrative or distributive), and (d) elapsed time on the negotiation task.

The final counteroffer was the sum of the offer levels across all three contract issues: wages, health and welfare, and pension. If an agreement was reached, the agreed-to offer was used to compute the final counteroffer. For example, if the agreed-to offer was 105-50-15, the final counteroffer was 170. If an agreement was not reached, the final counteroffer was computed from the offer that the management team would propose in response to the last offer from union team. For example, if the last union offer was 120-80-40, in response to this offer, the management team would propose 0-0-0. Thus, the final counteroffer would be 0.

Agreement was defined as whether or not the union team reached an agreement with the management team. A union team accepting the management team's offer was considered the same as the management team accepting the union team's offer. Type of agreement was defined as integrative and distributive. An integrative agreement (a win/win) was defined as when the union team reached an agreement with a final counteroffer of more than 120 points; a distributive agreement (win/lose) was defined as when the union team reached an agreement equal to or less than 120 points.

Elapsed time was defined as the time between the start of the negotiation and the termination of the negotiation. This time does not include the tutorial or question-answering training activities.

***Individual and Team-Level Teamwork Process Measures.*** The individual and team-level process measures attempted to measure the teamwork process (i.e., adaptability, coordination, decision making, interpersonal skills, leadership, and communication) for each individual and team during the negotiation. Individual team process was computed by calculating the number of messages sent in each team process category. For example, if an individual sent 5 messages from the coordination category, the individual's coordination score was 5. Communication score was computed by taking the total number of messages sent by an individual minus the number of "I sent the wrong message" (error) messages. Team-level teamwork processes were calculated by adding all messages sent by the three group members for each category. For example, if all three group members sent 5 messages each in the coordination category, then the team-level coordination performance score was 15.

## Participants

Participants in Study 1 included members of a local union–industry training program ($N$ = 153) and temporary agency workers ($N$ = 108). These participants had completed a teamwork skills questionnaire (see O'Neil et al., 1998, for detail on this questionnaire).

## Procedure

For union participants, the study was conducted at their training center. Researchers transported nine Macintosh PowerBook computers (2 PB 1400s, 3 PB 3400s, 4 PB 5300s), ethernet hubs, ethernet cables, RJ-45 couplers, mice, mouse pads, and power supplies for each data collection effort. Network setup time was approximately 3 hours.

There were three tables with three computers on each table in the room in which we conducted the study. Participants were randomly assigned to a computer, a team, and a role (leader vs. member). The researchers introduced themselves, informing the participants that they were part of a study examining teamwork skills. Participants were asked to fill out the teamwork skills questionnaire. After participants completed the questionnaire, one researcher gave a brief overview of the task in 10 minutes. A brief introduction on how to do the computer task was then provided. This was an interactive instructional period during which participants could ask questions about the task.

For the participants from the two temporary agencies, CRESST's computer lab was used to collect the data. Twelve Power Macintosh 6100 computers with 17" monitors were used. Participants were informed about the study and asked to fill out the teamwork skills questionnaire. After all participants completed the questionnaire, the researcher explained the team scenario and how to do the computer task, using a projector and transparencies.

Although there were a few differences in the data collection procedures for union participants and for temporary agency participants, most procedures were identical for both sets of participants. The following activities in the data collection were the same for both sets of participants.

***Team Scenario.*** Participants were informed that they were part of a three-person team consisting of a leader and two members. Only the leader could send offers for the union team and accept offers from the management team. The major responsibility of the other two union team members was to give advice and suggestions to the leader.

***Computer Use.*** Participants were given computer demonstrations on how to carry out critical functions such as sending messages. All participants went through a negotiation tutorial on the computer. The tutorial explained in detail the negotiation scenario, the negotiation issues and priorities, the team scenario, and how to operate the computer. Time required for the participants to complete the tutorial ranged from 13 to 22 minutes.

When participants finished the computer tutorial, they were instructed to study the message handout while waiting for everyone else to finish. After all participants completed the tutorial, they began the negotiation task. After finishing the negotiation, participants were given feedback regarding their performance. Participants were shown measures of their individual performance score—individual team process score (number of messages sent for each message category); they were also shown their team performance—team-level team process score (total number of messages sent by all team members for each message category) and negotiation performance (whether or not an agreement was reached), type of agreement (integrative or distributive, if an agreement was reached), and elapsed time for the negotiation task.

***Data Collection Problems.*** There were a few participants who had difficulty in using the computers or who left early during the negotiation task. As a result, three union participant groups were dropped from the data analysis. For the temporary agency sample, two groups were dropped due to missing computer data. A third group, which misunderstood the computer task and accepted the first offer, was also dropped. A total of 33 groups ($N = 99$) of temporary agency workers was included in our data analysis for the networked negotiation task.

## Results

***Individual-level Measures.*** The frequency of messages used by participants is shown in Table 12.2. The use of messages by individuals varied from 39 to 707 for union participants and from 25 to 674 for temporary agency workers. The results indicated that the frequency distributions were highly skewed positively, with most messages having a mode of zero, which was consistent with our previous study (O'Neil et al., 1996). In order to place each message on a common metric (0–1) and modify the underlying distribution, O'Neil, Chung, and Brown (1997) suggested that before aggregating the messages to generate team process scales, the message counts should be transformed from raw score frequency of usage for each individual to a percentage of all message usage for that individual relative to all participants. For example, if an individual used six instances of message 1, which had a total usage of 60 by all individuals, this individual's score on message 1 was transformed from 6 to .10 (6/60). The individual's transformed score for every message was used to compute the individual's team process scores.

For each participant, individual scores for adaptability, coordination, decision making, leadership, and interpersonal teamwork processes were computed by aggregating the messages from each category. Also, an overall com-

TABLE 12.2
Frequency Count of Messages for Both Samples on the Networked Negotiation Task

| *Message Number* | *Union (N = 144)* | *Temporary Agency (N = 99)* |
|---|---|---|
| 1 | 707 | 674 |
| 2 | 499 | 353 |
| 3 | 152 | 105 |
| 4 | 109 | 71 |
| 5 | 80 | 64 |
| 6 | 39 | 149 |
| 7 | 69 | 82 |
| 8 | 73 | 49 |
| 9 | 121 | 136 |
| 10 | 131 | 106 |
| 11 | 86 | 75 |
| 12 | 56 | 82 |
| 13 | 182 | 144 |
| 14 | 71 | 75 |
| 15 | 67 | 99 |
| 16 | 60 | 71 |
| 17 | 71 | 165 |
| 18 | 673 | 487 |
| 19 | 188 | 249 |
| 20 | 155 | 133 |
| 21 | 129 | 70 |
| 22 | 103 | 47 |
| 23 | 219 | 119 |
| 24 | 82 | 25 |
| 25 | 548 | 637 |
| 26 | 164 | 126 |
| 27 | 520 | 494 |
| 28 | 316 | 290 |
| 29 | 165 | 124 |
| 30 | 114 | 141 |
| 31 | 51 | 77 |

munication score was calculated by adding the individual's five teamwork process scores minus the total number of error messages the individual sent. Means and standard deviations for each scale for both samples are presented in Table 12.3. The means of all team processes were higher for temporary agency workers than for union participants, because union participants had higher total usage for approximately two thirds of all messages. In other words, the higher the total usage of messages, the lower the team process scores, because each team process score was calculated by its corresponding messages (proportion scores), which were transformed by division of the total usage of these messages.

***Team Process Reliability.*** The reliability of the five computer-based team process scales was tested using item raw scores (counts of times utilized) and transformed item scores (proportion scores). The results indi-

TABLE 12.3
Individual-Level Teamwork Process Scales on the Networked Negotiation Task

| | *Union (N = 144)* | | *Temporary Agency (N = 99)* | |
|---|---|---|---|---|
| Adaptability | .04 | .04 | .06 | .05 |
| Coordination | .04 | .05 | .06 | .07 |
| Decision making | .04 | .04 | .07 | .07 |
| Leadership | .05 | .04 | .07 | .07 |
| Interpersonal | .03 | .04 | .05 | .06 |
| Communication | .20 | .14 | .29 | .21 |

cated that the reliability of proportion scores was better than raw scores for all scales for temporary agency workers, whereas proportion scores were more reliable than raw scores only in the scales for decision making and leadership for union participants (see Table 12.4). The reliability of the decision-making scale for both raw score and proportion score for union participants was the lowest among all team processes. In general, these reliabilities are too low, and the message set should be revised to improve the reliabilities.

***Team-level Measures: Outcome Measures.*** There were four team-level outcome measures in the negotiation task: whether an agreement was reached, type of agreement reached, team performance scores, and time spent on negotiation. The outcome measure of agreement was coded dichotomously (1 = agreement; 0 = no agreement). The results indicated that about 70% of teams reached agreement for both samples (see Table 12.5).

Agreement type was determined by the value of performance score. If a team reached an agreement with a team performance score greater than 120, this team reached an integrative agreement. On the other hand, if a team reached an agreement with a team performance score equal to or less than 120, the team reached a distributive agreement. The type of agreement was also coded dichotomously (2 = integrative; 1 = distributive). Fifty percent of teams reached an integrative agreement for both samples (see Table 12.6).

For teams reaching agreement, team performance scores were computed as the total value of the agreed-on offer. For teams not reaching an agreement, team performance scores were calculated by taking the total value of the inverse of the last offer sent by the union team. The results indicated performance scores ranging from 0 to 150 for union participants, and from 10 to 165 for temporary agency workers. Time spent in

TABLE 12.4
Summary of Reliability Analysis for Teamwork Process Scales for the Networked Negotiation Task

| *Teamwork Process Scale* | *Message Numbers* | *Adjustment* | *Union Alpha* | *Temporary Agency Alpha* |
|---|---|---|---|---|
| Adaptability | m25,m26,m27,m28,m29 | use raw value | .60 | .50 |
| | m25,m26,m27,m28,m29 | use proportion scores | .60 | .52 |
| Coordination | m5,m6,m7,m8,m9,m12 | use raw value | .62 | .67 |
| | m5,m6,m7,m8,m9,m12 | use proportion scores | .61 | .73 |
| Decision making | m1,m13,m14,m15,m16,m17 | use raw value | .33 | .56 |
| | m1,m13,m14,m15,m16,m17 | use proportion scores | .44 | .59 |
| Interpersonal | m20,m21,m22,m23,m24 | use raw value | .69 | .51 |
| | m20,m21,m22,m23,m24 | use proportion scores | .66 | .58 |
| Leadership | m2,m3,m4,m10,m11,m18,m19 | use raw value | .47 | .49 |
| | m2,m3,m4,m10,m11,m18,m19 | use proportion scores | .54 | .62 |

TABLE 12.5
Percentage of Agreement for the Networked Negotiation Task

| | *Union* | | *Temporary Agency* | |
|---|---|---|---|---|
| | *Group (N)* | *Percentage (%)* | *Group (N)* | *Percentage (%)* |
| Agreement | 34 | 70.8 | 24 | 72.7 |
| No agreement | 14 | 29.2 | 9 | 27.3 |

TABLE 12.6
Type of Agreement for the Networked Negotiation Task

| | *Union* | | *Temporary Agency* | |
|---|---|---|---|---|
| | *Group (N)* | *Percentage (%)* | *Group (N)* | *Percentage (%)* |
| Integrative | 17 | 50 | 12 | 50 |
| Distributive | 17 | 50 | 12 | 50 |

the negotiation was obtained by the program for each team. Descriptive statistics for each of the outcome measures for these two samples are shown in Table 12.7.

The results of outcome measures were similar for both samples. The mean performance score was 111.25 for union participants, 105.45 for temporary workers. In addition, about 70% of teams reached agreement, and half of these teams reached an integrative agreement, as previously mentioned. The average time for teams to complete the negotiation was 22 minutes for union participants and 25 minutes for temporary agency workers.

***Tests of Hypothesis.*** There would be a positive relationship between networked team processes of decision making and outcome measures.

The results generally indicated that, with few exceptions, there were no significant relationships between team processes and outcome measures, except for time in negotiation, for both samples (see Table 12.8). The correlations in Table 12.8 are nonparametric. The reader may recall that the reliability of the teamwork process measures was low and thus may have influenced these results. Further, the number of groups is also fewer than desired. With respect to outcome, the results indicated that decision mak-

TABLE 12.7
Descriptive Statistics for Outcome Measures for Union and Temporary Agency Participants on the Networked Negotiation Task

| *Variable* | *Mean* | *SD* | *Minimum* | *Maximum* |
|---|---|---|---|---|
| Union participants | | | | |
| Time | 21.89 | 8.65 | 6.50 | 46.86 |
| Agreement | .71 | .46 | .00 | 1.00 |
| Performance score | 111.25 | 39.04 | .00 | 150.00 |
| Temporary agency participants | | | | |
| Time | 25.56 | 9.85 | 10.26 | 46.90 |
| Agreement | .73 | .45 | .00 | 1.00 |
| Performance score | 105.45 | 36.77 | 10.00 | 165.00 |

TABLE 12.8
Correlation Between Teamwork Process Scales and Outcome Measures for Union and Temporary Agency Participants on the Networked Negotiation Task

| *Team Process* | *Performance Score* | *Agreement* | *Agreement Type* | *Time in Negotiations* |
|---|---|---|---|---|
| Union participants | | | | |
| Adaptability | .26 | .09 | .20 | .53* |
| Coordination | .05 | -.10 | -.14 | .38* |
| Decision making | .22 | .06 | .35* | .54* |
| Leadership | .14 | .14 | .02 | .54* |
| Interpersonal | -.02 | .00 | -.10 | -.28* |
| Communication | .17 | .04 | .07 | .61* |
| Temporary agency participants | | | | |
| Adaptability | .17 | -.28 | .08 | .48* |
| Coordination | -.25 | -.17 | -.26 | .41* |
| Decision making | .14 | .03 | .23 | .53* |
| Leadership | -.13 | .03 | -.12 | .42* |
| Interpersonal | -.28 | .03 | -.12 | .44* |
| Communication | -.10 | -.11 | -.04 | .59* |

*$p < .05$.

ing was significantly related to integrative agreement for union participants, which was consistent with our previous study. It seemed that teams that engaged in more decision-making processes were more likely to reach agreement with higher value. However, although in the right direction, there was no significant relationship between decision making and quality of agreement for temporary agency workers.

Consistent with previous research by O'Neil, Chung, and Brown (1997), the results showed that all team processes were significantly related to time in negotiations. As previously mentioned, team processes were calculated using the messages sent by individuals. It was reasonable to assume that the more messages individuals sent, the longer it took to complete the negotiation. As a result, all team processes related positively to time in negotiations.

## STUDY 2: NETWORKED KNOWLEDGE-MAPPING TASK

Our previous research suggested that the networked knowledge-mapping task was feasible in measuring real-time teamwork skills (Chung et al., in press). Specifically, participants ranging from elementary school children to graduate students at university were able to jointly construct a knowledge map using our networked knowledge-mapping system with messages.

A knowledge (or concept) map is a node–link–node representation of content, where nodes represent concepts and links represent relationships between linked concepts (Dansereau, 1995; Jonassen, Beissner, & Yacci, 1993). Our present networked knowledge-mapping task was designed for a context of group knowledge-mapping activities, where the activities took place over networked computers. The networked knowledge-mapping task evaluated team processes in three-person teams as the teams constructed knowledge maps on topics in environmental science.

Our system provided to group members a set of predefined messages that matched our taxonomy of teamwork processes, such as adaptability, coordination, decision making, interpersonal skills, and leadership. Participants used these messages to communicate with their team members when they constructed their knowledge maps. The messages and their associated categories are discussed in more detail later.

Consistent with the networked negotiation task (Study 1), this work also focused on assessing team processes and on how team processes affected team performance. Similar to the Study 1 hypothesis, the following hypothesis was examined for the networked knowledge-mapping task: Participants' team processes will relate to their teamwork performance (outcome).

### Networked Knowledge Mapping System

The system was developed on the Macintosh with HyperCard 2.3. We used the built-in networking capabilities of the operating system and HyperCard to implement a basic peer-to-peer system. Every member's screen was updated as changes occurred (e.g., when leaders made a change in linking the concept). In other words, all team members' computers were synchronized with each other. Only the leader could add, delete, or move knowledge map components around the screen to make changes to the knowledge map. Each time a map event occurred, the event was dispatched to the other members' computers, and their maps were automatically updated. All map events were logged by the computer.

The software was designed to be simple and easy to use. Participants sent messages by clicking on number buttons. Concepts were added to the knowledge map via menu selections, and links were created by connecting two concepts and then selecting the desired link from a pop-up menu. A more extensive description of this task is provided by Chung et al. (in press).

***User Interface.*** Figure 12.2 shows the user interface to the system. The display is divided into three sections. The top fifth is reserved for knowledge mapping. The lower left section of the screen displays the

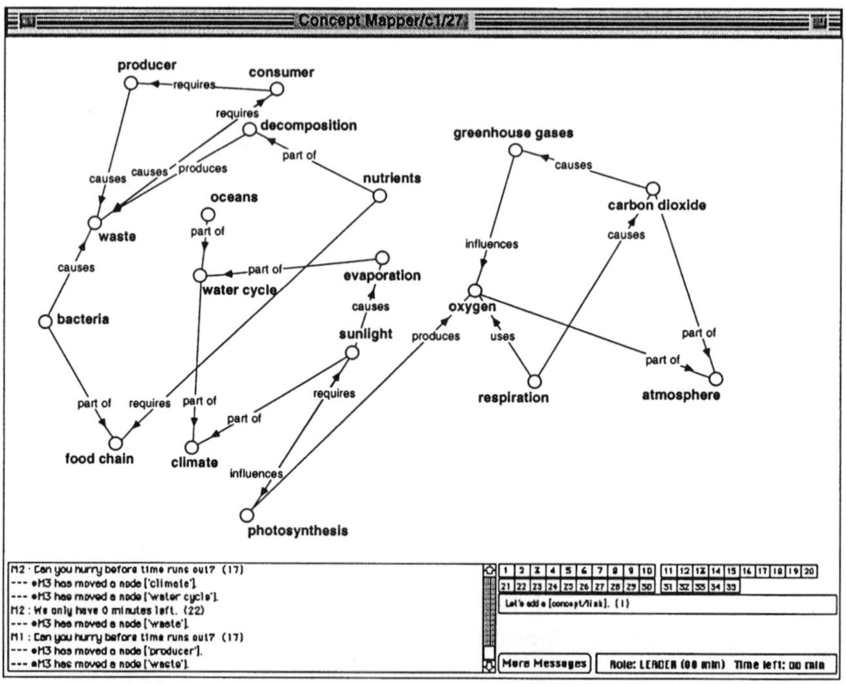


FIG. 12.2. User interface for the system in the networked knowledge mapping task.

messages. All messages sent are listed in the order sent. Thirty-five numbered buttons are provided to team members. Each participant is given a handout that lists all the messages, and the messages are numbered to correspond to the buttons on the computer screen. To send a message, a participant clicks on a button and the corresponding message is sent to all team members' computers. Other information displayed on the lower part of the screen is the remaining time for the entire task and the remaining time as a leader or nonleader. As indicated in Fig. 12.2, the remaining time for the current team member as a leader is 2 minutes, and the time remaining for the entire task is 32 minutes.

The task provided the team with 15 terms (atmosphere, bacteria, climate, CO2, decomposition, evaporation, food resources, oxygen, precipitation, respiration, sun light, vegetation, photosynthesis, waste, water) and 10 links (contributes to, causes, leads to, part of, result of, similar to, produced by, influences, type of, prior to; see Fig. 12.2).

***Message Types.*** Most messages could be sent without typing. Participants could send a message by clicking on a numbered button. For example, Fig. 12.3 shows how message 1 (Let's add a [concept/link]) was handled. The participant simply chooses the desired option (concept or link) and the message reflecting the choice is sent.

A second type of message required participants to type text. For example, the intent in Fig. 12.4 was for participants to complete the message stem.

There were a total of 37 messages used for the networked knowledge-mapping task. Thirty-four messages in five teamwork categories were used to measure team processes (i.e., adaptability, coordination, decision making, interpersonal skills, and leadership). Two messages ("yes" and "no") were part of the communication category. One message was provided to signal a previous message sent in error and was excluded from the categories. A sample message from each category is given below.

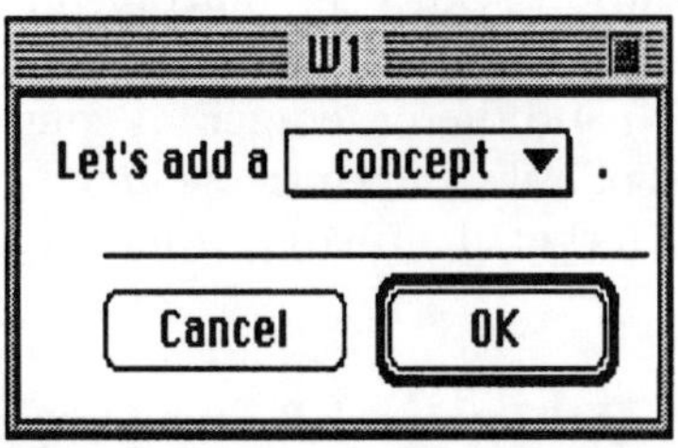


FIG. 12.3. Message requiring user input: selecting values for the networked knowledge mapping task.

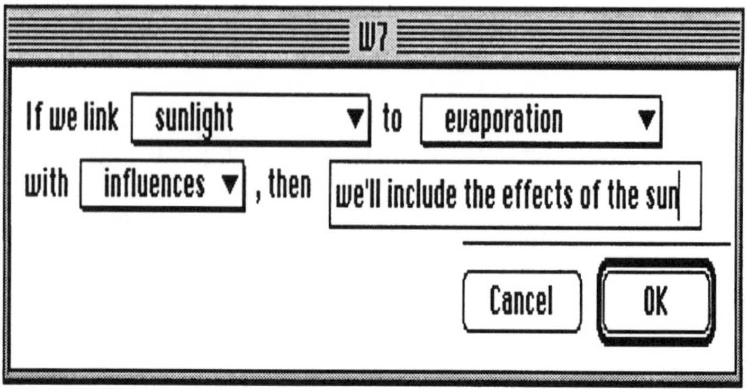


FIG. 12.4. Message requiring user typed text for the networked knowledge mapping task.

- Adaptability—recognizing problems and responding appropriately. This category consisted of 7 messages (e.g., What do you think, M1?).
- Coordination—organizing team activities to complete a task on time. This category consisted of 7 messages (e.g., We only have 5 minutes left).
- Decision making—using available information, make decisions. This category consisted of 6 messages (e.g., What if we add 'carbon dioxide'?).
- Interpersonal skills—interacting cooperatively with other team members. This category consisted of 7 messages (e.g., I need to hear from all of you).
- Leadership—providing direction for the team. This category consisted of 7 messages (e.g., Let's link 'carbon dioxide' to 'producer').

## Measures

***Team Outcome Measures.*** Our team outcome measures were computed by comparing a team's knowledge map against a set of four experts' maps (Herl et al., 1996) and then averaged. A content score, a structure score, and the numbers of links and concepts in the group map were used as outcome measures. Herl et al. (1996) provided a discussion of all scoring algorithms.

***Individual and Team-level Teamwork Process Measures.*** Consistent with the networked negotiation task, individual and team-level process measures assessed the degree to which individuals and their team engaged in each of the team processes (adaptability, cooperation, decision making, interpersonal skills, leadership, and overall communication). Individual team process was computed by calculating the number of messages sent in each team process category. This method was used for all team process categories except communication. Communication was measured by taking the total number of messages sent by an individual minus the number of "I sent the wrong message" messages. Team-level teamwork processes were calculated by adding all messages sent by all three group members for each category. For example, if all three group members sent 5 messages each in the leadership category, then the team-level leadership performance score was 15.

***Participants.*** One hundred seventeen people from two temporary agencies in southern California participated in this study. Many of these participants completed both the networked negotiation task (Study 1) and the knowledge-mapping task (Study 2). One hundred three participants also completed the original 23-item teamwork questionnaire before they started the networked knowledge-mapping tasks (see O'Neil et al., 1998).

***Procedure.*** The researcher informed the participants that they were part of a study examining teamwork skills and gave a brief overview of the activities to follow. A simple example of constructing a knowledge map was demonstrated on a white board at the front of the computer lab. The example used three concepts (birds, eagle, and wings) and three links (part of, type of, and uses). Participants were shown how to construct an actual knowledge map on the computer. The researcher also explained how to add or delete concepts and how to create links between concepts. In addition, message handouts for this task were passed out to all participants, and they were given a few minutes to study the messages. The researcher explained and showed how to use two different types of messages to communicate with other team members. Participants then began their networked knowledge-mapping task.

***Team Scenario.*** Participants were informed that they were part of a three-person team comprised of a leader and two members. Only the leader could manipulate the knowledge map (e.g., add concepts and links). Nonleaders could only advise by sending messages. Leadership rotated every 6 minutes, giving each participant the opportunity to control the knowledge map twice. Teams were given 36 minutes to complete the knowledge-mapping task.

***Data Collection Problems.*** One participant clicked on the OK button before she was told to, and we had to restart the computer and the network. She made the mistake a second time, and we had to restart the computer again. The task was delayed by 20 minutes. This team was dropped.

## Results

***Individual-level Measures.*** The frequency of messages used by participants is shown in Table 12.9. The use of messages by individual varied from 7 to 178.

Consistent with the networked negotiation task, frequency distributions were highly skewed positively, with most messages having a mode of zero. The message count was therefore transformed from raw score frequency of usage for each individual to percent of all message usage for that individual relative to all participants. For example, if an individual used 7 instances of message 1, which had a total usage count of 70, this individual's process score on message 1 was transformed from 7 to .10 (7/70).

For each participant, individual scores for adaptability, coordination, decision making, leadership, and interpersonal teamwork processes were computed by aggregating the messages from each category. A communication score was calculated by summing each of the five teamwork process scores for each individual, plus the number of "yes" and "no" messages sent, minus the total number of error messages for that individual. Means and standard deviations for each scale are shown in Table 12.10.

TABLE 12.9
Frequency Count of Messages for the Networked Knowledge-Mapping Task

| *Message Number* | *Frequency Count* | *Message Number* | *Frequency Count* |
|---|---|---|---|
| 1 | 121 | 21 | 29 |
| 2 | 65 | 22 | 47 |
| 3 | 79 | 23 | 41 |
| 5 | 49 | 24 | 10 |
| 6 | 178 | 25 | 16 |
| 7 | 32 | 26 | 20 |
| 8 | 62 | 27 | 29 |
| 9 | 32 | 28 | 106 |
| 10 | 7 | 29 | 68 |
| 11 | 50 | 30 | 87 |
| 12 | 66 | 31 | 85 |
| 13 | 16 | 32 | 159 |
| 14 | 26 | 33 | 74 |
| 15 | 13 | 34 | 46 |
| 16 | 35 | 35 | 19 |
| 17 | 43 | 36 | 74 |
| 18 | 23 | 37 | 29 |
| 19 | 42 | 38 | 24 |
| 20 | 23 | | |

*Note.* No message 4 for the task.

TABLE 12.10
Individual-Level Teamwork Process Scales for the Networked Knowledge-Mapping Task

| | *Mean* | *SD* | *Minimum* | *Maximum* |
|---|---|---|---|---|
| Adaptability | .06 | .06 | .00 | .30 |
| Coordination | .06 | .12 | .00 | .88 |
| Decision making | .05 | .09 | .00 | .61 |
| Interpersonal | .06 | .09 | .00 | .45 |
| Leadership | .06 | .06 | .00 | .29 |
| Communication | .31 | .26 | .01 | 1.30 |

***Team Process Reliability.*** The reliability of item raw scores and transformed item scores (proportion scores) for the five scales is shown in Table 12.11. The results indicate that the reliability of proportion scores is higher than raw value in the scales for decision making and interpersonal skills, whereas the reliability of raw value is better in the scales for adaptability, coordination, and leadership. However, as is seen in Table 12.11, the reliability for the scales is unacceptably low. With the exception of the coordination scale, all scales should be revised. In general, the reliability of these team process measures is lower than that of similar scales in the union–management task. The latter task had revised the message set based on a reliability analysis.

***Team-level Measures: Outcome Measures.*** There were four outcome measures for the networked knowledge-mapping task. The content score was the most representative of team performance on the knowledge-mapping task. The results indicated that team performance (content score) on the knowledge-mapping task varied from 0 to 16.75 (see Table 12.12). The average content score was 4.33, indicating low levels of team knowledge. Outcome measures were computed by comparing a team's knowledge map against the maps of a set of experts. The following are the agreement results for the knowledge-mapping task.

***Team-level Measures: Agreement Results for Knowledge Mapping Task.*** Exact agreement analyses were performed on the pairwise agreement using experts' maps to score participants' maps. The links in the experts' maps were used to score links in participants' knowledge maps. A pair of experts was considered to either agree or disagree on the correctness or incorrectness of participants' links. There are three possibilities for each participant-constructed link: (a) both experts construct links; (b) neither expert constructs a link; and (c) only one expert constructs a link. A link is scored as incorrect when it does not match either of the experts' links

TABLE 12.11
Summary of Reliability Analysis for Teamwork Process Scales for the Networked Knowledge-Mapping Task

| *Teamwork Process Scale* | *Message Numbers* | *Adjustment* | *Alpha* |
|---|---|---|---|
| Adaptability | m5,m7,m9,m29,m30,m31,m36 | use raw value | .39 |
| | m5,m7,m9,m29,m30,m31mm36 | use proportion scores | .32 |
| Coordination | m13,m17,m18,m19,m20,m21,m22 | use raw value | .74 |
| | m13,m17,m18,m19,m20,m21,m22 | use proportion scores | .73 |
| Decision making | m10,m11,m14,m15,m16,m37 | use raw value | .36 |
| | m10,m11,m14,m15,m16,m37 | use proportion scores | .45 |
| Interpersonal | m23,m24,m25,m26,m27,m28,m32 | use raw value | .41 |
| | m23,m24,m25,m26,m27,m28,m32 | use proportion scores | .59 |
| Leadership | m1,m2,m3,m6,m8,m12,m38 | use raw value | .23 |
| | m1,m2,m3,m6,m8,m12,m38 | use proportion scores | .17 |

TABLE 12.12
Descriptive Statistics for Outcome Measures for the Networked Knowledge-Mapping Task

| | *Mean* | *SD* | *Minimum* | *Maximum* |
|---|---|---|---|---|
| Content score | 4.33 | 3.89 | .00 | 16.75 |
| Structure score | .16 | .09 | .03 | .38 |
| Number of links used | 14.08 | 8.99 | 3.00 | 34.00 |
| Number of terms used | 11.15 | 4.40 | 4.00 | 18.00 |

(a), or whenever both experts do not construct links (c). The possibility of an agreement exists for (b) even though only one expert constructs a link. In this case, both the presence of one expert's link and the absence of another expert's link may determine the participant's link to be scored as incorrect. This type of agreement is not as strong as the case where agreement is reached either by the presence or absence of links in both experts' maps.

In Table 12.13, both types of agreement are reported. The number of links in each participant map represents the total number of comparisons made for that particular map. Exact agreement is defined as the number of agreements divided by the total number of comparisons. The interexpert agreement results using expert maps as criteria to score participant maps are presented in Table 12.13 for all pairwise combinations of experts. The exact agreement percentages are reasonably high (in the mid 80% range). However, it should be noted that if the more strict definition of agreement is used—that is, when experts are considered to disagree when only one of a pair constructs a link—then, even though the participant's link would be scored 0 by both experts, the exact percentages of agreement would range from 71% to 80%.

TABLE 12.13
Interexpert Pairwise Agreement Percentages on Group Map Scores

| *Rater pair* | | *Strict* | *Exact* |
|---|---|---|---|
| 2 | 1 | 79 | 83 |
| 3 | 1 | 76 | 86 |
| 3 | 2 | 74 | 85 |
| 4 | 1 | 71 | 80 |
| 4 | 2 | 80 | 87 |
| 4 | 3 | 77 | 86 |

*Note.* $n = 39$.

***Team Process Measures.*** Team-level process measures were computed by aggregating the three team members' scores on each team process. Means and standard deviations for each team process are provided in Table 12.14.

***Tests of Hypothesis.*** Participants' team process will relate to their teamwork performance (outcome).

As indicated in Table 12.15, only the coordination process was significantly related to outcome measures; unexpectedly, it was negative. In other words, the more coordination process teams engaged in, the lower the content and structure scores they attained. The reader may recall that only the coordination scale had acceptable reliability. Communication process was also significantly related to the structure score and the number of terms used in constructing the knowledge map.

Compared with our previous study using knowledge maps (Chung et al., in press), both studies showed that the networked team processes tended to have a negative correlation with outcome measures, although few of

TABLE 12.14
Descriptive Statistics for Team-Level Process Scales for the Networked Knowledge-Mapping Task

| | *Mean* | *SD* | *Minimum* | *Maximum* |
|---|---|---|---|---|
| Adaptability | .18 | .12 | .00 | .47 |
| Coordination | .18 | .20 | .00 | .94 |
| Decision making | .15 | .16 | .00 | .81 |
| Interpersonal | .18 | .14 | .00 | .47 |
| Leadership | .18 | .11 | .04 | .43 |
| Communication | .90 | .49 | .31 | 2.23 |

TABLE 12.15
Relationship Between Team Processes and Teamwork Outcome Measures for the Networked Knowledge-Mapping Task

| | *Content Score* | *Structure Score* | *Number of Terms Used* | *Number of Links Used* |
|---|---|---|---|---|
| Adaptability | -.02 | -.15 | -.13 | -.08 |
| Coordination | -.47** | -.45** | -.47** | -.47** |
| Decision making | -.22 | -.15 | -.19 | -.09 |
| Interpersonal | -.11 | -.10 | -.15 | -.18 |
| Leadership | -.11 | -.07 | -.02 | .00 |
| Communication | -.31 | -.33** | -.33** | -.29 |

*$p < .05$. **$p < .01$.

TABLE 12.16
Relationship Between Revised Scales Team Processes for the Networked Knowledge-Mapping Task

| | *Total* | *Cognitive Teamwork* | *Interpersonal Teamwork Number of Links Used* |
|---|---|---|---|
| Adaptability | -.16 | -.22 | .12 |
| Coordination | -.07 | -.02 | -.20 |
| Decision making | .29 | .30 | .06 |
| Interpersonal | -.01 | .02 | -.11 |
| Leadership | .11 | .01 | .35* |
| Communication | .04 | .04 | .01 |

$^{*}p < .05$.

them were significant. However, although this study indicated that coordination correlated significantly and negatively with content scores, our previous study indicated that communication significantly and positively correlated with content scores.

In general, Study 2 results indicated that team processes and the revised scales were not significantly related to outcome measures (e.g., content score; see Table 12.16). A possible explanation for these findings is that our task was highly knowledge dependent, and that as a result, participants who lacked the requisite knowledge were unable to engage each other at a substantive level. In other words, knowledge-mapping performance was more likely determined by the knowledge that teams had, rather than by their teamwork skills or team processes. Therefore, it was hard to find a relationship between knowledge-mapping performance and teamwork skills for this study. Further, the reliability of the team process measures was low. In addition, our sample size was relatively small.

## DISCUSSION

The purpose of the studies reported here was to measure team competencies of adults using our computer-based, networked teamwork assessment system. We were interested in how our teamwork measurement system performed with a diverse adult population and how the current results, specifically the relationships between team processes and team performance, compared with our prior work with middle and high school students. The tasks used in Study 1 and Study 2 were based on a common theoretical framework. The first study examined the team processes and performance of teams of local union–industry training program members and temporary agency workers negotiating a contract with a (simulated) management team. The second study examined the team processes and performance of teams of temporary agency workers jointly constructing a knowledge map on environmental science.

### Team Performance

In general, the performance results of the teams in our current set of studies were slightly different from the results in our previous studies, although not substantially. For the union–management negotiation task, our previous sample (see O'Neil, Chung, & Brown, 1997) performed highest on the main outcome measure—the value of the negotiated deal ($M$ = 121.7, $SD$ = 21.1)—followed by the union sample ($M$ = 111.3, $SD$ = 39.0) and the temporary agency sample ($M$ = 105.4, $SD$ = 36.8). Although the differences between the samples are not large, it is noteworthy to point them out. In our previous study, participants were drawn from a 10th-grade AP history class and a 12th-grade low-achievement class. The union sample was drawn from adult training classes, and the union participants were currently undergoing contract negotiations. The adult temporary agency workers were probably the furthest removed from the classroom setting. Although the temporary agency participants performed the lowest, the differences between all the samples do not appear substantial given the large standard deviations.

For the collaborative knowledge-mapping task, performance on all the knowledge-mapping outcome measures of the temporary agency adult sample was very similar to the performance of the pretest group of middle school students reported in Chung et al. (in press; mean = 19.3% correct). This result is not surprising and highlights the sensitivity and stability of the collaborative knowledge-mapping task to differences in prior content knowledge. Both the temporary agency worker adult sample and the pretest middle school student sample were presumed to have little prior knowledge in environmental science.

Thus, from the perspective of measurement of teamwork outcomes, the results for Study 1 and Study 2 suggest two interesting findings. First, the negotiation task seems to be more domain independent. Our data for three diverse samples show similar team performance. The finding that there was not a larger difference between the temporary agency worker sample and the other samples supports this claim, given that this sample was the most heterogeneous in terms of educational background, occupation, and age.

The second interesting finding was that the sensitivity of the collaborative knowledge-mapping task appears to be more domain dependent. Given that knowledge mapping is knowledge driven, this seems unsurprising. What is interesting is that this finding holds in a group context.

Although there were other differences between the sample in the present study (adults) and in our previous study (middle school students), our interpretation of the data is that the difference that mattered the most was the lack of knowledge about environmental science.

### Team Processes

Although the results for our team outcome measures appear reasonable and consistent in general with our prior results, the results for the relationship between team processes and team performance are less clear. In general, in Study 1 the correlations for the union and temporary agency samples differ for each team process in terms of magnitude. Further, these correlations differ from our prior work, although overall, the correlations for the union sample are more similar to our prior work.

When the correlations between teamwork processes and the main performance variables of the current studies are compared to our prior work, the pattern of correlations differs. In Study 1, the negotiation task, the union sample had correlations that were of lower magnitude but in the same direction. The temporary agency sample had correlations that were, in general, opposite in direction. One reason the pattern of correlations may be so diverse is that the samples were very different among the three studies. The union sample was comprised almost entirely of apprentices of the same trade. The temporary agency participants were sampled to vary in terms of age, gender, educational background, and occupational history. The high school classes sampled in our original study were comprised of AP history students and low-achieving students. Thus, it is not surprising that the alpha reliabilities and the correlations were found to vary so much.

Similarly, for Study 2, the correlations between team processes and the knowledge-mapping outcome measures are different from our prior work. In general, the difference is in the magnitudes of the correlations. As with our prior work, the correlations in Study 2 are in general negative. Overall, the more participants sent messages, the lower they performed on the knowledge-mapping task.

Although there are no clear explanations for the inconsistencies in the teamwork process analyses, we identified some contributing factors. First, the reliabilities of the teamwork process scales are poor for both Study 1 and Study 2. Cronbach's alpha for both samples was in the range of .3 to .6, which may have affected the correlations between team processes and team outcome measures. Poorer alpha reliabilities were obtained in Study 2, where the alpha ranged from .2 to .7. In this case, only coordination had an acceptable alpha of .7.

Another possible reason for the inconsistent correlations, which is related to the poor reliabilities, is impact of the interface on the task. As suggested by Chung et al. (in press), the user-interface of the collaborative knowledge-mapping task may have imposed too large a cognitive load on participants (see also Sweller, 1994). This may have contributed to the variation among participants in their selection of messages, contributing to the poor reliabilities and inconsistent correlations. Support for this explanation is seen in a comparison of the alpha reliabilities found in Study 1 (union–management negotiation task) and Study 2 (knowledge-mapping task). The user-interface of the negotiation task was much simpler and imposed a lower cognitive load on participants. Participants in the negotiation task needed only to focus on the message sheet and the messages sent by other team members. Further, the alpha reliabilities for the teamwork processes in the knowledge-mapping task were lower compared to the negotiation task.

Our results—relatively stable performance but unstable processes—raise an interesting issue warranting further investigation. Clearly, teams from different samples were able to achieve similar levels of performance. There is little doubt that participants were able to use the system, communicate with each other, and complete the tasks. There was apparently large variation in the way teams within a sample responded to the tasks, the user-interface, or both. What this suggests is that there are many paths to completing a group task, and that our measures may be too gross or too insensitive to these variations to measure them. These results also suggest that the interaction among individuals within a team and across task types is far more complex and less well understood than first thought.

From an assessment perspective, this offers an exciting and unique opportunity to refine our measures of teamwork processes. For the first time, we have sufficient data to assert that we have computer-based tasks that have (a) relatively stable performance across different samples, and (b) domain-independent (negotiation) and domain-dependent (knowledge-mapping) characteristics. Our next step is to disentangle the complex interaction between team tasks, user-interface design, knowledge requirements, and communication modes.

## ACKNOWLEDGMENTS

This work was supported in part by contract #85300-97-0056 from Statistics Canada, Government of Canada, to the National Center for Research on Evaluation, Standards, and Student Testing, University of California, Los Angeles, and in part under the Educational Research and Development Centers Program, PR/Award Number R305B60002, as administered by the Office of Educational Research and Improvement, U.S. Department of Education.

The findings and opinions expressed in this report do not reflect the positions or policies of Statistics Canada or the Government of Canada, nor do the findings and opinions expressed in this report reflect the positions or policies of the National Institute on Student Achievement, Curriculum, and Assessment, the Office of Educational Research and Improvement, or the U.S. Department of Education.

## REFERENCES

Burke, C. S., Volpe, C., Cannon-Bowers, J. A., & Salas, E. (1993, March). *So what is teamwork anyway? A synthesis of the team process literature*. Paper presented at the 39th annual meeting of the Southeastern Psychological Association, Atlanta, GA.

Chung, G. K. W. K., O'Neil, H. F. Jr., & Herl, H. (in press). Reliability and validity of a web-based assessment of collaborative problem solving. *Computers in Human Behavior*.

Dansereau, D. F. (1995). Derived structural schemas and the transfer of knowledge. In A. McKeough, J. Lupart, & A. Marini (Eds.), *Teaching for transfer: Fostering generalization in learning* (pp. 93–121). Mahwah, NJ: Lawrence Erlbaum Associates.

Fletcher, D. (in press). Use of distributed interactive simulation for assessment of team tactical skills (Problem solving in simulated combat) [Special Issue]. *Computers in Human Behavior*.

Herl, H. E., Baker, E. L., & Niemi, D. (1996). Construct validation of an approach to modeling cognitive structure of U.S. history knowledge. *Journal of Educational Research, 89*, 206–219.

Herl, H. E., O'Neil, H. F. Jr., Schacter, J., & Chung, G. K. W. K. (1998). *Assessment of CAETI STS1 technologies* (Deliverable to ISX). Los Angeles: University of California, National Center for Research on Evaluation, Standards, and Student Testing.

Jonassen, D. H., Beissner, K., & Yacci, M. (1993). *Structural knowledge: Techniques for representing, conveying, and acquiring structural knowledge*. Hillsdale, NJ: Lawrence Erlbaum Associates.

Morgan, B. B., Jr., Salas, E., & Glickman, A. S. (1993). An analysis of team evolution and maturation. *Journal of General Psychology, 120*, 277–291.

O'Neil, H. F. Jr. (Ed.). (1997). *Workforce readiness: Competencies and assessment*. Mahwah, NJ: Lawrence Erlbaum Associates.

O'Neil, H. F. Jr., Allred, K., & Dennis, R. A. (1997a). Use of computer simulation for assessing the interpersonal skill of negotiation. In H. F. O'Neil, Jr. (Ed.), *Workforce readiness: Competencies and assessment* (pp. 205–228). Mahwah, NJ: Lawrence Erlbaum Associates.

O'Neil, H. F. Jr., Allred, K., & Dennis, R. A. (1997b). Validation of a computer simulation for assessment of interpersonal skills. In H. F. O'Neil, Jr. (Ed.), *Workforce readiness: Competencies and assessment* (pp. 229–254). Mahwah, NJ: Lawrence Erlbaum Associates.

O'Neil, H. F. Jr., Baker, E. L., & Kazlauskas, E. J. (1992). Assessment of team performance. In R. W. Swezey & E. Salas (Eds.), *Teams: Their training and performance* (pp. 153–175). Norwood, NJ: Ablex.

O'Neil, H. F. Jr., Brown, R., & Chung, G. K. W. K. (1996, April). *Using dichotomous factor analysis to measure teamwork skills*. Paper presented at the annual meeting of the American Educational Research Association, New York.

O'Neil, H. F. Jr., Chung, G. K. W. K., & Brown, R. (1997). Use of networked simulations as a context to measure team competencies. In H. F. O'Neil, Jr. (Ed.), *Workforce readiness: Competencies and assessment* (pp. 411–452). Mahwah, NJ: Lawrence Erlbaum Associates.

O'Neil, H. F. Jr., Wang, S., Chung, G. K. W. K., & Herl, H. (1998). *Final report for validation of teamwork skills questionnaire using computer-based teamwork simulations*. Los Angeles: University of California, National Center for Research on Evaluation, Standards, and Student Testing.

Salas, E., Dickinson, T. L., Converse, S. A., & Tannenbaum, S. I. (1992). Toward an understanding of team performance and training. In R. W. Swezey & E. Salas (Eds.), *Teams: Their training and performance* (pp. 3–30). Norwood, NJ: Ablex.

Sweller, J. (1994). Cognitive load theory, learning difficulty, and instructional design. *Learning and Instruction, 4*, 295–312.

Webb, N. (1993). *Collaborative group versus individual assessment in mathematics: Processes and outcomes* (CSE Tech. Rep. No. 352). Los Angeles: University of California, National Center for Research on Evaluation, Standards, and Student Testing.

Webb, N., & Palincsar, A. S. (1996). Group processes in the classroom. In D. C. Berliner & R. Calfee (Eds.), *Handbook of educational psychology* (pp. 841–873). New York: Prentice Hall.

Chapter 13

# Assessment of Visual Requirements for Flight Simulation

Keith K. Niall
*Defence and Civil Institute of Environmental Medicine, Toronto, Canada*

Byron Pierce
*Air Force Research Laboratory, Mesa, Arizona*

The assessment of visual requirements for flight simulation is a study both sophisticated and fraught with complexity. It is commonplace for authors to declare the immensity or the richness of their subject, but vision is an old topic where much has been forged already, and where much remains to be accomplished (see Warner & Casey, 1995). Some issues in vision are deceptive in their apparent simplicity: a favorite anecdote is that Marvin Minsky once assigned the "problem of vision" as an exercise to a summer graduate school in engineering at M.I.T., fully expecting that his students would solve the problem by term's end. Needless to say, they did not. A lot of effort can be saved by learning to *think* about vision clearly, by learning to gauge the scope of particular problems in vision research as no larger and no smaller than they are. "For in psychology there are experimental methods and *conceptual confusion*. The existence of the experimental method makes us think we have the means of solving the problems which trouble us; though problem and method pass one another by" (Wittgenstein, 1958, p. 232e).

There is a dilemma characteristic to the assessment of visual requirements for flight simulation. One must decide whether to provide either complete conditions or else necessary conditions for visually guided performance on a task. That is, one can decide to reproduce exactly the changing pattern of light that reaches an aviator's eyes during flight, or else one can decide to reproduce only those aspects or properties of the visual scene that can be shown to be crucial to performance. On first consideration this may seem to weigh realism against expediency. There is a great deal of cost and care involved in promoting optical rigor in displays for flight simulation. Indeed, one of our goals is to reproduce the changing pattern of light presented to an aviator—to ensure that the illuminated scene presented by the simulator is indistinguishable from what might be experienced in flight. Yet it is worthwhile to remember this is an ideal. Usually we express the point by saying that if more money were available for engineering development, or if our graphics computers were only much faster, then realism would improve.

This emphasis on complete realism forestalls another legitimate concern: that we don't need to reproduce everything an aviator sees in the cockpit. That is because another of our goals is to determine which visual properties or aspects of the simulation are necessary to the task at hand, for those tasks that are made difficult by their dependence on vision (e.g., Heikens, Gray, O'Neill, & Salisbury, 1999). From this knowledge, we may say in detail how it is the aviator accomplishes the task, that is, in a refueling task by attending to the rivets of the companion plane, or in a search task by the systematic control of eye movements in raster pattern. Our second goal, then, is to appreciate the role of learning and adaptation in aircrew training insofar as it is guided by vision. Incomplete simulations can be used to great training advantage, because aviators adapt with marked elasticity to many novel or impoverished conditions of illumination. Moreover, observers learn to interpret new kinds of displays with alacrity: night vision goggles (Rabin, 1993), infrared cameras, and other devices present visible information that is useful in identification tasks, but where the characteristic reflectances of familiar objects may be different than under ordinary daylight conditions of observation. One goal of flight simulation is to provide complete optical fidelity, but some other goals are to understand which aspects of a simulation are essential for a task, and to extend the sensory facility of the aviator by presenting information from new domains, or by presenting information in new ways.

## METRICS

The search for appropriate metrics is a recurring theme in the assessment of visual requirements. A *metric* is loosely characterized as a scheme or a strategy for measurement: mathematics is the source of these schemes or strategies. Then the search for a metric consists in the application of appropriate mathematics to the domain of our concern. The establishment—that is, the choice and the publicly enacted convention—of a metric enables us to answer how much and how well a flight simulation succeeds. How much, in that an appropriate metric taken from optics can be applied to gauge the completeness of optical fidelity that is achieved in the simulation (e.g., Beall & Loomis, 1997). How well, in that enumeration and measurement can be made of the optical properties that are known to play a part in performance of a task. Indices of performance can be compared to the number and measure of these properties, as well. Three temptations should be avoided in the search for a metric; these temptations are easy to describe and difficult to avoid. The first is the temptation to pursue mathematics for mathematics' sake: an application ought to be enlightening at least to those with a general education in mathematics, if not always enlightening in practise to many who lack such an education. The second is the temptation to use a wrench for a hammer. The astute choice of a metric is an act of ingenuity or creativity: no recipe book will help to guide any truly novel application. The third, most dangerous temptation is quietism or cynicism in the search for a relevant metric. To put the matter vividly, trying to specify visual requirements without an appropriate metric is like trying to dig a hole in the earth with one's bare hands: an ill-considered waste of time. Mathematics is the language of science, and understanding is not given to those who refuse even to communicate.

## PROPERTIES OF LIGHT, LAND, AND SKY

A primary demand for the visual display of a flight simulator is that it provide adequate resolution. What drives this requirement of resolution? There are specific tasks that demand a high-resolution display for their simulation: an example is identification of the aspect angle of an aircraft in the middle distance (in terms of kilometers). One could say that high-resolution displays are required because our eyes are capable of resolving very small subtenses of visual angle, but that begs the question. The important question is what there may be in the environment that requires an observer to resolve small differences.

There are two aspects of the air environment that determine the requirement for resolution, but that are so obvious as to be ignored: land and sky. That is, one determining factor is the difficulty of accurately depicting terrain complexity (see Kleiss, 1996), and another factor is the modeling of scattering and absorption characteristics of the atmosphere. Depiction of the terrain may be categorized as: (a) the depiction of landmarks such as cultural features and gross geographical forms, and (b) the depiction of texture, including local geology, vegetation, and coloring of the soil or surface. But such a distinction is hardly adequate: first, it makes *texture* a heading on the same footing as *miscellaneous*—texture may then include anything at all. Second, the distinction is blurred: what was texture a moment ago may become a vitally important feature when size scale changes as one continues a flight path. There are several ways in which observers use terrain: in way-finding, in visual search, and in the intuitive judgment of speed, orientation, and heading by the relative motion of elements in the scene due to perspective (i.e., by what has come to be known as *optic flow*: Cutting, Vishton, Flückiger, Baumberger, & Gerndt, 1997). Even these several uses of detail in the landscape will mandate that depiction of the terrain be as finely-grained as is possible with current display technology—and we do not yet know how to count all the uses an aviator makes of changes in terrain.

How then do characteristics of the atmosphere determine resolution requirements? The scattering of light by aerosol particles, and absorption contribute to an overall attenuation of light in the atmosphere, that is, they contribute to the reduced transmittance of the air. In a pure, dry atmosphere (this ideal is known as a Rayleigh atmosphere) that consists of a mixture of gases without dust, a black object subtending a small visual angle against a bright sky could have a visible range of more than 350 kilometers along the observer's line of sight. (Middleton, 1952, pp. 18 and 104). However, this capability does not contribute to the requirement of resolution in a visual display as much as does a photometric relation: the cosine law of illumination (Smith & Atchison, 1997, p. 277). As an observer's line of sight draws parallel to a flat surface, as occurs when flaring to land on a runway, the visual range of objects and textures that are seen on the flat surface changes dramatically. The measure of the solid angle they subtend changes with the cosine of the angle the line of sight makes to the flat surface. Features on the runway that appear well spaced at one moment are crowded together by the effect of scene compression in the next moment (the magnitude of this compression is given by the cosine law: Domini & Caudek, 1999; Niall, 1997). In flight simulation, the effect of compression of detail poses a requirement for increased resolution, and this effect becomes crucial during approach and during takeoff. Then the transmittance of light in air and the cosine law of illumination can be considered characteristics of the atmosphere that determine resolution requirements in flight simulation.

How do we quantify the resolution that we require in a flight simulator? There is a hack solution to this problem: that the device be able to resolve a pair of lines that are separated by two arc-minutes of visual angle, for an observer at a comfortable distance from the display screen. That number is meant to represent the separation in visual angle between a pair of lines that can just be resolved by the human eye: it is supposed to represent an "eye-limiting resolution" of film quality. Yet the specific forms that are presented to an observer influences the minimum angular subtense that the observer can resolve: in a task called vernier acuity (Fahle & Harris, 1998), line offsets can be distinguished by eye at angular subtenses of a fraction of an arc-minute—less than the postulated minimum resolution. The measurement of resolution may seem to pose a problem, because in flight we are often concerned with distance and with orientation, and we are accustomed to measuring in those terms. Measures of distance and orientation do not suffice to gauge resolution, because those geometric properties are not preserved in the propagation of light by which an image is formed. Measure of visual angle and the cross-section of solid angle are somewhat better: in a display image, these measures change in a regular manner as an object of fixed size is depicted to recede in distance from the observer. Another family of measures may render greater service in the future. It has been suggested that measures that do not change with the distance and orientation of objects in the environment can be used to gauge how closely observers resolve a variety of forms. This family of measures is outlined in Mundy and Zisserman (1992; also Niall, 1999).

## PHYSIOLOGICAL CONSTRAINTS

Not only the medium of propagation of light, but also the constitution of the visual system itself influences the perception of properties and surfaces we may wish to display in a flight simulator. Often the many contributions of the visual system are construed as a succession of effects, a causal chain that ends with sight; yet such a picture is misleading. Those contributions are better conceived as posing constraints or bounding conditions on sight: that is, as being impediments to sight. Some of these bounding conditions are familiar in terms of visual requirements for acuity: near-sightedness and far-sightedness can be understood as contributions of refraction in the resting state of the lens, whereas presbyopia can be understood as the loss of ability to change the accommodative state of the lens. It may also be remembered that the greater contribution to refraction in the eye is made by the cornea, not the lens (such components do interact: Valluri et al., 1999). Astigmatism is the condition in which the shape of the cornea departs appreciably from spherical. Other bounding conditions on sight are more subtle. Aircrew selection standards in many countries mention both visual acuity (usually in terms of discrimination on the standard Snellen or Landolt charts) and ability to discriminate colors in terms of hue. The utility of both these standards for aviation is in dispute. Changes in visual acuity with age are widely familiar (Fahle & Daum, 1997; Kikukawa, Yagura, & Akamatsu, 1999) and a source of chagrin to pilots in their regularly scheduled medical examinations. If the influence of the visual system on the perception of form and color is misconstrued as a chain of causes, then these constraints that lens and cornea impose on vision will be understood as only the first links in an indefinitely long series of causes, each of which may be considered for its contribution to visual requirements.

Yet the story ends sooner than that. Such "neural" factors distinct from the said optical constraints are seldom if ever elaborated independently. All these factors may and should be filed under the heading of *acuity* instead. The term *acuity* has been appropriated as a rubric for a variety of concerns, where finally acuity indicates success or degree of failure in discriminating form or discriminating color. Standards of performance in discrimination are amenable to better treatment, as when they are elevated into the useful construct known as the *standard observer*. Several versions of a normal observer have been specified for the discrimination of color (Trezona, 1998), although similar conventions have not yet been established for the visual discrimination of form (hence our continued reliance on eye charts).

## DISPLAY PROPERTIES

The optical, mechanical, and electronic characteristics of the components of flight simulators may also constrain the performance of vision-based tasks. Naturally some issues concerning specific devices quickly become yesterday's news, as one technology or corporation is superseded by another. Some issues do recur: the angular placement of a seam or an occluder between adjacent panels of a large display can interfere with the performance of a specific task (Bentz, 1980: e.g., in air-to-air refueling, or in a formation manoeuver). Eternal compromise seems to mark other issues: as much as practical, the field of view presented by a flight simulator should be unrestricted. An unrestricted field of view has never been either convenient or practical, if unrestricted is taken to mean the presentation of a panoramic view, a complete wrap-around scene that includes a significant vertical extent and that enables the observer to turn around and view the scene behind. A large field of view has been a difficult requirement to satisfy, partly because it conflicts with another requirement that all parts of a visual display should be presented at maximum resolution. But, from the time of this writing, it was not terribly long ago that the rendering of figures in perspective and the development of a hidden-surface algorithm were big news in graphics.

The display systems that are used in contemporary flight simulators have two main components: the image generator, and the associated image source, the latter including software, database, and display. An image generator can be any process or system by which a computer produces an image. The cost of these devices ranges from several thousands of dollars for a personal computer to well over a million dollars for the latest and greatest of high-end devices. The choice of an image generator involves two concerns: the update rate (the speed with which an image can be computed) and its database (the complexity of the scene to be computed). Several studies have been made of the requirements for image generator update rate and scene complexity in flight simulation. A basic result is that an image generator used in flight simulation must compute and display imagery at maximal complexity and at high update rate. In the simulation of flight at high speeds and low altitudes, a minimum update rate of 60 frames per second is indicated; this generally rules out the use of low-end image generator devices. Most image generator devices that meet this requirement combine a general computing processor with a specialized, high-speed graphics processor. In the simulation of other flight tasks, a lower update rate can sometimes be used.

There are a number of factors that mandate image generator capacity. Some considerations are:

1. the number of polygons that the graphics system will compute each second
2. the number of point-light sources to be displayed in a second
3. the number of picture elements (pixels) the system generates
4. the interlacing of the display (interlaced systems compute every other horizontal line during an update; noninterlaced systems compute every line)
5. database management considerations
6. the scheme by which the level of textural detail is altered as an observer approaches a surface
7. contingencies for processor overload when more polygons are specified to be written than can be written in real-time
8. the hidden-line or occultation algorithm that is used
9. smoothing or anti-aliasing techniques
10. the application of computer-generated or photorealistic texture to single polygons
11. the requirement for the database to be correlated with other databases in networked simulations or sensor simulations.

The image computed by the image generator should be presented to the observer with sufficient brightness, contrast, resolution, field of view, and with minimal geometric distortion. This is the job of the image source. Differences in image quality also depend on their characteristics in terms of timing accuracy and positional accuracy. Primary technologies include cathode-ray tubes (CRTs), back-lit liquid crystal displays (LCDs), light-valve liquid crystal displays, and laser-based technologies. For simulator displays with a large field of view, these technologies are paired with projection optics, as opposed to being viewed directly.

CRTs have a single electron gun that scans an image sequentially on a screen, in either an interlaced or noninterlaced manner. Current CRTs can address an image of up to 2500 × 2000 pixels. However, greater addressability does not always mean improved image quality. Back-lit liquid crystal displays project light through a liquid crystal panel. Like CRTs, typical liquid crystal displays execute a sequential scan, although this is no constraint: it is the characteristic of many image generators that are used with liquid crystal displays. The persistence of liquid crystal display elements in time is far greater than that of most CRT phosphors: this greater persistence induces a blurring or smearing of moving images. Advances are being made in liquid crystal display switching speed, and in the number of elements that can be addressed at once in a liquid crystal display. A negative characteristic of most liquid crystal display panels is the visible grid that is formed by the seams or interstices of adjacent pixels. The effect has been described as similar to looking through a mesh of wires, like a screen door. Liquid crystal *light valves* use a CRT coupled to a liquid crystal layer. A high-intensity light source is sent to one side of the liquid crystal layer. There the light is modulated, then reflected by a mirror through a projection lens and onto a screen. Since a CRT does the image scanning for a light-valve, the image is scanned in raster pattern.

Liquid crystal light valves have a high light output and exhibit no "screen-door" effect. Laser-based technologies incorporate new solid-state eye-safe laser light sources: no display phosphors or liquid crystal elements are involved. In laser-based systems, laser light is shaped into many beamlets, then modulated, scanned, and projected onto a viewing screen. Such devices may be able to provide very high resolution, increased contrast, greater brightness, and improved color gamut in dynamic imagery.

There is such a variety of large-format flight simulator displays that they are difficult to classify. A start might be to distinguish screen displays from refractive or reflective displays. (*Refractive* displays bend light a little, or transmit it directly; *reflective* displays reverse the direction of light transmission.) Refractive or reflective displays can be further distinguished as pupil-forming or nonpupil-forming. With pupil-forming displays, there is a definite volume of air within which an observer's eyes must be positioned. This volume lies about an axis of projection. Nonpupil-forming displays allow the observer to move a reasonable distance off-axis and still observe the projected image in focus. Decisions about displays are made on the basis of eye relief, field-of-view, the range of an observer's head movements, light transmission, optical distortion, contrast rendition, and resolution.

One solution to the continuing compromise about resolution is the development of area-of-interest simulations. There, a small portion of the display is presented at fine resolution, while the graphics for the remainder of the screen are rendered more closely. This may be thought to mimic the changes in acuity that occur from foveal vision to peripheral vision (Anstis, 1998), though there are other better descriptions of the difference between central and peripheral vision (Näsänen & O'Leary, 1998). Area-of-interest displays may be presented in different ways, as well. Certain aspects of the display may have high intrinsic value and so be chosen to be rendered in greater detail (one may value incoming aircraft over more cirrus clouds) or the area of enhanced resolution may be slaved to the direction in which an observer is looking. This may be as simple as turning off the parts of the display behind the observer's head, or it may represent a sophisticated and seamless blending of a central area with the larger background, where the central area glides over the display following the direction of gaze. In that case, the area to be rendered in detail must be calculated from eye- and head-tracking devices, in real-time.

Other contemporary issues emerge from the massive computation used to present visual imagery in flight simulators. (Of course, "massive" is a relative term, if you are used to modeling the evolution of galactic clusters, or suchlike.) The image quality of a flight simulator display can be affected by the manner and sequence in which pixels are written (that is: refreshed) on the display screen. The standard method of pixel-writing follows a raster pattern: pixels are written left to right in the top row of the screen, then left to right in the row next to top, and so on until the entire screen has been refreshed. It is possible to engineer multiple-line displays as well: one of these follows a pattern in which all rows of the display are written at the same time, beginning with the leftmost column of pixels and continuing to the rightmost column.

In all these methods, each pixel of the display screen is refreshed completely before any pixel is overwritten: each refresh cycle of an entire screen is called a frame. The frame rate of a display is known to affect its image quality. A frame rate of 60 Hz (60 Hertz is 60 frames per second) is generally accepted and used in North America for both simulation and broadcast. This frame rate has one advantage: it is well above the average critical flicker fusion frequency of the human eye, so that successive frames appear as a continuously changing picture rather than as a flashing series of still pictures. This effect has distinct consequences for the perception of raster-scanned displays. (Thomson & Saunders, 1997). Further increase in frame rate—from 60 Hz at least to 120 Hz—has been shown to enhance image quality.

It may not be clear how else the pattern in which pixels are written can affect the perceived image quality of a display until one considers temporal changes on small parts of the display screen (Gorea & Hammett, 1998; Lindholm, 1992). Events that occur simultaneously should appear simultaneously on the display screen, but this common-sense relation is difficult to maintain when the pixels of the display are refreshed at different times, albeit very quickly. Suppose that a disk is pictured to move quickly across a screen, and not all the pixels of a frame are being refreshed at once. Because the disk moves, it takes on a different position in successive frames. Most often only part of any frame is present on a screen, since the pixels on the screen are being refreshed continuously between one frame and the next. The difference in position of the disk between frames may then appear as a "tearing" or jaggedness at the edges of the disk. Similarly, point sources of light may appear to be doubled. In this interaction between spatial position and motion in time, the pattern in which pixels are written can affect image quality greatly. Effects of asynchrony also appear in what are known as transport delays in simulation. A simulation is not simply a movie, to be viewed passively; observers in the simulator are also active participants. Yet there may be an unnatural lag between the actuation of a device in the simulator cockpit and the depicted effect of that control on the screen image. Such lags can be on the order of hundreds of milliseconds, and these lags have been cited as a primary cause of simulator-induced nausea and headaches (called *simulator sickness* on analogy with motion sickness, but see van der Steen, 1998).

How can all these effects on image quality be assessed? There does exist a variety of methods with which to begin. Given a standard set of patterns, we can measure how well those patterns have been displayed by the simulator (one can also think of variations of the display over time in this context). Or better, we can ask how well that standard set of patterns can be perceived by an observer in the simulator. There are a number of standard sets of patterns (called *basis sets*) with which to begin. A basis set should have the property that an arbitrary pattern on the screen can be reproduced as a combination of the patterns from the set. Several basis sets are in common use: fuzzy bars of different frequencies, whose luminance changes in sinusoidal fashion from left to right (one-dimensional Fourier gratings); black-and-white checkerboards of different sizes (two-dimensional Hadamard or Haar patterns); and fuzzy checkerboards of different sizes, whose luminance changes in sinusoidal fashion from left to right, and up to down (two-dimensional Fourier patterns). Any of these sets may be used (see Papoulis, 1968), though a two-dimensional set is to be preferred where two-dimensional images are in question.

When such a basis set of patterns is displayed, how well are the patterns transmitted at a given luminance, or how well can those patterns be detected? Each pattern in the set is transmitted or detected more or less well at one luminance: very high frequency patterns may not be detected at all in some cases. The function described by this quantitative assessment of transmission or detection is known as a *modulation transfer function* (Jennings & Charman, 1997; López-Gil, Iglesias, & Artal, 1998). It provides a descriptive and graphical illustration of changes in image quality. Image statistics can be calculated on the modulation transfer function or directly on changes in the patterns themselves. There is a particular statistic or figure of merit that is suited to the comparison of two-dimensional images: the crosscorrelation coefficient (Collet & Quinquis, 1994; Thomas, Hanson, & Oliensis, 1994). It provides a measure of the similarity of two-dimensional images. Together, these methods are a first step toward a larger field of inquiry (Young & Calvert, 1974)—the application of image-processing techniques to problems of image quality in flight simulation.

## TENTATIVE CONCLUSIONS

*Slowly we are learning*
*We at least know this much,*
*That we have to unlearn*
*Much that we were taught,*
*And are growing chary*
*Of emphatic dogmas;*
*[Sight] like Matter is much*
*Odder than we thought.*
*—Auden (1976, p. 207)*

One function of a flight simulator is to present the same pattern of light to an observer's eyes as would be present to those of an aviator during flight. Yet an important study of training effectiveness has a narrower focus. There we seek to know which parts of or variations in the visual display affect performance. Sometimes "the experimenter should not hope, and does not need, to display all the information in an ambient array. He is not trying to simulate reality. . . . What is required is only that the essential invariant be isolated and set forth" (Gibson, 1979, p. 305; cf. Startchik, Milanese, & Pun, 1998). The first goal is one of exact reproduction, whereas the second goal is one of analysis. We are far from realizing either goal. It would be futile to make predictions for the direction of future research. One of the characteristics of research, distinct from the applications of technology, is that in research we are widening the circle of our knowledge, and we simply do not know what lies beyond the circumference. According to one philosopher, the best way of identifying someone who has real genius is that we never discover exactly where it was they lost their way.

However, there are some things we can say about current directions of research: that the notion of resolution or acuity has been used as a grab-bag notion that it would be worthwhile to unpack, and explicate in detail; that a close and astute application of metrics is a welcome contribution; and that there has not been sufficient consideration of the "environmental" properties of light, including the structure of landscape terrain and the influence of the atmosphere on aerial scenes. These problems are recommended to the reader for a solution.

## REFERENCES


Anstis, S. (1998). Picturing peripheral acuity. *Perception, 27*, 817–825.

Auden, W. H. (1976). *Collected poems*. E. Mendelson (Ed.), New York: Random House.

Beall, A. C., & Loomis, J. M. (1997). Optic flow and visual analysis of the base-to-final turn. *International Journal of Aviation Psychology*, 7(3), 201–223.

Bentz, J. L. (1980). Joining techniques for optically combined visual display systems. *Proceedings of the Second Interservice/Industry Training Equipment Conference and Exhibition* (pp. 46–59). Arlington, VA: National Security Industrial Association.

Collet, C., & Quinquis, A. (1994). Motion estimation methods and noisy phenomena. *Machine Vision and Applications*, 7, 247–258.

Cutting, J. E., Vishton, P. M., Flückiger, M., Baumberger, B., & Gerndt, J. D. (1997). Heading and path information from retinal flow in naturalistic environments. *Perception & Psychophysics, 59*(3), 426–441.

Domini, F., & Caudek, C. (1999). Perceiving surface slant from deformation of optic flow. *Journal of Experimental Psychology: Human Perception and Performance, 25*(2), 426–444.

Fahle, M., & Daum, I. (1997). Visual learning and memory as functions of age. *Neuropsychologia, 35*(12), 1583–1589.

Fahle, M., & Harris, J. P. (1998). The use of different orientation cues in vernier acuity. *Perception & Psychophysics, 60*(3), 405–426.

Gibson, J. J. (1979). *The ecological approach to visual perception*. Boston, MA: Houghton Mifflin Company.

Gorea, A., & Hammett, S. T. (1998). Spatio-temporal vernier acuity. *Spatial Vision, 11*(3), 295–313.

Heikens, M. F., Gray, G. W., O'Neill, H. J., & Salisbury, D. A. (1999). *1997 Canadian Forces Air Operations Vision Survey. Section I: Operational Vision Requirements*. Toronto, Canada: Technical Report 1999-019 of the Defence and Civil Institute of Environmental Medicine, Department of National Defence.

Jennings, J. A. M., & Charman, W. N. (1997). Analytic approximation of the off-axis modulation transfer function of the eye. *Vision Research, 37*(6), 697–704.

Kikukawa, A., Yagura, S., & Akamatsu, T. (1999). A 25-year prospective study of visual acuity in the Japan Air Self Defense Force personnel. *Aviation, Space, and Environmental Medicine, 70*(5), 447–450.

Kleiss, J. A. (1996). Assessing the influence of operational factors on the perceived structure of real-world scenes viewed during low-altitude, high-speed flight. *International Journal of Aviation Psychology, 6*(4), 335–358.

Lindholm, J. M. (1992). Perceptual effects of spatiotemporal sampling. In M. A. Karim (Ed.), *Electro-optical displays* (pp. 787–807). New York: Marcel Dekker.

López-Gil, N., Iglesias, I., & Artal, P. (1998). Retinal image quality in the human eye as a function of accomodation. *Vision Research, 38*, 2897–2907.

Middleton, W. E. K. (1952). *Vision through the atmosphere*. Toronto, Ontario: University of Toronto Press.

Mundy, J. L., & Zisserman, A. (1992). *Geometric invariance in computer vision*. Cambridge, MA: MIT Press.

Näsänen, R., & O'Leary, C. (1998). Recognition of band-pass filtered hand-written numerals in foveal and peripheral vision. *Vision Research, 38*, 3691–3701.

Niall, K. K. (1997). "Mental rotation," pictured rotation, and tandem rotation in depth. *Acta Psychologica, 95*, 31–83.

Niall, K. K. (1999). Estimates of shape by eye: or, The little invariant that could. *Acta Psychologica, 100*, 291–320.

Papoulis, A. (1968). *Systems and transforms with applications in optics*. New York: McGraw-Hill.

Rabin, J. (1993). Vernier acuity through night vision goggles. *Optometry and Vision Science, 70*(8), 689–691.

Smith, G., & Atchison, D. A. (1997). *The eye and visual optical instruments*. Cambridge, England: Cambridge University Press.

Startchik, S., Milanese, R., & Pun, T. (1998). Projective and illumination invariant representation of disjoint shapes. *Image and Vision Computing, 16*, 713–723.

Thomas, J. I., Hanson, A., & Oliensis, J. (1994). Refining 3D reconstructions: A theoretical and experimental study of the effect of cross-correlations. *Computer Vision, Graphics, and Image Processing: Image Understanding, 60*(3), 359–370.

Thomson, W. D., & Saunders, J. (1997). The perception of flicker on raster-scanned displays. *Human Factors, 39*(1), 48–66.

Trezona, P. W. (1998). Theoretical aspects of mesopic photometry and their implication in data assessment and investigation planning. *Color Research and Application, 23*(5), 264–273.

Valluri, S., Minkovitz, J. B., Budak, K., Essary, L. R., Walker, R. S., Chansue, E., Cabrera, G. M., Koch, D. D., & Pepose, J. S. (1999). Comparative corneal topography and refractive variables in monozygotic and dizygotic twins. *American Journal of Ophthalmology, 127*(2), 158–163.

van der Steen, F. A. M. (1998). An earth-stationary perceived visual scene during roll and yaw motions in a flight simulator. *Journal of Vestibular Research, 8*(6), 411–425.

Warner, H. D., & Casey, E. P. (1995). *Flight simulator visual system research and development: Bibliography of support provided by the Aircrew Training Research Division* (Tech. Report AL/HR-TR-1995-0107). Mesa, Arizona: Air Force Research Laboratory, Warfighter Training Research Division.

Wittgenstein, L. (1958). *Philosophical investigations* (2nd ed.). Oxford: Basil Blackwell & Mott, Ltd.

Young, T. Y., & Calvert, T. W. (1974). *Classification, estimation, and pattern recognition*. New York: American Elsevier.

Chapter 14

# Benefit–Cost and Cost-Effectiveness Systems Analyses for Evaluating Simulation for Aircrew Training Systems

William C. Moor
*Arizona State University, Tempe, AZ*

Dee H. Andrews
Robert T. Nullmeyer
*Air Force Research Laboratory, Mesa, AZ*

Burke Burright
*Air Force Research Laboratory, Brooks AFB, TX*

The aircrew training community has a long tradition in evaluating training systems. It takes the scientific experiment as the appropriate paradigms for evaluation and focuses on very disaggregate systems. Training researchers in this tradition sometimes addressed only the impact of a single change in a single course (Sassone, 1985; Kamsickas, 1991).

Unfortunately, methods for evaluating training effectiveness developed at this microlevel do not scale up easily to large, expensive training systems with complex interactions. They have not proved useful for evaluating extensive and fundamental changes in larger, complex training systems (Kamsickas, 1991; Fast, Dunson, Wortman, Stone, & Engquist, 1992). Moreover, these traditional methods were developed to evaluate training processes that were in place. So they do not provide much information pertinent to deciding whether to make major investments in new training systems.

Benefit–cost analysis and cost-effectiveness analyses are methods used frequently to evaluate investment projects proposed by government agencies. They provide evaluation paradigms that overcome the limitations of the one traditionally used by aircrew training researchers. This chapter delineates the general steps involved in a benefit–cost analysis and provides an example for a large, complex aircrew training system. Then, it goes on to discuss when cost-effectiveness provides a more appropriate paradigm and gives an example of its application to another aircrew training system.

## BENEFIT–COST ANALYSIS

Benefit–cost analysis is based on the ideas of quantifying benefits, to whomsoever they accrue, provided by each alternative, and of expressing them in dollar values. First, we summarize the general steps involved in doing a benefit–cost analysis for any kind of governmental investment (Fabrycky, Thuesen, & Verma, 1998; Hanley & Spash, 1993; Londero, 1996; Maciariello, 1975; Oxenfeldt, 1979; Park, 1993; Pearce, 1983). Then, we describe how one of us (Moor) implemented those steps in an analysis of alternatives for enhancing pilot training.

### Overview of the Process

The process of doing a benefit–cost analysis involves a sequence of steps.

***Step 1: Specifying the Goals.*** This step involves specifying the functional goals that the alternatives must achieve. A benefit–cost analysis should provide a framework for comparing any alternative that could achieve the given goals. It is up to the analyst to construct a set of alternatives that meet the goals.

***Step 2: Defining the Alternatives.*** This step involves conceptualizing the alternatives that are proposed as ways for reaching these goals. The analyst must lay out each alternative in sufficient detail to allow himself to estimate its impacts. He should construct each alternative so that it is mutually exclusive with respect to each other alternative. However, each alternative may include one or more subcomponents that may not be exclusive between alternatives.

***Step 3: Identifying the Impacts of Each Alternative.*** This step involves identifying and estimating the impacts that each alternative would have. Although each alternative should have the same functional goals, each would achieve them at different levels. The analyst must find credible methods for estimating such differential effects.

***Step 4: Quantifying the Benefits.*** Benefits include "all the positive effects (less the negative benefits) resulting from the realization of the project, regardless to whom they fall" (Hanspeter, 1973, p. 19). Each benefit must be defined in such a way that its nature, magnitude, and timing are understood; that the groups receiving the benefits are clearly known; and that the calculations for estimating the benefits in dollars are clear stated. There are no agreed-on methods of constructing metrics for evaluating benefits (Canada, Sullivan, & White, 1996; DeGarmo, Sullivan, Bontadelli, & Wicks, 1997; Maciariello, 1975; McDonald, Broeder, & Cutak, 1989; Smith, 1986); indeed, the major challenge posed by many benefit–cost analyses lies in devising credible ways for estimating quantifying benefits that all involved parties will accept. However benefits are measured and estimated, the methods must be acceptable to all parties involved and must be defined in terms of the specific objectives to be met (Bowman, 1999; Frost, 1975; Gostling & Jackson, 1986; Park, 1997; Ray, 1984).

***Step 5: Identifying and Estimating Costs.*** All equipment acquisition, space utilization, operating, maintenance, and personnel requirements of each alternative must be clearly identified. Moreover, assumptions regarding when the various costs would be incurred must be stated clearly. Then, the analyst must build or find credible methods for estimating each element of cost. They should be developed that follow as closely as possible all current Department of Defense (DOD) requirements for cost estimation and documentation (Byrns, Corban, & Ingalls, 1995; Dept. of the Air Force, 1988a; Ortwerth, 1990; Tanner & Neathammer, 1991).

***Step 6: Discounting of Benefits and Costs.*** In general, the benefits and costs of each alternative will appear at different points of time in the future. But, a dollar's worth of benefits or costs in a distant future period is not as valuable as a dollar's worth of benefits or costs tomorrow. Discounting of all benefits and costs to an appropriate, common period is necessary to put all the alternatives on the same footing. Therefore, the benefits and the costs associated with each alternative should be discounted to the present values in the common, comparison period. This approach uses dollar estimates stated in terms of a single point in time (present dollars) and therefore removes the effect of inflation (by correcting future estimates of actual cash flows to their present equivalencies).

***Step 7: Identifying the Superior Alternative.*** The superior alternative provides the greatest net benefits. Once all benefits and costs are discounted to a common zero period, net benefits can be calculated for each alternative by subtracting the sum of its discounted costs from the sum of its discounted benefits. Then, the analyst can compare the net benefits provided by each alternative to create a final ranking of the alternatives for the decision-making process.

Incremental benefit–cost (B/C) ratios provide an equally valid way of expressing the results when the alternatives can be lined up from smaller to larger. Then one can calculate the incremental benefits–cost ratio associated with moving from one alternative to the next largest alternative by dividing the change in discounted total benefits by the change in discounted total costs. It will benefit society to continue to increase the size of the project until the incremental benefit–cost ratio falls below one.

***Step 8: Analyzing the Sensitivity of the Results.*** A sensitivity analysis provides a way to learn something about the robustness of the results. It provides a way of learning how much our conclusion about which alternative is superior depends on the values used to compute key parameters. It requires that the factors that drive the estimates of the most important benefits and costs be identified. Once identified, each driver or set of drivers must be varied systematically to find out at what points, if any, the choice among alternatives would change. In other words, sensitivity analysis can tell the decision-makers over what sets of conditions the alternative is superior. Clearly, an alternative that proves superior under a wide range of assumptions about future conditions and parameter values is less risky.

## An Example Application: Multiship Pilot Training

Here we turn to an example application that illustrates these eight steps. It is based on an analysis performed several years ago by one of us. (Moor & Andrews, 1992). This analysis focused on alternative simulation systems for multiship pilot training.

***Step 1: Specifying the Goals.*** USAF doctrine is that warplanes fly and fight in pairs or in multiples of pairs. Most aircrew training simulation systems have not been able to train multiperson (or multiship) interaction tasks. However, with the rapid technological improvements made in microcomputers and vision systems, it has become possible to construct aircrew training simulation systems that are able to train these multiperson interaction tasks. Today's technology allows such systems to be configured in several alternative ways. The analysis issue then became which alternative system configuration would provide the greatest benefit in training at a cost within budgetary limits.

***Step 2: Defining the Alternative.*** The alternatives generated for this study reflect criteria concerning multiship simulation developed through a literature search (Amdor, Isley, & Pierce, 1988; Graham, Menton, & Schwoerer, 1981; Institute for Simulation & Training, 1990; Lethert, 1985; McDonald et al., 1989). They also reflect discussions with experts in the field of simulation training (Bell, Geltmacher, & Andrews, personal conversation; Genet, 1988; Houck, Thomas, & Bell, 1990). Then, using the general criteria for organizational alternatives and the conditions established for these alternatives, four alternatives were constructed from the available component technologies. They form a representative set (not an exhaustive set) of ways multiship simulation training could be provided to operational pilots.

- Alternative I: A pair of Combat Engagement Trainers (CETs) with visual display helmets with head tracking equipment in each squadron. However, no provision is provided for communicating with simulators located at other squadrons. The master computer at the focus squadron would be capable of generating up to two "enemy" aircraft to be flown against. There would be a direct (download) link with a data base development center that would generate the data bases representing different flying environments. Ground control would be provided by simulation personnel. This configuration would permit element simulation with a human opponent (each "flying" one of the CETs or a pair of aircraft flying against up to two computer-generated opponents.
- Alternative II: A pair of CETs with visual display helmets with head tracking equipment in each squadron. All the simulators in a squadron would be able to communicate to other simulators in squadrons within a specific wing, but no other communication would be possible. There would be a direct (download) link with a data base development center that would generate the data bases representing different flying environments. Ground control would be provided by simulation personnel. This configuration would permit up to 4 vs. 2 simulation with human opponents. That is, flights of up to four simulators could be made with an "enemy" composed of up to two opponents "flying" in two other simulators.
- Alternative III: Pair of CETs with visual display helmets in each squadron and a pair of Display for Advanced Training (DART) domes with visual display helmets and head tracking systems at a central or wing location. All the elements within the wing would be capable of communicating with all other elements. There would be a direct (download) link with a data base development center that would generate the data bases representing different flying environments. Ground control would be provided by simulation personnel. This configuration would permit up to 4 vs. 2 simulation with human opponents.
- Alternative IV: A pair of CETs with visual display helmets in each squadron and four regional centers, each center with four DART domes with Fiber Optic Helmet Mounted Display (FOHMD) with head tracking equipment. Pilots would spend part of their training time in a regional center. There would be the capability of full intercommunication with any one of the set of four squadrons (there are a total of 16 squadrons) associ-

ated with a specific regional center. There would be a direct (download) link with a data base development center that would generate the data bases representing different flying environments. This configuration would permit up to 4 vs. 4 simulation with human opponents.

***Steps 3 and 4: Identifying the Impacts and Quantifying the Benefits.*** For this particular training system, it was felt that the impact linked directly to the interpretation of benefits in dollar terms. Therefore, these two steps were combined. These steps involved the construction of an analytic model for estimating benefits and costs. It started with a set of possible multiship training simulation alternatives like the ones defined earlier in Step 2. The analytic model links performance areas with simulation environments. A performance area is an operational activity for which a sortie would be scheduled so that a pilot could acquire, improve, or maintain the skills necessary to carry out that specific activity. A simulation environment is the environment (inside the simulator) of the pilot who is practicing the performance area. The model finally developed used seven performance areas and four simulation alternatives.

The following are the names of each of the performance areas used for this model and a brief definition for each of them:

1. Radar Employment/Sorting: Tactics used for radar search and the sorting of enemy formations and individual formation members.
2. Multibogey (4 or More): Tactical employment against multiple (4 or more) enemy air threats.
3. Tactical Intercept: An intercept using specific single or multiple ship tactics, using either ground control radar or radar on the aircraft.
4. Electronic Identification: Using electronic systems to determine the identity of another aircraft.
5. Tactical Electronic Warfare System (TEWS) Assessment: Use of the TEWS system built into the aircraft to detect potential threats, primarily via the radar warning receiver.
6. Radar Lookout: A briefed responsibility of each flight member as to where he is primarily to look for threats. More time is spent looking at the radar than outside the aircraft.
7. Beyond Visual Range (BVR) Employment: Tactics designed to operate in a BVR environment, where radar and radar missile capabilities must be considered.

The analytic model includes a simulation benefit factor (SBEN) for each of the 28 combinations of performance areas and simulation environments. The SBEN factor reflects both the degrees to which the simulation environment emulates the actual cockpit experience and the numbers of times a performance area can be practiced in the simulator versus the aircraft. The SBEN factors used in the analytic model are shown in Table 14.1.

The numeric values in Table 14.1 are computed by multiplying the imputed capability of the simulation environment to represent the actual aircraft environment (measured on a scale from 0.00 to 1.00 and derived from expert opinion) times the ratio of the number of times a specific performance area could be practiced in a specific simulator divided by the number of times the performance area could be practiced during an actual aircraft sortie.

The direct economic impact of each simulator reflects the opportunity costs of using aircraft and weaponry, which are approximated by the marginal costs of this equipment. They directly reflect the training/practice benefit accruing due to the use of simulation:

1. The marginal (incremental) cost of flying the aircraft on a per sortie basis (or per hour), corrected for performance area if appropriate. The Marginal Aircraft Cost is continuous and is identified as MAC$.
2. The weaponry cost of using ammunition, weaponry, or other consumable expended per aircraft sortie for each performance area. The WEAPonry cost is identified as WEAP$.

The indirectly measured impacts reflect potential losses of pilots and aircraft used in flying sorties. They are indirect in the sense that they are based on sortie time rather than a measure of training benefit. They are a measure of risk rather than training:

TABLE 14.1
Simulation Benefit Factors

| | *Simulation Alternatives* | | | |
|---|---|---|---|---|
| *Peformance Areas* | *1* | 2 | *3* | *4* |
| 1 | 0.0250 | 0.0500 | 0.1167 | 0.1667 |
| 2 | 0.0400 | 0.0800 | 0.1867 | 0.2667 |
| 3 | 0.0400 | 0.0800 | 0.1867 | 0.2667 |
| 4 | 0.0400 | 0.0800 | 0.1867 | 0.2667 |
| 5 | 0.0250 | 0.0500 | 0.1167 | 0.1667 |
| 6 | 0.0400 | 0.0800 | 0.1867 | 0.2667 |
| 7 | 0.0400 | 0.0800 | 0.1867 | 0.2667 |

*Note.* From Moor and Andrews (1992).

1. The cost of loss of the aircraft as a function of its use in flying sorties of the specific performance area. This is a probability-based measure computed by (cost of an aircraft)*(probability of loss per sortie). The AIRcraft loss Cost is identified as AIRC$.
2. Cost of losing a pilot due to training accident as a function of his exposure to risk in specific performance areas. This is a probability-based measure computed by (cost of the pilot)*(probability of loss per sortie). The PILot death Cost is identified as PILC$.

The values for estimating the benefit elements were derived from the 1988 and 1989 Air Force Budgets (Department of the Air Force, 1987; 1988a; 1988b). Documents that delineated estimates for the value of a pilot's life (General Accounting Office, 1987), costs of operating an F-15 (Institute for Simulation & Training, 1990), and costs of various weapons (Department of Defense, 1989) were also used. All cost figures were corrected using Consumer Price Index (CPI) correction values (Directorate of Engineering and Services, 1988, p. 6) applied to the reference year of the cost elements to appropriate 1990 values. The monetary conversion factors resulting from this analysis are shown in Table 14.2.

The number of simulation sorties available per year (NUM) for each of the alternatives was computed based on expected operating hours and an equal usage of the simulation systems for each of the seven performance areas. There were 7200 sorties for each performance area available for Alternative 1, 7200 for Alternative 2, 2400 for Alternative 3, and 2600 for Alternative 4. These values were then multiplied to determine a monetary benefit estimate for each alternative and each performance. Then, the monetary elements were summed across all seven performance areas.

TABLE 14.2
Benefit Conversion Factors

| | *Aircraft Use* | *Weapons Use* | *Loss of Aircraft* | *Loss of Pilot* |
|---|---|---|---|---|
| *Peformance Areas* | *(MAC$)* | *(WEAP$)* | *(AIRC$)* | *PILC$)* |
| 1 | $6,300.00 | $2,000.00 | $266.00 | $116.76 |
| 2 | $5,850.00 | $2,000.00 | $247.00 | $108.42 |
| 3 | $5,850.00 | $2,000.00 | $247.00 | $108.42 |
| 4 | $5,850.00 | $2,000.00 | $247.00 | $108.42 |
| 5 | $6,300.00 | $2,000.00 | $266.00 | $116.76 |
| 6 | $5,850.00 | $2,000.00 | $247.00 | $108.42 |
| 7 | $5,850.00 | $2,000.00 | $247.00 | $108.42 |

*Note.* From Moor and Andrews (1992).

These procedures cumulated in the following equation for estimating the benefits for each alternative per year:

$$\text{Benefits} = \text{NUM} \times \text{SBEN} \times \{\text{MAC\$} + \text{WEAP\$}\} + \text{NUM} \times \{\text{AIRC\$} + \text{PILC\$}\}$$

This equation says that the number of sorties available for a given simulation alternative is multiplied times the estimated benefit factor for the alternative, and the result is multiplied times the cost of the marginal (shadow) cost of the aircraft plus the cost of the weapons the aircraft would use during an average sortie. This number is then added to a number computed by multiplying the number of sorties available for a given simulator times the sum of the average cost of losing the aircraft plus the average cost of losing the pilot if an accident should occur in an actual training mission. Table 14.3 gives the estimated annual benefits for each alternative. These values have been computed using this equation.

These values are the benefits values for the final benefit cost evaluation. Many of the variables used for benefits computation are qualitative and can only be seen as best estimates. This is a common factor in any benefit–cost analysis.

***Step 5: Identifying and Estimating Costs.*** The cost model was constructed to follow, as closely as possible, the Air Force's rules and procedures. See AF REGULATION 173-15, March 1988, "Economic Analysis and Program Evaluation for Resource Management, Cost Analysis" (Department of the Air Force, 1988a). This regulation identifies the key cost elements, which must be considered in constructing estimates of life-cycle cost of acquiring, operating, maintaining, and disposing of a system alternative. So, the process of constructing the cost model for this analysis became a problem in selecting the germane cost elements and finding credible information for estimating their values. The data for the cost model were derived from a number of sources. They included interviews with personnel who were very familiar with the commercial simulation device mar-

TABLE 14.3
Monetary Benefits Values Imputed to Alternatives (per Year)

| | |
|---|---|
| Alternative 1 | $32,598,864 |
| Alternative 2 | $46,890,864 |
| Alternative 3 | $75,225,152 |
| Alternative 4 | $87,908,342 |

*Note.* From Moor and Andrews (1992).

ket (Thomas, personal conversation), Air Force cost estimating manuals (Directorate of Engineering and Services, 1988; Grant & Thornley, 1988), and civil service and military pay rates (Air Force Association, 1990).

The cost estimates also involved several assumptions. It was assumed that each alternative would be manned by military and civil service personnel and that all buildings and equipment would be acquired new. It was further assumed that all computer-based equipment would have a 3-year economic life and that all other capital equipment would have a 9-year economic life. The cost evaluation therefore has a life cycle of 9 years. All cost cash flows were stated in monetary terms at the point in time when they were expected to occur. These are not uniform cash flows, and their conversion to uniform annual equivalencies (to facilitate comparisons) was undertaken during the financial discounting step.

***Step 6: Discounting of Benefits and Costs.*** When this analysis was done, AF REGULATION 173-15 (Department of the Air Force, 1988a) specified that an appropriate discounting rate was 10%, which the analysis used. All cash flows of benefits and costs were discounted to the base year and then expressed as annual equivalencies.

***Step 7: Identifying the Superior Alternative.*** As the reader can see, in Table 14.4, the analysis points to Alternative 4 as providing the greatest benefits.

Alternatively, this results can be expressed in terms of incremental benefit–cost (Canada et al., 1996; DeGarmo et al., 1997). This approach requires that the alternatives be ranked in order of increasing initial costs. The increment (*not* the total value) of benefits to costs is computed. Any alternatives with an incremental value greater than 1.0 represent worthwhile investments. In Table 14.5, the alternatives are already ranked in order of first cost.

The benefit–cost ratio approach requires that any alternative with a ratio of less than 1.0 be rejected and the increment for the next alternative be constructed ignoring the values for the rejected alternative. Any increment that is greater than 1.00 shows that the alternative for which the increment is computed is desirable.

The approach does *not* allow for computing incremental ratios and merely selecting the highest incremental value. If this were to be done, the incorrect alternative could be accepted. The highest incremental benefit–cost ratio in a set of incremental ratios does not indicate the preferred alternative. In the example case, because all increments are greater than 1.00, Alternative 4 is the preferred alternative, which compares to the decision made based on the data in Table 14.4.

TABLE 14.4
Benefit–Cost Comparison Elements

| *Alternative* | *Annual Benefits* | *Annual Costs* | *Annual Net Benefits* |
|---|---|---|---|
| 1 | $32,598 | $12,919 | $19,679 |
| 2 | $46,890 | $13,397 | $33,493 |
| 3 | $75,225 | $21,698 | $53,527 |
| 4 | $87,908 | $29,671 | $58,237 |

*Note.* All cash flows in 000's. From Moor and Andrews (1992).

***Step 8: Analyzing the Sensitivity of the Results.*** To carry out a sensitivity analysis, the benefit–cost model discussed in Steps 5 and 6 had to be capable of showing the impacts of changes in the values of any of the key variables on the ranking of the alternatives. For this analysis, the capabilities of the alternatives, the shadow prices, the marginal costs of flying the aircraft some additional number of hours over those currently being flown, cost of the aircraft, the number of sorties, flights of the aircraft that could be scheduled, availability, and the discount rate could be varied to determine whether the change would impact the superiority of the alternatives.

## Results

In the study on which this example is based, a spreadsheet was developed to provide the benefit–cost analysis, which is linked to the spreadsheets for benefits and cost. All computed values in the spreadsheet are based on the

TABLE 14.5
Benefit–Cost Ratio Approach

| | *Incremental* | |
|---|---|---|
| *Alternative* | *Benefit–Cost Ratio* | *Decision* |
| 1 | $32,598/$12,919 = 2.52 | Accept Alternative 1 |
| 2 | $14,292/$477 = 29.92 | Accept Alternative 2 |
| 3 | $28,334/$8,300 = 3.41 | Accept Alternative 3 |
| 4 | $12,683/$7,973 = 1.59 | Accept Alternative 4 |

*Note.* From Moor and Andrews (1992).

equations presented in this chapter. These values are relatively few in number, and the spreadsheets are arranged so that the analyst can easily change key values and observe its change on the decision criterion. With this spreadsheet approach, a great many sensitivity analyses may be easily performed, varying any of the parameters built into the models. (More details and modifications in this approach may be found in Moor, 1991, 1993; Moor and Andrews, 1992, 1993.)

### Conclusions

This project applied, in complete detail, a method for the benefit–cost analysis to the acquisition of simulators that are designed for implementation at the operational (squadron or wing) level. Table 14.5 demonstrates the clarity with which the results of this can be presented to a decision maker choosing between or among competing simulators. It should be repeated that, although the numeric figures presented for the benefits of each alternative are regarded as valid for purposes of comparison, they should not be interpreted as indicating the absolute dollar value of a given simulation system.

This analysis shows the alternative that is preferred by a considerable margin over its nearest alternative. It does not necessarily show that the alternative is to be preferred over other, as yet, unanalyzed simulation systems.

## COST-EFFECTIVENESS ANALYSIS

Cost-effectiveness analysis can provide useful guidance to decision makers when the information needed for a full benefit–cost analysis is not available. As we saw, benefit–cost analysis aims to identify the alternative that would provide the largest net benefits to the Air Force. This aim means that a benefit–cost analysis needs to include credible estimates of the dollar values of the major benefits provided by each alternative. Cost-effectiveness analysis never can tell us which alternative would make us better off. Its goal is more limited but still useful. Cost-effectiveness analysis aims to identify the alternative that would achieve a set of goals at the lowest total cost. In other words, it aims to identify the most cost-efficient way to get something done.

Cost-effectiveness demands less information. It only requires an analyst to understand the relationships between the inputs and outputs of the alternative systems and the dollar cost of the inputs. It does not require an analyst to try to estimate dollar values for the systems' outputs. Estimating the dollar values of some aircrew training system benefit may be impossible. For instance, it is difficult to determine precisely the value of having one more person trained and ready to fly a particular mission.

When such estimating problems are pervasive, cost-effectiveness analysis can provide a useful and workable framework.

### Cost-Effectiveness Analysis for Evaluating Aircrew Training

Both benefit–cost analysis and cost-effectiveness analysis provide frameworks for evaluating alternative ways for configuring systems with similar goals. Of course, the term *system* has come to mean different things to different people. For a cost-effectiveness analysis, a system includes the resources, processes, and standards used to provide an output or set of outputs. It also includes the interactions among all its elements and the key connections between the system and its environment.

Analysts most commonly apply cost-effectiveness analysis to do ex-ante evaluations of alternative ways to configure systems. This kind of application provides information for the individuals who must decide which alternative is best. However, cost-effectiveness analysis can also provide a framework for ex-post evaluation of systems that have already become operational. In this kind of application, it provides a framework for evaluating the comparative efficiency of the new training system.

### An Example Application: the C-130 Aircrew Training System

The analysis (Derrick, Tomczak, Nullmeyer, & Burright, 1992) described next demonstrates a way of solving this problem. The primary goal of the analysis was to determine if contractor-supported training was more cost-effective than an all-military supported system. It shows that cost-effectiveness analysis can provide a framework for evaluating the effectiveness of extensive and fundamental changes in large, complex training systems. In doing so, it had to address and overcome several methodological problems particular to the use of cost-effectiveness analysis as a framework for the ex-post evaluation.

***The C-130 Aircrew Training System.*** The C-130 is a four-engine, turbo-prop, tactical airlift aircraft. It can be used to air-drop personnel and cargo and can operate from austere air strips. We defined the C-130 aircrew training system to include the resources, processes, and requirements that were in place to enable the flight-related courses that comprised instruction for C-130 crews throughout their flying careers. The system output is measured in terms of the numbers of crewmembers who were successfully trained to specified levels of skill and knowledge. The crew positions are aircraft commander, copilot, navigator, flight engineer, and loadmaster (the person who is responsible for passengers and cargo). The system covers 15 courses taught at the formal school at Little Rock Air Force Base, AR, where crews receive initial training and become qualified to fly the C-130 aircraft. It also included 29 continuing training courses delivered at operating bases throughout the United States.

The analysis compared two alternative training system configurations. In the original system, the Air Force was responsible for all facets of aircrew training. In the newer hybrid system, training became a joint Air Force and contractor activity with each element having defined functions. A training contractor performed ground training and support functions. This included validating training requirements; designing, developing, and maintaining courseware; delivering academic and simulator training; maintaining all training devices; and managing ground training including student scheduling. In this new hybrid system, the Air Force provided in-flight instruction, contract monitoring, and facilities for the contractor's portions of the system. Final evaluation of student skills and knowledge was an Air Force responsibility.

Aircrew training media in the earlier, all-Air Force system included: classrooms; the C-130 aircraft; a full-crew, full-mission C-130 weapon system trainer (a system that could simulate everything that a C-130 normally would do); lower fidelity C-130 part-task trainers and cockpit procedures trainers (trainers that would simulate only defined aspects of the total operating environment of the C-130); a satellite navigation station for navigator training; and a fuselage trainer (just the fuselage portion of the C-130 aircraft) for loadmaster instruction. The newer hybrid Air Force and contractor system retained these media, but added computer-based training and reduced classroom instruction.

Originally, the Air Force had sole responsibility for the operation and maintenance of the training devices. In the newer hybrid system, the contractor operated and maintained all media except the aircraft itself. In the old system, ground and flight training management functions, such as resource scheduling and tracking, were Air Force functions. Crewmembers, fully certified by the Air Force as trained for their specific jobs, did them manually. In the new system, student scheduling and tracking for academic and simulator instruction became contractor responsibilities. The contractor added computerized scheduling and tracking capabilities to assist in these training management functions. The addition of automated management tools resulted in a reduction in the number of people assigned to these management tasks.

***Equating the Output of the Alternative Systems.*** When using cost-effectiveness as a framework for the ex-post evaluation of aircrew training systems, one cannot assume automatically that the systems are equally effective. Rather, an analyst must treat *equal-effectiveness* as a tentative hypothesis requiring testing and confirmation. Our first step was to compare the training practices and formal training objectives of the two systems. Virtually all the formal training objectives of either system could be translated into the objectives of the other system. An examination of course flows revealed a similar pattern of commonality. The sequence of training activities, as well as the tasks covered in simulator scenarios and flight profiles, were very similar in the two systems. Only one major difference in training practice was found: Some instruction that was accomplished in the classroom in the old system was delivered using computer-based training in the new system.

Our next step was to compare the characteristics of graduates from the two systems. Approximately 4,500 formal school student-training folders, student records of the training and evaluation of the students, from the old and new systems were reviewed. Student proficiency ratings were nearly identical in the two systems. Interviews were then conducted with Air Force personnel both at the formal school and at training units who had worked with students from both ground-training programs. They revealed a strong consensus among Air Force instructors that the original Air Force ground training and the new contractor-supported ground training provided virtually identical students going into the flying phase. (A small proportion of flight instructors, often loadmaster instructors, felt that students were better prepared for flight training in the old system, but that any differences were made up with no impact on resource consumption. They were attributed, by these instructors, to changes that had been made in prerequisites for students entering the loadmaster career field.) In short, evaluation standards and methods remained unchanged, and the two systems produced equivalent "pass" rates for all crew positions.

Two externally driven differences between the old and new systems were discovered that required adjustments for a meaningful cost comparison. First, an overseas base closed at about the time the new system was implemented. To compensate for this change, the old system was adjusted to reflect the new base configuration. Second, a reduction in the use of actual flying hours for training was made to the new system based on a budget decision that would have equally affected the old system. Therefore, the flying hour cut was applied to the old system as well.

### Results

A well-done cost-effectiveness analysis should provide new insights into how large, complex systems work and into what drives their costs. It should provide a new way of seeing the systems. Several important insights emerged from this analysis of the C-130 aircrew training systems.

Moving from an all-Air Force training system to a mixed system for C-130 aircrews provided relatively small but significant efficiencies at historical levels of student throughput. Over 10 years of operations, the mixed system would yield cumulative savings (in constant 1992 dollars) of $39 million. Until capital equipment would have to be replaced (in about 2010), annual savings would amount to about 3.2% of overall operating costs.

The savings in the costs of Air Force instructors would offset the cost of the contract. In addition, the Air Force gained efficiencies through 3% to 4% reductions in the costs of both student days and flying hours.

Shifting to the mixed system means moving to an aircrew training system with higher fixed costs, investments made in simulation systems and components, and lower variable costs, costs attributable to instructor hours and hours using the system. A 20% fall in student throughput would wipe out the savings from the mixed system. Larger falls in student throughput would make the mixed system more expensive than the all-Air Force system. However, the greater economies of scale associated of the mixed system mean that a 20% increase in student throughput would triple the savings provided by the mixed system.

## Conclusions

This cost-effectiveness analysis validated an assumption made at the beginning of the program: contractor support for C-130 aircrew training could reduce training costs without sacrificing training quality. The source of the savings came almost entirely from a reduced need for personnel to accomplish training tasks. More specifically, computer-based instruction allowed training to be accomplished with fewer instructors. Similarly, computer-based training management tools reduced the number of people needed in support functions.

Perhaps the most surprising result was the importance of student loads for determining which alternative was most cost-effective. With today's rapid downsizing of the military, decision makers should give serious consideration to how changes in training systems will impact their capabilities to shed costs as student loads fall.

This effort also demonstrated that a cost-effectiveness analysis framework provides a powerful tool for ex-post evaluation of changes to large, complex training systems. Lessons learned from this analysis have been applied to other training system acquisition programs.

Finally, the effort provided an analytic model that the Air Mobility Command (AMC) continues to use and expand. It still uses the model to guide decision making regarding C-130 instruction. Moreover, the Deputy Commander of Operations for Training also had the model expanded to capture the costs and dynamics of AMC's other aircrew training systems. AMC's continued and expanded use of the analytic model may provide the strongest validation of the value of the cost-effectiveness analysis framework for evaluating large, complex training systems.

## SUMMARY AND CONCLUSIONS

The analysis performed indicated that the Aircrew Training System (ATS) is cost-effective relative to pre-ATS training in the configuration under then current contract prices and historical levels of student throughput in all the courses. The steady-state operational cost of approximately $167 million per year for combined Air Force and contractor training with the ATS was some $5.4 million less than the pre-ATS cost for the same number of graduates. Assuming a constant student flow, this $5.4 million savings (constant 1992 dollars) will hold until some of the major capital equipment must be replaced, at which time the difference (in favor of the ATS system) will increase.

Although the benefit–cost model presented in this chapter was not fully implemented by the Air Force, numerous presentations were made to Air Force decision makers and, based on comments received by the authors, an impact was perceived. Notably, this impact appears to have resulted in an increased use of sensitivity analysis for decision making and an increased willingness to consider the use of transfer ratios as an input for benefit–cost computations.

The chief benefit of the project was demonstrating, in complete detail, a method for the benefit–cost analysis to the acquisition of simulators that are designed for implementation at the operational (squadron or wing) level. This approach demonstrates the clarity with which the results of this can be presented to a decision maker choosing between or among competing simulators. It should be repeated that, although the numeric figures presented for the benefits of each alternative are regarded as valid for purposes of comparison, they should not be interpreted as indicating the absolute value of a given simulation system.

## ACKNOWLEDGMENT

This work is drawn from research performed at the Aircrew Training Research Division of the Armstrong Laboratory from 1990 through 1995 and supported by the U.S. Air Force (Derrick, Tomczak, Nullmeyer, & Burright, 1992; Moor, 1991, 1993; Moor & Andrews, 1992, 1993).

## REFERENCES


Air Force Association. (1990). USAF Almanac 1990: USAF in facts and figures. *AIR FORCE Magazine 73*(5), 37–56.

Amdor, S. L., Isley, F. W., & Pierce, B. J. (1988). *Computer-based instruction/simulator program for fighter lead-in training: feasibility research*. The BDM Corp. Research Report AFHRL-TP-87-64, Air Force Systems Command, Brooks AFB.

Bowman, M. S. (1999). *Applied economic analysis for technologists, engineers, and managers*, New York: Prentice-Hall.

Byrns, Jr., E. V., Corban, E., & Ingalls, S. A. (1995). A novel cost–benefit analysis for evaluation of complex military systems. *Acquisition Review Quarterly*, Winter, 1–20.

Canada, J. R., Sullivan, W. G., & White, J. A. (1996). *Capital investment analysis for engineering and management*, (2nd ed.). New York: Prentice-Hall.

Degarmo, E. P., Sullivan, W. G., Bontadelli, J. A., & Wicks, E. M. (1997). *Engineering economy*, (10th ed.). New York: Prentice-Hall.

Department of the Air Force. (1987). Justification of Estimates for Fiscal Years 1988/1989: Operation and Maintenance, Air Force, Volume 1. *Biennial Budget Estimates*, Submitted to Congress. Washington, DC. AD A180522.

Department of the Air Force (1988a). *AF Regulation 173-15*, Headquarters US Air Force. Washington, DC.

Department of the Air Force. (1988b). Justification of Amended Fiscal Years 1988/1989: Aircraft Procurement, Air Force, Volume 1. *Biennial Budget Estimates*. Submitted to Congress. Washington, DC. AD A198259.

Department of Defense. (1989). Program Acquisition Costs by Weapon System. *Department of Defense Budget for Fiscal Years 1990 and 1991*. Washington, DC. AD A204415.

Derrick, W., Tomczak, W., Nullmeyer, R., & Burright, B. (1992). Cost and training effectiveness impacts of contracted C-130 aircrew training. *Proceedings of the 14th Interservice/Industry Training Systems and Education Conference*, 550–562.

Directorate of Engineering and Services. (1988). *Annual construction pricing guide for FY 90 program*. Department of the Air Force, Headquarters United States Air Force (HQ USAF/LEECD, Pentagon, Room 5D483). Washington, DC.

Fabrycky, W. J., Thuesen, G. J., & Verma, D. (1998). *Economic decision analysis* (3rd ed.). New York: Prentice-Hall.

Fast, J. C., Dunson, B. H., Wortman, R., Stone, B. M., & Engquist, S. K. (1992). *Measuring benefits of manpower, personnel, and training (MPT) research and development*. Armstong Laboratory, Human Resources Directorate, Manpower and Personnel Research Division (Tech. Report No. AL-CR-1991-0007). Brooks AFB, TX.

Frost, M. J. (1975). *How to use cost benefit analysis in project appraisal*. New York: John Wiley.

General Accounting Office. (1987). *Air force pilots; developing and sustaining a stable, combat-ready force, briefing report*. GAO/NSIAD-88-49BR, Washington, DC.

Genet, R. M. (1988). *The WARNET Papers*. Report No. AFHRL-TP-87-63, Human Resources Laboratory, AF Systems Command. Brooks AFB, TX.

Gostling, J. J., & Jackson, L. B. (1986). Getting the most out of benefit–cost analysis: Application in the Wisconsin department of transportation. *Government Finance Review*, 33–39.

Graham, G. L., Menton, A. F., & Schwoerer, R. R. (1981). Training effectiveness—A total system perspective. *Proceedings of the Interservice/Industry Training Systems Conference*, 179–186.

Grant, R. J., & Thornley, A. (Eds.). (1988). *Means square foot costs: Residential, commercial, industrial, institutional, 1988*. 9th Annual Edition. R. S. Means Company, Inc., Kingston, MA.

Hanley, N., & Spash, C. L. (1993). *Cost–benefit analysis and the environment*. Cheltenham, Glos, UK: Edward Elgar Publishing.

Hanspeter, Georgi. (1973). *Cost-benefit analysis and public investment in transport: A survey*. London: Butterworths.

Houck, M. R., Thomas, G. S., & Bell, H. H. (1990). *Requirements analysis for multiplayer air combat simulation*. Unpublished paper, Human Resources Laboratory, U.S.A.F. Williams AFB, AZ.

Institute for Simulation & Training. (1990). *Report of industry task force, navy training—2000*. University of Central Florida, Orlando, FL.

Kamsickas, G. M. (1991). An objective look at the modularization and standardization of training systems. *Proceedings of the 13th Interservice/Industry Training Systems Conference*, 357–367.

Lethert, J. F. (1985). F-16 Simulators—What Have We Learned? *Proceedings, Interservice/ Industry Training Systems Conference*, 323–337.

Londero, E. (1996). *Benefits and beneficiaries: An introduction to estimating distributional effects in cost–benefit analysis* (2nd ed.). Washington, DC: Johns Hopkins University Press.

Maciariello, J. A. (1975). *Dynamic benefit–cost analysis*. Lexington, MA: D. C. Heath and Co.

McDonald, G. W., Broeder, R. F., & Cutak, R. J. (1989). Multi-ship air combat simulation. *Proceedings of Interservice/Industry Training Systems Conference*, 148–158.

Moor, W. C. (1991). Benefit–cost evaluation of simulator based multiship training alternatives. *United States Air Force Summer Faculty Research Program: Program Technical Report*, Rodney Darrah, editor, Universal Energy Systems, Inc., Dayton, OH. AD A244517.

Moor, W. C. (1993). A method for comparison of alternative multiship aircraft simulation systems utilizing benefit estimation. *United States Air Force Summer Faculty Research Program: Program Technical Report, Vol. 2*, Research & Development Laboratories, Culver City, CA. AD A261989.

Moor, W. C., & Andrews, D. H. (1992). *Benefit–cost model for the evaluation of simulator-based multiship training alternatives*. Armstrong Laboratory, Human Resources Directorate, Aircrew Training Research Division, Williams AFB, AZ. AL-TP-1992-0023.

Moor, W. C., & Andrews, D. H. (1993). *Benefits estimation for simulation systems used for aircrew training in a multi-ship environment* (Final Tech. Report, AL/HR-TR-1993-0158). Human Resources Directorate, Aircrew Training Research Division, Williams AFB, AZ.

Ortwerth, M. S. (1990). *Fundamental Concepts of Cost–Benefit Analysis*. Cost Memorandum 90-01, U. S. Army Troop Support Command, Directorate for Resource Management, Cost Analysis Division. St. Louis, MO. AD-A232933

Oxenfeldt, A. R. (1979). *Cost–benefit analysis for executive decision making*. New York: AMACOM.

Park, C. S. (1993). *Contemporary engineering economics*. New York: Addison-Wesley.

Park, C. S. (1997). *Contemporary engineering economics* (2nd ed.). New York: Addison-Wesley.

Pearce, D. W. (1983). *Cost–benefit analysis* (2nd ed.). London: MacMillan.

Ray, A. (1984). *Cost–benefit analysis: Issues and methodologies*. Baltimore: Johns Hopkins University Press.

Sassone, P. G. (1985). *Training extension course research: Review of the literature on cost and training effectiveness*. Research Note 85-74, Mellonics Systems Development Division, Litton Systems, Inc., U. S. Army, Research Institute for the Behavioral and Social Sciences, Alexandria, VA. AD A160632

Smith, V. K. (1986). A Conceptual Overview of the Foundations of Benefit–Cost Analysis. In J. D. Bentkover, V. T. Covello, & J. Mumpower (Eds.), *Benefits assessment: The state of the art* (pp. 13–34). Dordrecht, Holland: D. Reidel.

Tanner, P., & Neathammer, R. D. (1991). *Cost Estimating Guide for Performing Economic Analyses in the Military Construction, Army (MCA) Process*. U.S. Army Corps of Engineers, USACERL Technical Report P-91/08. AD-A232 873

Chapter **15**

# Synthetic Team Training Environments: Application to Command and Control Aircrews

Samuel G. Schiflett
*Air Force Research Laboratory, Brooks AFB, TX*

Linda R. Elliott
*Veridian Engineering, San Antonio, TX*

This chapter describes our general approach to integrating basic research in cognitive science and advances in training technology to the domain of command and control team performance. Integration is an ongoing goal, as military research demands that scientific rigor be focused on operational problem solving. This necessitates continual interplay to balance the requirements and payoffs of basic research to those of our operational customers. Here, we describe our efforts to develop synthetic team training environments, using PC-based platforms, as analogues of actual operational performance domains.

It has been noted (Salas, Cannon-Bowers, & Kozlowski, 1997) that computer-based training has yet to be fully exploited for the advancement of the science and practice of training. Salas and his colleagues provide compelling arguments as to how to improve the science and practice of training. They, and others (Fletcher, 1994; Ford, Kozlowski, Kraiger, Salas, & Teachout, 1997; Swezey & Salas, 1992), have discussed opportunities for improvement, centering around the incorporation of more recently developed propositions regarding presentation formats, media characteristics, and insights into the nature of expertise, as opposed to the use of technology as a proxy for training improvements.

It is clear that progress is being carried forward with regard to issues such as cognitive task analyses (MacMillan et al., 1998; McNeese, Zaff, Citera, Brown, & Whitaker, 1995; Rasmussen, 1983; Rogers, Maurer, Salas, & Fisk, 1997; Shute, Torreano, & Willis, 1998) individual differences and motivational processes (Kozlowski & Salas, 1997; Mathieu & Martineau, 1997; Mathieu, Tannenbaum, & Salas, 1992; McIntyre & Salas, 1995), and teamwork training (Ford et al., 1997; Franken & O'Neil, 1994; Kozlowski, Gully, McHugh, Alas, & Cannon-Bowers, 1996; Salas, Bowers, & Cannon-Bowers, 1995; Salas, Cannon-Bowers, & Johnston, 1997; Swezey, Meltzer, & Salas, 1994). In fact, several theoretical frameworks have been offered to predict and explain the interplay between training method and context, individual factors, and organizational context, and to provide a structured environment for the development of testable propositions (Kozlowski & Salas, 1997; Mathieu et al., 1992; Salas, Cannon-Bowers, & Blickensderfer, 1997; Swezey & Salas, 1992). These frameworks should guide the selection and evaluation of advanced capabilities as they enhance training systems.

Advances in technology offer compelling new capabilities to researchers and trainers. Recently, there has been a great investment in highly realistic simulations of battlespace scenarios, involving military operational personnel in the roles of aircraft pilots and command, control, and communications ($C^3$) intelligence personnel. This emphasis results from prior success in demonstrating the efficacy of simulation technology (Alluisi, 1991) and greater focus on cost reduction, including conserving the use of aging aircraft. Still, increased funding naturally focuses attention of stakeholders requesting proof of the efficiency and effectiveness of these highly realistic simulation systems (Clark, 1997). At the same time, realistic and complex scenarios previously restricted to the high-end simulators, can now be executed using desktop PC-based computers, at much lower cost. The challenge is to identify and address trade-offs among these systems, which range from extremely realistic to task scenarios that claim psychological fidelity based on cognitive demands (Salas, Cannon-Bowers, & Blickensderfer, 1997; Salas, Cannon-Bowers, & Kozlowski, 1997).

Technological advances have reduced costs and enhanced capabilities in generating highly realistic, complex, multiforce simulators. In addition, artificial intelligence technology enables further advancement in scenario realism and data capture. "Intelligent" decision-making agents can be used for purposes such as decision aiding, intelligent tutoring, and the emulation of "synthetic" team members (MacMillan et al., 1998; Marcella & Johnson, 1998; Mahan, 1998; Tecuci & Keeling, 1998; Towns, Fitzgerald, & Lester, 1998). These decision-making agents are programmed with some degree of intelligence to model the performance of friendly or enemy personnel. Friendly agents can range in expertise and purpose, from novice fellow team members to trusted senior "advisor." Agents can also be programmed for particular types of decision error, based on characteristics such as level of risktaking, biases (such as central tendency in assessing threat), and/or strategies (such as "elimination by aspects" or "similarity") that do not take into account baysian probabilities. These agents afford the training researcher a more finely tuned exploration of learning and decision making and greater control in assessing the performance of individuals in a team context. They allow investigation of individual skill acquisition and decision making in a team context, while controlling for the actions and level of expertise of other team members. Use of these agents also allows investigation of teams differing in type and degree of expertise among team members. Agents can also be used to enhance performance monitoring/feedback. For example, agents can serve to monitor the workload of team members, providing feedback and/or recommendations to the team leader or members. Enhanced data collection capabilities provide more detailed interpretation of individual and team-decision processes with online performance feedback capabilities. Many other enhancements to particular streams of research can be derived from these fundamental capabilities.

In our program of research, prototype agents are being developed for application in command and control team task environments. This overall effort is augmented by agencies such as the Air Force Office of Scientific Research, Space Systems Command, the National Defense University, and the office for small business innovative research. In addition, several universities have central and/or supporting roles, including the University of Georgia (Robert Mahan), Michigan State University (John Hollenbeck, Daniel Ilgen, Steve Kozlowski), Penn State University (John Mathieu), George Mason University (Richard Klimoski, Steve Zaccaro), and University of South Florida (Michael Coovert). The foundation of this effort is based on the development of configurable synthetic team task environments for research and training in team performance. Building on existing commercial and Department of Defense (DOD) technology, these task environments will include technological improvements such as authoring systems for scenario generation, online performance assessment/feedback, automated data collection, and internet capability.

It is evident that opportunities for productive research in training are abundant, as long as one maintains a systematic theory-based approach. As noted previously, this chapter describes our general approach to integrating basic research in training theory and advances in training technology to the domain of command and control team performance. We begin with a description of the performance domain, followed by discussion of issues to consider in the development of these systems. Two issues are emphasized, that of degree and type of fidelity, and performance measurement and feedback.

## GENERAL APPROACH: LINKING TO THE C³ PERFORMANCE DOMAIN

Our approach takes into account trade-offs inherent in research conducted at differing levels of operational fidelity; that is, degree of realism with regard to our performance domain of interest. Trade-offs exist pertaining to type and degree of different aspects of "validity," such as internal, construct, content, and external validity. These issues are embedded in our program of research, which utilizes research platforms that span the dimension of fidelity from psychological fidelity to operational fidelity. Our goal is to coordinate basic and applied research and facilitate multidisciplinary investigation of operational command and control performance. Basic research in military contexts must also strive for some indication of external validity. Budget reductions for DOD expenses further increase the need to underline the relevance of research to operational problem solving. Research is far more likely to be supported when demonstrated in an environment generalizable to operational performance under "battlespace" conditions.

We have extended our existing program of research in command and control environments to include development of STEs as research and training platforms. Changes in our program of research is in response to the opportunities afforded by computer-based technological advances, recent emphasis on high-fidelity distributed mission training, fundamental importance of experimental approaches based on theory-driven propositions, and need to transition research findings to benefit our customers in operational settings. These STEs will serve as an underlying layer of experimental platforms for multidisciplinary research, such that increments in performance effects can be compared across interventions, whether they are focused on training, individual differences, human-factors engineering, or team structure/process. They serve the purpose of bridging the gap between experimental tasks commonly used in university basic research and the high-fidelity environments, which emulate real-world complex tasks. In this chapter, we discuss research issues and approaches in the development of a STE that captures core characteristics of an Airborne Warning and Control System (AWACS) team of Weapons Directors (WD).

Prototype intelligent agents within an AWACS-driven STE were recently developed and demonstrated in a command and control task domain as part of our program of research at the Air Force Research Laboratory, Brooks AFB, TX, under partial support from the Air Force Office of Scientific Research (Chiara & Stoyen, 1997; Stoyen et al., 1999). This integrated program will generate configurable synthetic team task environments for research and training in team performance (MacMillan et al., 1998).

To ensure that our research is relevant to issues of command and control performance, we designed and developed a research platform that emulates the core task characteristics of AWACS aircrews. The crew stations and scenarios that simulate the tactical air power $C^3$ mission of the E-3 AWACS aircraft are referred to as Command, Control, and Communication Simulation Training And Research Systems ($C^3$STARS). Our research does not focus on a particular method or intervention (e.g., display enhancement, training methodology, etc.). Instead, we have focused our research on a task performance domain (i.e., command and control performance). Our vision entails the complete understanding of a domain in order to bring an array of tools to bear on the enhancement of performance. As a result, we have striven to focus our efforts toward meeting the training objectives of operational "customers," such as the AWACS Air Combat Wing's requirement for measures of training effectiveness. Other command and control customer-based Air Force organizations we serve are the Air Intelligence Agency (research in information warfare), Air Combat Command (participation in distributed multisimulator research and training exercises), Air Education and Training Command (development and transfer of training technologies), Space and Missile Command (use of intelligent agents as decision-aiding technology), and the Electronic Systems Command (validation of new AWACS display enhancements).

The basic design of the $C^3$STARS research facility allows investigators to record and capture virtually every keystroke and utterance of each team member. From this generic capability, investigators can tailor measures specific to their theoretical framework and research goal. Display features, equipment controls, and scenarios are highly realistic. Hostile threats (e.g., enemy aircraft and surface-to-air-missile sites) are computer generated. Support participants in a battle scenario, such as friendly pilots and ground-based forces, are "role played" by experienced personnel, augmented by speech synthesizers (Dalrymple & Schiflett, 1997). In addition, the participants can be geographically dispersed using other distributed training simulators. $C^3$STARS is connected to other Air Force simulators, located elsewhere, through the DOD Advanced Distributed Simulation (ADS) internet (formerly the Defense Simulation Internet). The ADS is a secure broadband internet that allows large-scale computer exchange between simulators at various DOD facilities, enabling multiforce simulation of battlespace scenarios (Platt & Crane, 1993). The $C^3$STARS facility has supported Distributed Mission Training (DMT) exercises to evaluate situation assessment and decision making in pilots (Waag & Bell, 1997) under various levels of air control. Recently, team decision-making skills required for aircrew coordination were identified and assessed between AWACS air weapon controllers and F-16, F-15, and A-10 pilots in several mission contexts. The bottom-line value in all these DMT exercises is it provides a real world "train like you fight" assessment of multiship coordination skills required to plan, execute, and debrief in a naturalistic team environment.

## SYNTHETIC TASK ENVIRONMENTS: PAYOFFS AND TRADE-OFFS

$C^3$ teams are a case study in naturalistic decision making, as they are typified by complex interdependencies, high-stakes decision making under conditions of stress, ambiguity, and time pressure (Elliott, Schiflett, & Dalrymple, 1998). In addition, optimal tactics and strategies are a function of the overall mission plan and of dynamic circumstance. How do we choose and develop scenarios that capture these characteristics without reverting to full high fidelity? How do we choose the display characteristics that must be emulated? Questions arise with regard to validation of the synthetic environment. How do we know we captured the core characteristics and scenarios at the "proper" level of fidelity that represent core operational demands?

Validation of high-fidelity systems relies on content validity and on the performance of subject-matter experts. Subjects known as high-performers in the operational environment should perform best in the high-fidelity task and vice versa. Is this strategy sufficient for a STE that will be presented to novice trainees or research participants? Is there a set of tools or methods that can be applied to a performance domain to identify core characteristics of a synthetic task? Can recent methods developed for cognitive task analysis ensure that critical characteristics are identified? In our opinion, there is no single optimal methodology. Rather, this is related to the question of what type of validity and for what purpose.

Psychological validity is not usually an issue to researchers who are familiar with highly realistic simulations. High fidelity simulations rely on realism to ensure the highest degree of extrapolation of findings to operational environments. Theories driving this strategy strive to enhance training transfer through incorporation of identical stimulus cues and general rules, as well as manipulation of conditions of practice (Kozlowski & Salas, 1997). Operational needs drive the requirement for realistic external validity, and of utmost importance, the generalizability of findings to and from specific operational contexts. For example, in a past study, the Office of the Surgeon General wanted an evaluation of the effect of various antihistamines on AWACS WD performance. This request was driven by a "need to know" operational question: Would nondrowsiness antihistamines that may alleviate sinus pain exacerbated by flight degrade performance? To answer this question, we assessed performance at many different levels: (a) at a basic individual skills level on a cognitive task battery, (b) individual team member performance on their assigned tasks, under low and high workload, (c) overall team performance, and (d) mission outcome. Although statistically significant effects were found by Nesthus, Schiflett, Eddy, and Whitmore (1991) for individual cognitive tasks, they were not detrimental to individual or team operational performance (Dalrymple, Eddy, & Schiflett, 1995; Eddy, Dalrymple, & Schiflett, 1992). This may be due to a number of factors, such as behavioral automaticity typical of high expertise performance or lack of sensitivity of measures. Thus, we found that performance on these very basic cognitive tasks did not generalize to the more operational performance tasks. We concluded that the multilevel measurement approach was necessary to model effects at different levels. It also demonstrated to us the need for more immediate measures of performance that can bridge the gap in generalizing performance on basic cognitive skill task batteries to operational performance. What is needed is a synthetic team task environment that measures individual and team participation.

Compelling arguments have been stated for the relevance of STE systems based on psychological fidelity, as well as the absolute need for internal validity for the advancement of scientific knowledge (Berkowitz & Donnerstein, 1982; Cook & Campbell, 1979; Cook, Campbell, & Peracchio, 1990; Mook, 1983). Measures must be reflections of constructs, and constructs must be driven by theory and research by theoretical propositions; otherwise we suffer from a theoretical "dustbowl" of empirical predictions. The result will be no underlying understanding, and no generalizable propositions to offer to a different domain.

Advances in theory must guide task and scenario characteristics. This is fully understood. Even so, answers are not clear cut. As an example, if we were interested in human performance as a function of interface symbology, the synthetic task would need to emulate and manipulate a range of display characteristics. On the other hand, investigation of types of teamwork and tactical strategy (e.g., content, formulation, implementation, reformulation) would require similar scenarios, but not necessarily the knowledge of complex symbologies and switch actions also inherent in the operational task. However, the task cannot become too simplistic or it will lose core characteristics of the task as a whole and extrapolation will suffer. For example, one cannot make all communications computer-mediated if in the actual task they are very heavily verbal. Research would then be embedded in a task that may be too dissimilar to the operational setting to allow any degree of extrapolation of findings to training strategies for high-fidelity simulations.

Figure 15.1 presents trade-offs inherent in research based on tasks differing in level of fidelity to operational task demands. High fidelity envi-

**Level of Operational Fidelity (i.e. "realism")**

| **"Synthetic Task Environment"** | **"High-Fidelity Environment"** |
|---|---|
| Purpose: Generalizability | Purpose: Extrapolation to specific context |
| Psychological fidelity | Operational fidelity |
| "Naïve" Subjects | "Expert" Subjects |
| University context | Operational context |
| "Basic" Research | "Applied" Research |
| Internal validity<br>Construct validity | External validity<br>Content validity |
| Hypotheses re "cause-effect"<br>(e.g. training/skill acquisition) | Hypotheses re "ecological validity"<br>(e.g. expert recognition responses) |

FIG. 15.1. Advantages of lower versus higher levels of operational fidelity.

ronment (HFE) systems with full operational capabilities have the highest degree of realism and thus are most generalizable to the specific subject and task domain. They provide task-specific findings with the most accurate extrapolation for performance models. In contrast, tasks derived from cognitive task analysis may be more generic in nature, particularly when the focus is on the cognitive task demands and not the emulation of a specific operational environment. When the focus is on the task demands, it enables a greater generalizability to other operational environments with similar demands. However, as tasks become more generic, there is the risk of losing accuracy in terms of predicting performance in a specific environment. As Weick (1979) and others have pointed out, research cannot be general, simple, and accurate all at once. Two out of three is the most one can hope for. If research is highly simple and/or general, it will not be as accurate in a particular context. In turn, the more specific and accurate the predictions, the less generalizable it is to other situations. In this case, highly realistic and complex simulation environments maximize accuracy of predictions to a particular context, whereas task environments based on psychological and construct fidelity (cognitive task demands) are more systematically generalized to environments with similar features.

STE systems based on psychological fidelity are usually simpler, with more highly controlled scenario/task features and more naïve and homogenous subjects (O'Neil, Chung, & Brown, 1997). Higher numbers of subjects can be utilized in a more fully controlled and/or randomized research design, thus providing greater statistical power (Cohen, 1992). This approach can also be used to investigate skill acquisition issues, whereas high fidelity environments require subject expertise for participation (Driskell & Salas, 1992). As a result, the synthetic task approach can more easily demonstrate construct and internal validity, but extrapolation to a particular operational environment is lower. Highly realistic environments enable performance to be investigated from a systems engineering perspective: Performance is "modeled," and any deviations between predicted and actual performance becomes incorporated in subsequent models (Coovert & McNelis, 1992). This more empirical modeling approach maximizes predictive accuracy, but presents challenges with regard to cost, subject matter expert availability, and the array of confounding variables that arise from the complex and dynamic scenarios and individual differences in experience, training, and expertise (Andrews, Waag, & Bell, 1992). See Kleinman, Luh, Pattipati, and Serfaty (1992) for an approach that balances the two perspectives. Further discussions of the characteristics and assessment of the aspects of *validity* can be found elsewhere (Berkowitz & Donnerstein, 1982; Cook & Campbell, 1979; Cook et al., 1990; Cronbach, 1989; Mook, 1983; Nunnally, 1978; Shavelson & Webb, 1991).

Our overall strategy was founded on a high level of operational fidelity to maximize predictive accuracy and now is extended to simpler synthetic environments based on psychological fidelity. We are developing a distributed research platform for Synthetic Task Environments for Assessment and Modeling (SynTEAM) that will emulate core characteristics of individual and team $C^3$ decision-making constructs. Research findings from the STE systems would guide high-fidelity research, along with feedback from the operational community, and this in turn provides feedback to the basic researchers.

Several universities and companies are assisting in this evolving process of developing a SynTEAM network. The growing list of universities include Michigan State, Penn State, George Mason, Georgia, South Florida, New Mexico State, and Carnegie Mellon. The distributed STE will be the underlying infrastructure that ties the laboratories together and represents the core command and control elements in the distributed high-fidelity simulation network. For example, our high-fidelity $C^3$STARS facility, representing the AWACS E-3 aircraft functional capabilities is connected to high-fidelity simulations of fighter aircraft located at the Warfighter Training Research Division, Mesa, AZ, and enemy integrated air defense systems at the Information Warfare Center, Kelly AFB, TX, such that participants are immersed in a common battlespace.

Figure 15.2 captures our vision: To develop analogue STE and HFE systems such that linkages can be demonstrated from insights gained from basic theoretical research and application-oriented research for an operational payoff. At this time, we have basic researchers in universities generating and investigating an array of theories and methods. We also have

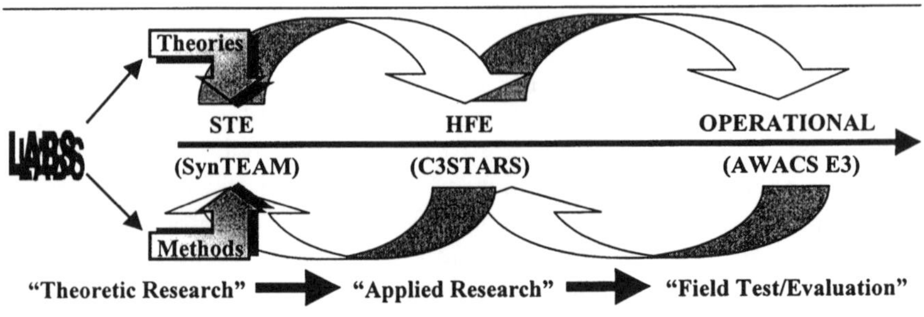


FIG. 15.2. Establish interactive relationships among basic and applied research programs for an operational command and control domain.

applied researchers within the Air Force Research Laboratory using High-Fidelity Environment (HFE) systems. However, there is a gap between them, often attributed to the lack of generalizability from university-based findings to the operational context or to the lack of theory driven focus on internal validity among military researchers. We cannot simply demand that universities utilize HFE systems; neither can we expect HFE-based research to meet the higher levels of control, statistical power, and domains of investigation typical of universities.

## STE TASK CHARACTERISTICS: IDENTIFICATION AND VALIDATION

Consider the characteristics of a $C^3$ team, for example, a team of AWACS Weapons Directors. AWACS Weapons Directors coordinate communications received from a number of sources, both within a team (e.g., other weapons directors and air surveillance operators) and external teams (e.g., friendly aircraft and ground-based operations). To accomplish this, AWACS team members must exchange, interpret, and effectively weigh information to optimize resource allocation decisions across team members over time and under stress. These decisions are often regarding dispersion of shared resources, such as strike forces (e.g., fighter/bomber aircraft), surface-to-air missile sites, reconnaissance, jamming, refueling, and search-and-rescue aircraft. Relevant information must be distributed to appropriate personnel and updated over time, in dynamic conditions that may require shift changes in personnel. Information is often complex, uncertain, verbal, and may be missing, degraded, passed along from unfamiliar sources, or misinterpreted by others. Also, information may be communicated or interpreted by individuals with only partially overlapping awareness of the battlespace. Add to this the distinguishing characteristics of war scenarios, where hostile intelligence forces are at work, with aversive and disastrous consequences of error. The result is the environment described as naturalistic and difficult to replicate in university settings. What is required is a practical approach that uses available methods and tools that can facilitate the determination of appropriate core characteristics to establish validity in a command and control task domain such as just described.

## APPROACH

We first determined the type and level of performance that will be of primary importance by identifying the core characteristics of the SynTEAM STE that functionally represents the task domain of the AWACS WDs. Because the STE will serve to bridge the generalizability gap between existing university tasks and more high-fidelity tasks, we decided to focus on more complex individual and team-level tasks that most fully distinguish operational performance from performance demands of simpler tasks. The fundamental platform that evolves from this orientation should be capable iterative refinements to meet specific research and/or training needs.

### Cognitive Task Analyses

Several cognitive task analyses have been performed on AWACS operational personnel. First, analyses were performed for AWACS individual tasks with a primary focus on display enhancements (Klinger et al., 1993). In addition, Fahey and his associates (Fahey, Rowe, Dunlap, & deBoom, 1997), investigated the AWACS task domain with a special focus on AWACS team tasks, utilizing a procedural knowledge elicitation technique (Hall, Gott, & Pokorny, 1995) and critical incidents techniques. The Fahey et al. (1997) report provides a comprehensive description of the AWACS roles, and the general finding was that critical incidents technique was more useful for the description of AWACS individual tasks and AWACS team tasks in particular.

In addition, we initiated a follow-up analysis of AWACS tasks, with a primary focus on team functions (MacMillan et al., 1998). This was a preliminary investigation, to test a team-based approach to cognitive task analysis using the critical decision method for eliciting knowledge (Klein, 1993). Results were very informative, providing frameworks and constructs within an information-requirements approach to task descriptions. Further data collection using this approach will be accomplished over the next year on a larger, diverse set of AWACS air controllers and surveillance operators.

As data from these sources accumulate, they will guide the development of additional STE characteristics. They will also provide the framework for the development of SynTEAM AWACS-based scenarios to represent pertinent cognitive task demands, training goals, and training methods. We will incorporate other relevant taxonomic approaches to command and control teamwork and team training. As opportunities for data collection arise, we plan to investigate other approaches to the analysis of team task demands. There are many taxonomies of interest that offer opportunities for accumulated and comprehensive understanding of the $C^3$ performance domain, not only of teamwork requirements, but of varying dimensions of individual task demands (Coovert & Craiger, 1997; Eggemeier, Fisk, Robbins, & Lawless, 1988; Fleishman & Zaccaro, 1992; Goldsmith & Kraiger, 1997; Klein, Calderwood, & MacGregor, 1989; Rogers et al., 1997).

### Identification of Team-Level Performance Tasks

Much progress has been made with regard to the conceptualization, measurement and interpretation of team-level phenomenon, primarily from the fields of industrial–organizational psychology, management and psychometrics (Hollenbeck, Ilgen, & Sego, 1994; Hollenbeck et al., 1995; Jones, 1974; Rousseau, 1985). Researchers have clearly demonstrated the importance of levels of analysis (i.e., individual, team, and organizational levels), particularly with regard to measurement and interpretation. For example, take the concept of team coordination. Coordination is a team-level construct conceptualized and measured at the team level. Individuals may perform in a manner that facilitates or impedes coordination, but the concept of coordination requires more than an individual. Also, the measurement of coordination may be more or less than the sum or average of individual performance. Steiner (1972) anticipated some of these measurement issues of team-level phenomenon when characterizing the types of team tasks as a function of the (a) average or sum of team member performance (e.g., the productivity of a typing pool), (b) highest-performing member (e.g., a group highly dependent on an energetic leader), or (c) least-performing member (e.g., a group highly dependent on a chain of actions, such as serial exchange of information).

It is widely believed that team-level phenomena can be a function of the type and degree of interdependence among team members (Saavedra, Earley, & Van Dyne, 1993; Swezey & Salas, 1992). In fact, we have developed an initial taxonomy of team-level functioning based on type and degree of interdependence among team members. The basis of the taxonomy rests upon the core conceptualization of teams and teamwork. Teams are distinguished from groups in general by a common purpose or goal, performed by interdependent team members (Salas, Dickinson, Converse, & Tannenbaum, 1992). From this definition we derived a core definition of teamwork, as contrasted with individual taskwork: *The fundamental function of teamwork is the effective managing of interdependencies to accomplish a team goal.* This definition allows specification of different profiles of teamwork behaviors for different type of teams. A particular team performance domain should be analyzed to identify first the type and degree of interdependencies, prior to specification of particular behaviors (e.g., communication, monitoring others, etc.) or knowledge requirements (e.g., distributed, overlapping, shared knowledge or expertise).

The essence of teamwork is thus conceptualized as the management of team member interdependencies. From this definition, we can identify the fundamental processes by which interdependencies are managed. We generated a list of functions we believe captures the degree to which a team relies on teamwork (i.e., the degree to which they are interdependent), the core functions by which teamwork is accomplished, and the type or level of complexity, based on cognitive demand, for each teamwork function. This would then allow teams to be classified as to type and degree of teamwork required, and it provides the foundation by which effective training techniques can be identified.

One fundamental distinction can be made in terms of how teams can differ in the nature of their interdependencies. Assuming interdependencies exist, teams will vary in the degree to which these interdependencies can be articulated in advance. For example, a symphony can explicate its interdependencies among team members to very specific terms, by articulating synchronized procedural strategies (who must do what, with whom, and when), given a score of music. Interdependencies among team members can be characterized and measured by quantitative methods, when the tasks which need to be coordinated can be specified in advance.

On the other hand there are situations where coordination cannot be specified in advance, or when plans change during performance. This requires additional knowledge and training in contingency planning, problem recognition, problem solving, and/or conflict resolution. This we refer to as dynamic replanning, and it is expected to be a fundamental aspect of teamwork and independent from the concept of coordination complexity.

***Coordination Complexity.*** Taxonomies of teamwork have referred to the degree of interdependence among teams, distinguishing teams based on the complexity of coordination requirements. Coordination requirements are minimal when tasks are simply pooled and become more complex when tasks must be sequenced. Although these coordination sequences may become very complex and demanding, quantitative approaches can be applied to represent and describe complexity.

Assuming the coordination plan is explicit, quantitative delineation of the coordination flow would yield an estimate of coordination complexity. Evaluation of team coordination can then be based on adherance to the plan to provide a measure of coordination efficiency. Many approaches exist which can provide indices of both coordination efficiency (of the plan) and coordination performance, as determined by execution of the plan (Bowers, Baker, & Salas, 1994; Coovert & McNelis, 1992; Coovert & Craiger, 1997; Coovert, Craiger, Cannon-Bowers, Driskell, & Salas, 1995). Entire disciplines have focused on the quantification of efficiency. Highly sophisticated mathematical modeling techniques are arising from numerous quantitative disciplines, and are becoming increasingly user-friendly through advances in PC technology. Many of these models have been applied to the performance domain of command and control, to ascertain the effectiveness of various team task architectures and/or to generate more effective structures (Benson, Kemple, Kleinman, Porter, & Serfaty, 1998; Brehmer, 1992, 1998; Curry, Kleinman, & Pattipatti, 1998; Kleinman, 1998; Krackhardt & Carley, 1998; Handley, Zaidi, & Levis, 1998; Lee & Carley, 1998; Levchuk, Pattipatti, Krackhardt, & Carley, 1998; Levchuk, Pattipatti, & Kleinman, 1998; Levchuk, Lou, Levchuk, Pattipati, & Kleinman, 1999; Serfaty, 1996; Paley, Levchuk, Serfaty, & MacMillan, 1999). Thus, the opportunity exists to model and evaluate the efficiency and execution of coordination plans of very complex interactions.

***Dynamic Replanning.*** Whereas coordination complexity is focused on efficient coordination planning and execution, dynamic replanning refers to the degree to which team members must adapt to dynamic circumstances. Teams may be very low on coordination complexity as defined by the complexity of a pre-specified plan, and yet still have a high requirement for adaptive teamwork—to rework the plan, and reallocate tasks and/or resources. Given a dynamic environment, teams must be able to problem solve and replan "on the fly." It then becomes difficult, and perhaps unfair, to evaluate the team through adherence to a prespecified coordination of action. Teams in general can be distinguished with regard to the degree to which the team/leader must generate coordination strategies during performance execution as opposed to following a predetermined and static plan. Certainly, execution of a complex reciprocal coordination plan can be challenging even if determined beforehand, but the task requirement is different when plans must be revised or improvised during performance execution. The contrast is obvious between orchestras vs. jazz bands, or theater acts vs. performance art. However, most teams will vary in more moderate levels for both static and dynamic aspects of coordination. For example, most sports teams have a reportoire of "strategies," and also a set of contingencies for switching strategies if the current strategy is not working well.

## Application to AWACS Teams

The constructs described above were investigated as part of a strategy to identify team-level constructs. Other taxonomies and frameworks also served a large role in our effort (Cannon-Bowers, Oser, & Flanagan, 1992; Fleishman & Zaccarro, 1992; MacMillan et al., 1998; Smith-Jentsch, Johnston, & Payne, 1998) along with military doctrine (rules of engagement), and information pertaining to general training requirements.

Discussion with focal groups comprised of AWACS subject matter experts led to refinement of the team functional taxonomy as applied to the AWACS task domain. Review of the taxonomy, along with alternatives, led to the identification of four broad AWACS-based teamwork functions: (a) mission planning, (b) communication content/timing, (c) adherence to communication protocol (clear, concise, correct), and (d) dynamic replanning. Subsequently, we presented this revised taxonomy to 38 AWACS operational personnel (weapons directors and instructors), who supported this delineation of teamwork and provided critical incidents that would demand a high level of teamwork.

***Mission Planning: Formulation of "Contracts."*** AWACS performance can be divided into three phases: Premission planning, task execution, and debriefing (discussion after task execution). Mission planning impacts AWACS teamwork by establishing roles and responsibilities, and through contingency planning. As discussed by Fahey et al. (1997) and MacMillan et al. (1998), AWACS weapons directors explicate roles, responsibilities, and strategies to manage team member interdependencies through establishment of "contracts." These contracts are made among AWACS team members (internal contracts), and to the "external" team (i.e., friendly assets).

***Communication: Adherence to Protocol.*** AWACS weapons director tasks are predominately based on verbal information exchange. Communications are heavily standardized, in terms of content (jargon) and process. This aspect of communication effectiveness refers to the degree to which an individual follows guidelines for communication exchange. In addition to proper jargon and syntax, communications must be clear, concise, and correct.

***Communication to Support Situation Awareness.*** Communications may be clear, concise, correct, and delivered according to proper protocol, but still be superfluous. The maintenance of situation awareness also requires that pertinent information be exchanged to the right person, at the right time. In such a communication-rich environment, too much communication can impede performance, when unnecessary information "steps over" other, more urgent communication. Indeed, part of the proper timing of AWACS communications to pilots involves knowledge of when to keep quiet.

***Dynamic Replanning.*** AWACS weapons directors had no problem identifying the relevance and importance of this aspect of teamwork, although most had a different term to describe it (e.g., "making it up as we go along," "winging it"). All agreed it was required on a regular basis due to poor planning, unexpected events, and/or equipment breakdowns. They also agreed that it was an aspect that distinguished novices from experts, and is usually improved with experience. One reason may be that it, along with teamwork in general, is not formally taught in training, but rather is expected to be acquired on the job.

## IDENTIFICATION OF PERFORMANCE MEASURES

A fundamental challenge in the development of STE tasks is that of performance measures and the strategy by which the STE environment can be evaluated for training effectiveness. Issues arise that are particular to the measurement of team performance as opposed to individual performance (Baker & Salas, 1992; Bowers, Baker, & Salas, 1994; Bowers, Salas, Prince, & Brannick, 1992; Brannick, Roach, & Salas, 1993; Coovert, Craiger, Cannon-Bowers, Driskell, & Salas, 1995; Eddy, 1989; Fowlkes, Lane, Salas, Franz, & Oser, 1994; Guzzo & Salas, 1995; Hollenbeck, Colquitt, Ilgen, LePine, & Hedlund, 1998; Salas, Prince, Baker, & Shrestha, 1995; Salas et al., 1995; Salas et al., 1997; Smith-Jentsch, Johnson, & Payne, 1998; Steiner, 1972; Stout, Cannon-Bowers, & Salas, in press). In our approach we identify five levels of performance measures, as exemplified in Fig. 15.3 and discussed next (Schiflett, Strome, Eddy, & Dalrymple, 1990).

Our approach relies upon systematic conceptualization and measurement of constructs, in order to establish convergent validity. We do this by use of several taxonomies, regarding (a) levels of performance, (b) individual and team performance constructs, and (c) methods of measurement. Ultimately, measures can be visualized within an overall multitrait, multimethod matrix, to enable investigation of reliability and validity of performance measures.

Figure 15.3 illustrates five broad categories of performance measurement, and some tradeoffs with regard to interpretability (performance as it relates to the mission), sensitivity (performance measures as functions of the performer as opposed to context), and generality (relationships to other teams and/or situations). At the tip of the triangle are measures related to overall mission effectiveness; those measures that indicate overall performance of the team in light of mission objectives. These relatively straightforward measures are driven by the context and goals of the team and are most easily interpreted when assessing team performance in relationship to mission outcomes. These measures are less straightforward with regard to assessing individual or teamwork effectiveness. Mission outcomes may be due to the effect of other variables, such as overwhelming number and tempo of hostile forces. Thus, the higher up on the triangle, the more effects can be influenced by exogenous variables. The team may have coordinated information as efficiently as possible, but still fail in overall outcomes. How then do we differentiate "good" teams from "bad" teams when all "failed" or "succeeded" in mission outcome?

Below mission outcome measures are measures of the organization. Missions are often accomplished by the coordination of several teams. When the mission is a function of more than one team, the structure and/or performance of the overall system can and should be examined.

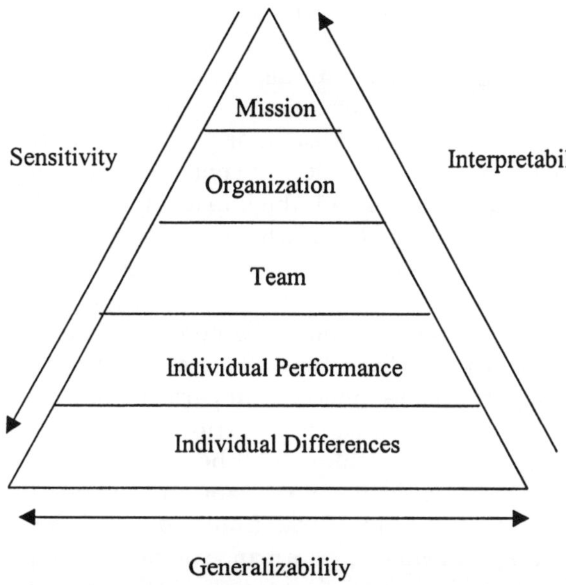


FIG. 15.3. Performance measurement hierarchy.

Next are measures of team performance as defined by the degree the team managed interdependencies among team members. This level of measurement would include dimensions such as planning, coordination execution, and dynamic replanning, as described earlier. These measures must be investigated within the context of factors such as workload tempo and duration. There may not be a consistently positive relationship between a teamwork measure, such as information exchange and mission outcome. For example, full information exchange under conditions of high time pressure can have a detrimental effect on mission effectiveness (Elliott, Dalrymple, & Neville, 1997; Elliott, Neville, Dalrymple, & Tower, 1997).

Generalization of relationships among measures of team performance and mission effectiveness should be restricted to contexts having similar core characteristics of the mission context. For example, in current $C^3$ investigations, measures selected to represent mission effectiveness may only be appropriate to AWACS teams, and even then, to particular types of scenarios. In one scenario we generated a defensive counter-air mission, where destruction of hostile aircraft is necessary to prevent loss of friendly assets. In other missions, the measure of effectiveness may be more a function of penetration into hostile territory or destruction of non-airborne hostile assets. In addition, research findings regarding effects of predictor variables or study manipulations (i.e., ambiguity, communication effectiveness) are expected to generalize to other military command and control performance domains with similar task demands, but not to team contexts which have a different profile of task characteristics.

At the bottom of the triangle are the measures of individual task performance, and of individual differences in capabilities. One should note that relationships among these measures and mission effectiveness would be most attenuated, due to the variety of contextual variables, which can affect mission effectiveness. Instead, the impact of individual capabilities should be related to individual performance, which are then related to measures of team, organization, and mission effectiveness. Team performance is usually expected to be related to measures of individual performance on a task; however, it should be noted that the relationship is reciprocal. That is, team performance will affect individual performance, just as individual performance affects team performance.

Our goal is to generate measures of performance across these levels such that the impact of interventions will be identified with dependent variables that are most sensitive and relevant. At the same time, impacts can be modeled with regard to effects, from individual performance on basic cognitive tasks, individual performance on task, team performance, and mission outcomes. This modeling approach to performance provides greater understanding of the impact of predictors on performance.

***Types of Measures.*** In addition to categories of performance, we also can construct a list of categories of measurement approaches. These approaches include embedded measures—those measures that are automatically and consistently captured by the simulator (e.g., number of hostile aircraft destroyed, number of friendly assets lost, any type of keystroke/action/event). They also include scenario-driven measures—those measures which are systematically generated as event-driven measures (e.g., reactions to communication/decision-making events to assess team information exchange). In addition there is observational data—those measures that are captured by subject matter expert (SME) observation of behaviors (e.g., checklists), and more subjective evaluative data—those measures based on evaluation of attributes (e.g., SME ratings, peer rating, self ratings), and questionnaire data (e.g., measures of knowledge or recall). Ideally, we will have at least two measures of each construct, and utilize a range of measurement methodology.

Criterion measures should be appropriate to the predictors of interest. For example, in assessing the impact of a particular drug on performance, it is apparent that the greatest impact would be on measures of individual capabilities and performance and lowest for mission effectiveness (Dalrymple, Eddy, & Schiflett, 1995). The data collected in this study was also generated to be consistent with individual and team-level variables within the Multi-Level Theory of Team Decisionmaking (Hollenbeck et al., 1995). It is critical to address methodological issues with regard to conceptualization, measurement, and analysis of data when investigating relationships between individual and team-level variables. The realism and complexity of the scenario characteristics will directly affect the choice of performance measures. For example, if teamwork performance is of interest, scenarios must be generated that are both (a) typical for the domain and (b) reliant upon teamwork for mission success. In a high-fidelity AWACS environment, we manipulated scenario characteristics in order to investigate team communication and decision-making performance, by directing information required by one team member to a different team member. That team member then had to send misdirected information to the correct team member, an action that is common to AWACS teams (Elliott et al., 1997; Elliott et al., 1998).

## SUMMARY

Our goal is to develop a STE environment that facilitates multidisciplinary investigation of complex and dynamic teamwork. In this chapter, we described our approach to the development of an STE system based on the tasks and responsibilities of weapons director teams. First, we described the characteristics of this command and control performance domain, based on several cognitive task analysis methods. We also discussed tradeoffs resulting from research performed at different levels of fidelity. We then described a generic approach to delineation of teamwork performance, based on the type and degree of interdependency among team members. From this we described the identification of teamwork aspects particular to this command and control performance domain. We also discussed characteristics of performance measures which will enable and enhance performance modeling.

This approach was applied to the development of several STE systems based on the same performance domain, for different purposes. Validation of these systems is underway. We will also pursue improvements which will further enhance research capabilities. These are features that will make the STE platform more "user-friendly" to the researcher. Questions include: How can scenarios be more easily generated and modified? How can data collection and analysis be facilitated? Characteristics of interest include features such as authoring systems for scenarios, menu-driven manipulation of characteristics, measurement of subject awareness and interpretation of environmental cues. Also, measurement of team decision processes, online performance assessment and feedback, and automatic generation of data files are necessary in evaluating training objectives in a timely manner.

As discussed previously, development and validation of these STE tasks is somewhat problematic. Indeed, one STE cannot be all things to all researchers. Again, research cannot be simple, general, and accurate all at once. However, it is a compelling vision, to augment existing basic and high-fidelity tasks with a research platform that can serve to help bridge that gap and enhance generalizability of university research to more challenging, operational environments.

## REFERENCES

Alluisi, E. A. (1991). The development of technology for collective training: SIMNET, a case history. *Human Factors, 33,* 343–362.

Andrews, D., Waag, W., & Bell, H. (1992). Training technologies applied to team training: Military examples. In R. Swezey & E. Salas (Eds.), *Teams: Their training and performance* (pp. 283–328). Norwood, NJ: Ablex.

Baker, C., & Salas, E. (1992). Principles for measuring teamwork skills. *Human Factors, 34*(4), 469–475.

Berkowitz, L., & Donnerstein, E. (1982). External validity is more than skin deep. Some answers to criticisms of laboratory experiments. *American Psychologist, 37*(3), 245–257.

Bowers, C., Baker, D., & Salas, E. (1994). Measuring the importance of teamwork: The reliability and validity of job/task analysis indices for team-training design. *Military Psychology, 6*(4), 205–214.

Bowers, C., Salas, E., Prince, C., & Brannick, M. (1992). Games teams play: A method for investigating team coordination and performance. *Behavior Research Methods, Instruments, and Computers, 24*(4), 503–506.

Brannick, M., Roach, R., & Salas, E. (1993). Understanding team performance: A multimethod study. *Human Performance, 6*(4), 287–308.

Cannon-Bowers, J., Oser, R., & Flanagan, D. (1992). Work teams in industry: A selected review and proposed framework. In R. Swezey & E. Salas (Eds.), *Teams: Their training and performance* (pp. 355–378). Norwood, NJ: Ablex.

Chiara, J., & Stoyen, A. (1997). Intelligent, adaptive research and training platform for individual and team training, with web capabilities and applications. *Phase II SBIR Report AF971-018*. Brooks AFB, TX: Air Force Research Laboratory, Warfighter Training Research Division.

Clark, C. (1997). Simulation buying spree needs restraint. *Defense Week* (Vol. 18), June 9.

Cohen, J. (1992). A power primer. *Psychological Bulletin, 112*, 155–159.

Cook, T., & Campbell, D. (1979). *Quasi-experimentation: Design and analysis issues for field settings*. Boston: Houghton Mifflin.

Cook, T., Campbell, D., & Peracchio, A. (1990). Quasi-experimentation. In M. Dunnette & L. Hough (Eds.), *Handbook of industrial/organizational psychology* (Vol. 1, pp. 491–576). Palo Alto: Consulting Psychologists Press, Inc.

Coovert, M., & Craiger, J. (1997). Performance modeling for training effectiveness. In J. Ford, S. Kozlowski, K. Kraiger, E. Salas, & M. Teachout (Eds.), *Improving training effectiveness in work organizations* (pp. 47–72). Mahwah, NJ: Lawrence Erlbaum Associates.

Coovert, M., Craiger, J., Cannon-Bowers, J., Driskell, J., & Salas, E. (1995). Innovations in modeling and simulating team performance: Implications for decision making. In Guzzo & E. Salas (Eds.), *Team effectiveness and decision making in organizations* (pp. 149–203). San Francisco: Jossey-Bass.

Coovert, M., & McNelis, K. (1992). Team Decisionmaking and Performance: A review and proposed modeling approach employing Petri nets. In R. Swezey & E. Salas (Eds.), *Teams: Their training and performance* (pp. 247–280). Norwood, NJ: Ablex.

Cronbach, L. (1989). Construct validity after 30 years. In R. L. Linn (Ed.), *Intelligence: Measurement, theory and public policy*. Urbana and Chicago: University of Illinois Press.

Dalrymple, M., Eddy, D., & Schiflett, S. (1995). *Comparative effects of antihistamines on AWACS weapons directors performance under sustained operations* (AL-TR-1995). Brooks AFB, TX: Armstrong Laboratory.

Dalrymple, M., & Schiflett, S. (1997). Measuring situational awareness of AWACS weapons directors. In A. Schopper (Ed.), *Situational awareness in the tactical air environment* (SOAR 97-01). Proceedings of the Naval Air Warfare Center's First Annual Symposium at Patuxent River, MD. WPAFB, OH: CSERIAC.

Driskell, J., & Salas, E. (1992). Can you study real teams in contrived settings? The value of small group research to understanding teams. In R. Swezey & E. Salas (Eds.), *Teams: Their training and performance* (pp. 101–126). Norwood, NJ: Ablex.

Eddy, D. (1989). *Selected team performance measures in a C3 environment: An annotated bibliography* (USAFSAM-TR-87-25). Brooks AFB, TX: School of Aerospace Medicine.

Eddy, D., Dalrymple, M., & Schiflett, S. (1992). *Comparative effects of antihistamines on aircrew mission effectiveness under sustained operations* (AL-TR-1992-0018). Brooks AFB, TX: Armstrong Laboratory.

Eggemeier, F., Fisk, A., Robbins, R., & Lawless, M. (1988, September). Application of automatic/controlled processing theory to training tactical command and control skills: II. Evaluations of task-analytic methodology. *Proceedings of the Annual Meeting of the Human Factors Society*, Anaheim, CA.

Elliott, L., Dalrymple, M., & Neville, K. (1997, September). Assessing performance of AWACS command and control teams. *Proceedings of the Annual Meeting of the Human Factors Society*, Albuquerque, NM.

Elliott, L., Neville, K., Dalrymple, M., & Tower, S. (1997, April). Impact of Ambiguity on AWACS Team Communications and Decisionmaking. *Proceedings of the Ninth International Symposium on Aviation Psychology*, Columbus, OH.

Elliott, L., Schiflett, S., & Dalrymple, M. (1998, April). Modeling the decision making process and performance of Airborne Warning and Control System (AWACS) Weapons Directors. *Proceedings of the NATO Symposium on Collaborative Crew Performance in Complex Operational Systems*, Edinburgh, United Kingdom.

Fahey, R., Rowe, A., Dunlap, K., & deBoom, D. (1997). *Synthetic task design: Preliminary cognitive task analysis of AWACS weapons director teams*. Preliminary Report. Brooks AFB, TX.

Fleishman, E., & Zaccarro, S. (1992). Toward a taxonomy of team performance functions. In R. Swezey & E. Salas (Eds.), *Teams: Their training and performance* (pp. 31–56). Norwood, NJ: Ablex.

Fletcher, J. (1994). What networked simulation offers to the assessment of collectives. In H. O'Neil (Ed.), *Technology assessment in software applications*. Hillsdale, NJ: Lawrence Erlbaum Associates.

Ford, J., Kozlowski, S., Kraiger, K., Salas, E., & Teachout, M. (1997). *Improving training effectiveness in work organizations*. Mahwah, NJ: Lawrence Erlbaum Associates.

Fowlkes, J., Lane, N., Salas, E., Franz, T., & Oser, R. (1994). Improving the measurement of team performance: The TARGETS methodology. *Military Psychology, 6*, 47–61.

Franken, J., & O'Neil, H. (1994). Stress-Induced anxiety of individuals and teams in a simulator environment. In H. O'Neil & M. Drillings (Eds.), *Motivation: Theory and Research*. Hillsdale, NJ: Lawrence Erlbaum Associates.

Goldsmith, T., & Kraiger, K. (1997). Structural knowledge assessment and training evaluation. In J. Ford, S. Kozlowski, K. Kraiger, E. Salas, & M. Teachout (Eds.), *Improving training effectiveness in work organizations* (pp. 73–98). Mahwah, NJ: Lawrence Erlbaum Associates.

Guzzo, R., & Salas, E. (1995). *Team effectiveness and decision making in organizations*. San Francisco: Jossey-Bass.

Hall, E., Gott, S., & Pokorny, R. (1995). *A procedural guide to cognitive task analysis: The PARI methodology* (Air Force Tech. Rep. AL/HR-TR-1995-0108). Air Force Armstrong Laboratory, Human Resources Directorate, Manpower and Personnel Research Division, Brooks AFB, TX.

Hollenbeck, J., Colquitt, J., Ilgen, D., LePine, J., & Hedlund, J. (1998). Accuracy decomposition and team decision making: Testing theoretical boundary conditions. *Journal of Applied Psychology, 83*(3), 494–500.

Hollenbeck, J., Ilgen, D., & Sego, D. (1994). Repeated measures regression: Enhancing the power of leadership research. *Leadership Quarterly, 5*(1), 3–23.

Hollenbeck, J., Ilgen, D., Sego, D., Hedlund, J., Major, D., & Phillips, J. (1995). The multilevel theory of team decision-making: Decision performance in teams incorporating distributed expertise. *Journal of Applied Psychology, 80*(2), 292–316.

Jones, M. (1974). Regressing group on individual effectiveness. *Organizational Behavior and Human Performance, 11*, 426–451.

Klein, G. (1993). A recognition-primed decision (RPD) model of rapid decision making. In G. Klein, J. Orasanu, R. Calderwood, & E. Zsambok (Eds.), *Decision making in action: Models and methods* (pp. 138–147). Norwood, NJ: Ablex.

Klein, G., Calderwood, R., & MacGregor, D. (1989). Critical decision method for eliciting knowledge. *IEEE Transactions on Systems, Man, and Cybernetics, 19*, 462–472.

Kleinman, D., Luh, P., Pattipati, K., & Serfaty, D. (1992). Mathematical models of team performance: A distributed decision making approach. In R. Swezey & E. Salas (Eds.), *Teams: Their training and performance* (pp. 177–218). Norwood, NJ: Ablex.

Klinger, D., Andriole, S., Militello, L., Adelman, L., Klein, G., & Gomes, M. (1993). *Designing for performance: A cognitive systems engineering approach to modifying an AWACS human-computer interface* (AL/CF-TR-1993-0093). Wright-Patterson AFB, OH: Armstrong Laboratory.

Kozlowski, S. (1996). Training and developing adaptive teams: Theory, principles, and research. In J. A. Cannon-Bowers & E. Salas (Eds.), *Decision making under stress: Implications for training and simulation* (pp. 115–154). Washington, DC: APA Books.

Kozlowski, S., Gully, S., McHugh, P., Alas, E., & Cannon-Bowers, J. (1996). A dynamic theory of leadership and team effectiveness: Developmental and task contingent leader roles. *Research in Personnel and Human Resources Management, 14*, 153–305.

Kozlowski, S., & Salas, E. (1997). An organizational systems approach for the implementation and transfer of training. In J. Ford, S. Kozlowski, K. Kraiger, E. Salas, & M. Teachout (Eds.), *Improving training effectiveness in work organizations* (pp. 247–290). Mahwah, NJ: Lawrence Erlbaum Associates.

Levchuck, Y., Pattipatti, K., & Kleinman, D. (1998, June). Designing adaptive organizations to process a complex mission: Algorithms and applications. *Proceedings of the 1998 Command and Control Research and Technology Symposium*. Naval Postgraduate School, Monterey, CA.

MacMillan, J., Serfaty, D., Young, P., Klinger, D., Thordsen, M., Cohen, M., & Freeman, J. (1998). *A system to enhance team decision making performance: Phase 1 final report*. Brooks AFB: Air Force Research Laboratory, Warfighter Training Research Division.

Mahan, R. (1998). Air Force recruits a psychologist for crew training. *APA Monitor*, July, p. 45.

Marcella, S., & Johnson, W. (1998). An instructor's assistant for team training in dynamic multi-agent virtual worlds. In B. Goettl, H. Halff, C. Redfield, & V. Shute (Eds.), *Lecture notes in computer science: Intelligent tutoring systems*. Proceedings from the Fourth International Conference, ITS, San Antonio, TX: Springer.

Mathieu, J., & Martineau, J. (1997). Individual and situational influences in training motivation. In J. Ford, S. Kozlowski, K. Kraiger, E. Salas, & M. Teachout (Eds.), *Improving training effectiveness in work organizations* (pp. 193–222). Mahwah, NJ: Lawrence Erlbaum Associates.

Mathieu, J., Tannenbaum, S., & Salas, E. (1992). The influences of individual and situational characteristics on measures of training effectiveness. *Academy of Management Journal, 35*, 826–847.

McIntyre, R., & Salas, E. (1995). Measuring and managing for team performance: Emerging principles from complex environments. In R. Guzzo & E. Salas (Eds.), *Team effectiveness and decision making in organizations* (pp. 9–45). San Francisco: Jossey-Bass.

McNeese, M. D., Zaff, B. S., Citera, M., Brown, C. E., & Whitaker, R. (1995). AKADAM: Eliciting user knowledge to support participatory ergonomics. *International Journal of Industrial Ergonomics, 15*(5), 345–363.

Mook, D. G. (1983). In defense of external invalidity. *American Psychologist*, 379–387.

Nesthus, T., Schiflett, S., Eddy, D., & Whitmore, J. (1991). *Comparative effects of antihistamines on aircrew performance of simple and complex tasks under sustained operations* (AL-TR-1991-0104). Brooks, AFB, TX: Armstrong Laboratory.

Nunnally, J. (1978). *Psychometric Theory* (2nd ed.). New York: McGraw-Hill.

O'Neil, H., Chung, G., & Brown, R. (1997). Use of networked simulations as a context to measure team competencies. In H. O'Neil (Ed.), *Workforce readiness*. Mahwah, NJ: Lawrence Erlbaum Associates.

Platt, P., & Crane, P. (1993). Development, test, and evaluation of a multiship simulation system for air combat training. *Proceedings of the 15th Interservice/Industry Training Systems and Education Conference* (pp. 629–639). Orlando, FL: National Security Industrial Association.

Rasmussen, J. (1983). Skills, rules, and knowledge: Signals, signs, and symbols, and other distinctions in human performance models. *IEEE Transactions on Systems, Man, and Cybernetics* (Vol. SMC-13, No. 3), 257–266.

Rogers, W., Maurer, T., Salas, E., & Fisk, A. (1997). Task analysis and cognitive theory: Controlled and automatic processing task analytic methodology. In J. Ford, S. Kozlowski, K. Kraiger, E. Salas, & M. Teachout (Eds.), *Improving training effectiveness in work organizations* (pp. 19–46). Mahwah, NJ: Lawrence Erlbaum Associates.

Rousseau, D. M. (1985). Issues of level in organizational research: Multi-level and cross-level perspectives. In L. L. Cummings & B. Staw (Eds.), *Research in organizational behavior* (Vol. 7, pp. 1–38). Greenwich, CT: JAI Press.

Saavedra, R., Earley, P., & Van Dyne, L. (1993). Complex interdependence in task-performing groups. *Journal of Applied Psychology, 78*(1), 61–72.

Salas, E., Bowers, C., & Cannon-Bowers, J. (1995). Military team research: 10 years of progress. Special Issue: Team processes, training, and performance. *Military Psychology,* 7(2), 55–75.

Salas, E., Cannon-Bowers, J., & Blickensderfer, E. (1997). Enhancing reciprocity between training theory and training practice: Principles, guidelines, and specifications. In J. Ford, S. Kozlowski, K. Kraiger, E. Salas, & M. Teachout (Eds.), *Improving training effectiveness in work organizations* (pp. 291–322). Mahwah, NJ: Lawrence Erlbaum Associates.

Salas, E., Cannon-Bowers, J., & Johnston, J. (1997). How can you turn a team of experts into an expert team? Emerging training strategies. In C. Zsambok & G. Klein (Eds.), *Naturalistic decision making* (pp. 359–370). Hillsdale, NJ: Lawrence Erlbaum Associates.

Salas, E., Cannon-Bowers, J., & Kozlowski, S. (1997). The science and practice of training: current trends and emerging themes. In J. Ford, S. Kozlowski, K. Kraiger, E. Salas, & M. Teachout (Eds.), *Improving training effectiveness in work organizations* (pp. 357–369). Mahwah, NJ: Lawrence Erlbaum Associates.

Salas, E., Dickinson, T., Converse, S., & Tannenbaum, S. (1992). Toward and understanding of team performance and training. In R. Swezey & E. Salas (Eds.), *Teams: their training and performance* (pp. ). Norwood, NJ: Ablex.

Salas, E., Prince, C., Baker, D., & Shrestha, L. (1995). Situation awareness in team performance: Implications for measurement and training [Special Issue: Situation awareness]. *Human Factors, 37*(1), 123–136.

Schiflett, S., Strome, D., Eddy, D., & Dalrymple, M. (1990). *Aircrew Evaluation Sustained Operations Performance (AESOP): A triservice facility for technology transition* (USAFSAM-TP-90-26). Brooks AFB, TX: USAF School of Aerospace Medicine.

Shavelson, R., & Webb, N. (1991). *Generalizability theory: A primer*. Newbury Park: Sage.

Shute, V., Torreano, L., & Willis, R. (1998). DNA: Uncorking the bottleneck in knowledge elicitation and organization. In B. P. Goettl, H. Halff, C. Redfield, & V. Shute (Eds.), *Lecture notes in computer science: Intelligent tutoring systems*. Proceedings of the Fourth International Conference, ITS, San Antonio, Texas: Springer.

Smith-Jentsch, K., Johnston, J., & Payne, S. (1998). Measuring team-related expertise in complex environments. In J. A. Cannon-Bowers & E. Salas (Eds.), *Decision making under stress: Implications for individual and team training* (pp. 61–88). Washington, DC: APA.

Stout, R., Cannon-Bowers, J., & Salas, E. (in press). Team situational awareness: Cue recognition training. In D. Garland & M. Endsley (Eds.), *Situation awareness analysis and measurement*. Mahwah, NJ: Lawrence Erlbaum Associates.

Steiner, I. (1972). *Group process and productivity*. Orlando, FL: Academic Press.

Swezey, R., Meltzer, A., & Salas, E. (1994). Some issues involved in motivating teams. In H. O'Neil & M. Drillings (Eds.), *Motivation: Theory and research*. Hillsdale, NJ: Lawrence Erlbaum Associates.

Swezey, R., & Salas, E. (1992). Guidelines for use in team training development. In R. Swezey & E. Salas (Eds.), *Teams: Their training and performance* (pp. 219–246). Norwood, NJ: Ablex.

Tecuci, G., & Keeling, H. (1998). Developing intelligent educational agents with the disciple learning agent shell. In B. P. Goettl, H. Halff, C. Redfield, & V. Shute (Eds.), *Lecture notes in computer science: Intelligent tutoring systems*. Proceedings of the Fourth International Conference, ITS, San Antonio, Texas: Springer.

Towns, S. G., Fitzgerald, P., & Lester, J. (1998). Visual emotive communication in lifelike pedagogical agents. In B. P. Goettl, H. Halff, C. Redfield, & V. Shute (Eds.), *Lecture notes in computer science: Intelligent tutoring systems*. Proceedings of the Fourth International Conference, ITS, 1998. San Antonio, Texas: Springer.

Tziner, A., & Eden, D. (1985). Effects of crew composition on crew performance: Does the whole equal the sum of the parts? *Journal of Applied Psychology, 70*, 85–93.

Waag, W., & Bell, H. (1997). Situation asessment and decision making in skilled fighter pilots. In C. Zsambok & G. Klein (Eds.), *Natural decision making*. Mahwah, NJ: Lawrence Erlbaum Associates.

Weick, K. E. (1979). *The social psychology of organizing* (2nd ed.). New York: Random House.

# Author Index

C

H

N

Q-R



S

X-Y



Z

# Subject Index